CRYSTALLOGRAPHY

CRYSTALLOGRAPHY

By

Dr. A. Goel

DISCOVERY PUBLISHING HOUSE
NEW DELHI-110002

First Published-2006

ISBN 81-8356-170-5

Published by

DISCOVERY PUBLISHING HOUSE
4831/24, Ansari Road, Prahlad Street,
Darya Ganj, New Delhi-110002 (India)
Phone: 23279245 • Fax: 91-11-23253475
E-mail:dphtemp@indiatimes.com

Printed at:
Amit Enterprises

Preface

This book has been written for the students of under-graduate and post-graduate level of the various universities in India. A special feature of the book is that the text has been illustrated with a large number of line diagrams and the data presented in the form of numerous tables for reference and comparison. In the preparation of text standard works and review by renowned author have been freely consulted and the reference given chapter wise. At the end of the book will be found useful by those who wish to make a more detailed study of the topics discussed.

We are extremely grateful to our respected Managing Director Shri Tilak Wasan, Discovery Publishing House, for valuable and active cooperation. We humble request our students and chemistry teachers to send us their constructive criticism and suggestions which we shall be using in the publication of the next edition.

Author

Contents

Pages

Preface

1. **Chemical Crystallography and Liquid Crystals** 1

Introductory, Types of Solids, Distinction Between Crystalline and Amorphous Solids, What is a Crystal, Liquid Crystals or Mesomorphic States, The Glassy State, The Growth of Crystals, External Features of Crystals, Symmetry of Crystals, Point Groups, space Lattice or Lattice, The Unit Cell and Primitive Lattice Cell, Classification of Crystals, Bravais Lattices, Conduction in Solids, Separation Between Lattice Planes, Coordination Number, Calculation of Number of Atoms Per Unit Cell, Radius Ratio Rule, Isomorphism, Mitscherlisch's Law of isomorphism, Laws of Crystallography, Solid State Defects, Packing of uniform Spheres, The Closed Packed Structure and the Nature of Voids, Dimensions of the Holes, Close Packing of Spheres, Internal Structure of Crystals, Structure of Sylvine (KCl), Structure of Zinc Blende, Structure of Diamond, The Graphite Crystal, Cesium Chloride Structure, Structures of Ionic Solids of General Formula, AX_2, Electron Microscope.

2. **Molecular Spectrum and Atomic Spectrum** 113

Spectroscopy, Molecular Spectra, Microwave Spectroscopy (Rotational Spectroscopy), Instrumentation for Microwave Spectroscopy, Infrared Spectroscopy, Nuclear Magnetic Resonance, Electronic Spectra, The Fortrat Diagram, The Charge Transfer Spectra, Isotopic Effect in Molecular Spectra, Electron Spin Resonance Spectroscopy, Atomic Spectrum, Different Special Lines of Hydrogen, Failures of Bohr's Theory, The Compton

Effect, Vector Atom Model, Spectra of Alkali Metals, Explanation of the Doublet Structure, The Sommerfeld Model.

3. Magnetochemistry and Magnetic Properties of Substances **293**

Magnetochemistry, Magnetic Properties of Substances, Definitions and Units, Theories of Paramagnetism, Theories of Diamagnetism, Magnetic Susceptibility is an Additive and Substitutive Property, Theory of Ferromagnetism, Measurement of Magnetic Susceptibility, Applications of Magnetic Susceptibilities.

4. Electric Properties of Molecules and Dipole Moment **330**

Introduction, Dipole Moment, Induced or Distortion Polarisation, Orientation Polarisation, Total Molar Polarisation, Polarisability, Mosotti-Clausius Equation, Debye Equation : Orientation Polarisation, Dependence of Polarizability on Frequency, Determination of Dipole moment, Bond Moments and the Molecular Dipole Moment, Applications of Dipole Moment.

1

CHEMICAL CRYSTALLOGRAPHY AND LIQUID CRYSTALS

INTRODUCTORY

Crystallography is the science of crystals which is devoted to the study of their development and growth, their external form, internal structure, and physical properties. For a very long time *crystallography* was regarded as a part of mineralogy. At the end of the 19th century, however, it became an independent discipline for several reasons, the most important of which were the following :

1. With the progress of chemistry, it became evident that a tremendous number of substances, having nothing in common with minerals, had crystalline structures.
2. Crystals, or more precisely crystalline substances are very widespread. The over-whelming majority of bodies and objects surrounding us have a crystalline structure. The earth and the stones we walk on, the walls and roof of our houses, even wood all this have a totally or partly crystalline structure, often a very complicated one.
3. Crystals have highly diversified specific properties.

TYPES OF SOLIDS

When a substance is at such a low temperature that the thermal agitation of its molecules (or constituent particles) is not strong enough to move them all around, the inter-molecular forces tend to hold the molecules together in more or less fixed positions. The material acquires a shape and is then said to be in the solid state. The so-called "solid" material may be divided into two distinct classes:

(a) *Crystalline Solids :* In a number of solids we observe them to be, bounded by well-defined plane faces, having same geometry irrespective of the source from which they have been obtained. Such solids are known as crystals and are mainly characterised by a spatial regularity of the crystal lattice. The atoms are closely packed in a regular manner to form a characteristic three-dimensional pattern.

In a single-crystal the regularity of arrangement of the pattern extends throughout the solid and all points are completely equivalent. In other words, if the point under consideration is given any amount of displacement in any direction, the point finds itself in the same environment as that in which it was previously. Thus, the most important and the *essential feature of a crystal is the periodicity of the arrangement alongwith regularity.* Some examples are :

(i) If we examine grains of ordinary table salt under a magnifier, we shall find most of them to have the shape of regular cubes.

(ii) In many solids we may not clearly see the shape of the crystals because many small sized crystals are tightly packed together without any specific order. An ordinary copper wire does not look like a crystalline substance but when its surface is examined under a microscope, its crystalline nature becomes visible, thus, a copper wire is said to have a *microcrystalline structure.*

(b) *Amorphous Solids :* An amorphous solid differs from a crystalline substance in being without any shape of its own and has a completely random particle arrangement, *i.e.*, no regular arrangement.

Amorphous substances in many respects resemble liquids which flow very slowly at room temperature. Strictly speaking, *amorphous substances are regarded as supercooled liquids in which the force of attraction holding the molecules together is so great that the material is rigid but there is no regularity of the structure.*

Examples of amorphous substances are glass and plastics which appear to be solid in nature but never found in crystalline form.

DISTINCTION BETWEEN CRYSTALLINE AND AMORPHOUS SOLIDS

1. *Characteristic Geometry :* Crystalline solids have a regular arrangement of particles in typical geometrical forms, whereas amorphous solids have a complete random particle arrangement.

2. *Melting Point :* A crystalline substance has a very sharp melting point while an amorphous substance does not have. For example, as the temperature of glass is raised gradually, it softens and then starts flowing without undergoing any abrupt or sharp change from solid to liquid state.

3. *Cooling Curve :* Cooling curve for an amorphous substance is smooth, while the curve for a crystalline substance has two breaks (a and b) which correspond to the beginning and the end of the process of crystallization. Temperature remains constant during crystallization. The process of crystallization is accompanied by some liberation of energy, which compensates for the loss of heat and causes the temperature to remain constant (Figs. 1.1 and 1.2).

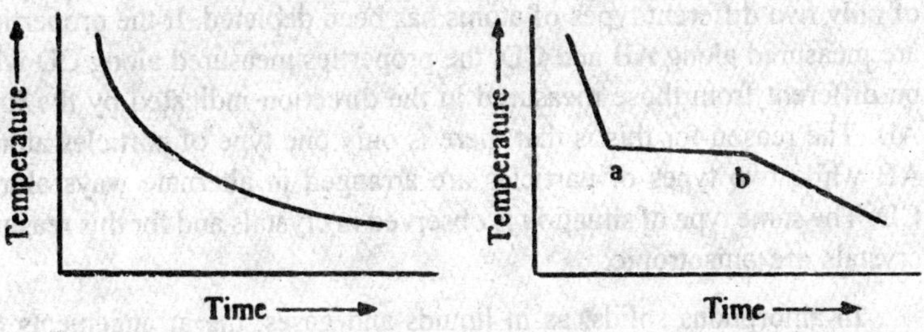

Fig. 1.1 : Amorphous substances. **Fig. 1.2 : Crystalline substances.**

4. *Isotropy and Anisotropy :* Amorphous substances are said to be isotropic, if their properties such as electrical conductivity, thermal conductivity, mechanical strength and refractive index are same in all directions. Crystalline solids, on the other hand, are anisotropic, *i.e.,* their physical properties are different in different directions.

For example, in a crystal of silver iodide the coefficient of thermal expansion is positive in one direction and negative in the other direction. Another example is that velocity of light passing through a crystal varies with the direction in which it is measured. It means that a ray of light passing through such a crystal may split up into two components; each will be following a different path and travelling with a different velocity The phenomenon is known as double refraction.

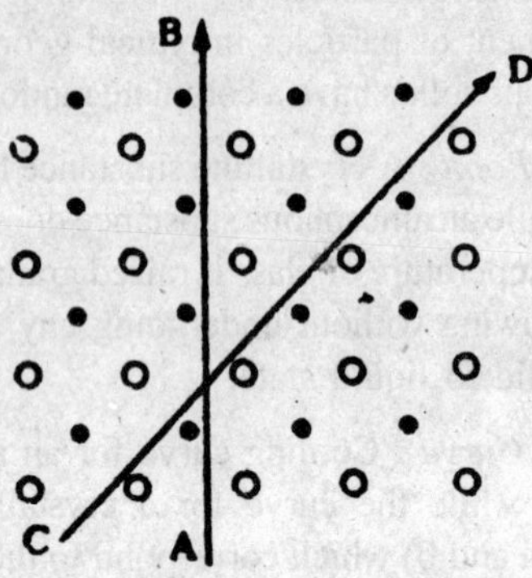

Fig. 1.3

The phenomenon of anisotropy offers a strong evidence for the presence of ordered molecular arrangements in crystals. This can be understood from Fig. 1.3, in which a simple two-dimensional arrangement of only two different types of atoms has been depicted. It the properties are measured along AB and CD, the properties measured along CD will be different from those measured in the direction indicated by the line AB. The reason for this is that there is only one type of particles along AB while two types of particles are arranged in alternate ways along CD. The same type of situation is observed in crystals and for this reason, crystals are anisotropic.

In amorphous solids, as in liquids and gases, the arrangements of particles are random and disordered. Due to this, all directions are equivalent and, therefore, all the properties of these amorphous substances remain same in all directions.

Discussion : The distinction between crystalline and amorphous solids, however, is not absolute. Fine precipitates, which we generally regard them as amorphous and which do not exhibit any crystalline structure even under the most powerful microscope, are found to possess definite crystal lattices of their own when examined by the X-ray methods.

The main points of difference between crystalline and amorphous solids may be summed as Table 1.1 follows :

Table 1.1

Crystalline solid	*Amorphous solid*
1. The constituent particles are arranged in a regular fashion containing short range as well as long range order.	1. The constituent particles are not arranged in any regular fashion. There may be at the most some short range order only.
2. They have sharp melting points.	2. They melt over a range of temperature.
3. They are *anisotropic i.e.*, properties like electrical conductivity, thermal expansion, etc. have different values in different directions.	3. They are *isotropic i.e.*, properties like electrical conductivity, thermal expansion, etc. have same in different directions (just as value in case of gases and liquids).
4. They undergo a clean cleavage.	4. They undergo an irregular cut.

WHAT IS A CRYSTAL

A crystalline solid consists of a large number of small units, called crystals, each of which possesses a definite geometric shape bounded by plane faces. The crystals of a given substance produced under a definite set of conditions are always of the same shape. On the change of conditions, one crystalline form of a given substance may change into another altogether different crystalline form.

LIQUID CRYSTALS OR MESOMORPHIC STATES

When a solid melts, the vibrations of its constituent particles acquire such magnitude that they lapse into translatory motions. However, in some cases molecular or ionic clusters do not lose their identity beyond the melting point of the solid. In such solids, the breakdown of clusters in some directions is very slow and they yield very viscous cloudy liquids at a more or less sharp characteristic temperature, known as the *transition point*. If the temperature is increased beyond the transition point, the cloudiness disappears again sharply at a new temperature, known as melting point of a solid. Beyond the melting point, the substance usually behaves as an ordinary liquid. Between the temperatures, *i.e.*, the transition point and *melting point* the cloudy liquid shows *double refraction*, a behaviour usually exhibited by crystals line solids. On this account the cloudy liquid state is sometimes called *mesomorphic state* or regarded to be made up of *liquid crystals*.

The term "liquid crystals" is made up of liquid and crystals. The word liquid is there because these tend to take the shape of the container and the word crystals is there because they still contain one or two dimensional arrays.

The term ."liquid crystals' is not satisfactory because the substances in this state do not have properties of crystalline state. Actually they are more like liquids in having properties such as mobility, surface tension, viscosity, etc.

To understand the liquid crystal, let us take an example. The compound 5-chloro-6-n-heptyloxy-2-naphthoic acid, exists in ordinary crystalline form. On heating to 438-5 K, the crystal fuses sharply yielding turbid liquid. If this is spread on a flat plate, one may see steps or ridges; it is birefringent, showing coloured areas under polarized light and having different optical properties parallel and perpendicular to the plate. Thus, it shows anisotropic behaviour (they have different physical properties in different directions). The turbid liquid is termed as *liquid crystal*.

Cl

O—*n*—C_7H_{15}

HOOC—

Types of Liquid Crystals. Liquid crystals have been classified into the following types :

(a) Nematic,

(b) Cholesteric, and

(c) Smectic.

These are obtained by heating the crystal phase. More than 500 organic compounds are known which exhibit one or the other of the three types of liquid crystals.

1. *Nematic Liquid Crystals :* These are closer to true anisotropic liquids than are the smectic phases. These liquid crystals have a low viscosity and flow readily. In polarised light they appear to have mobile thread-like structure. The molecules are

aggregated together in groups or swarms with their axis parallel to one another. These molecules move sideways or up and down along their length. Each molecule can rotate around its axis. Fig. 1.4 shows the arrangement of molecules in the nematic type liquid crystal.

In an electric or a magnetic field the swarms of nematic liquid crystals get oriented in the same direction. The turbidity in nematic phase is due to the scattering of light as we find in finely powdered glass.

With rise in temperature swarms diminish in size and at the m.p., these are too small to scatter light. The liquid becomes clear.

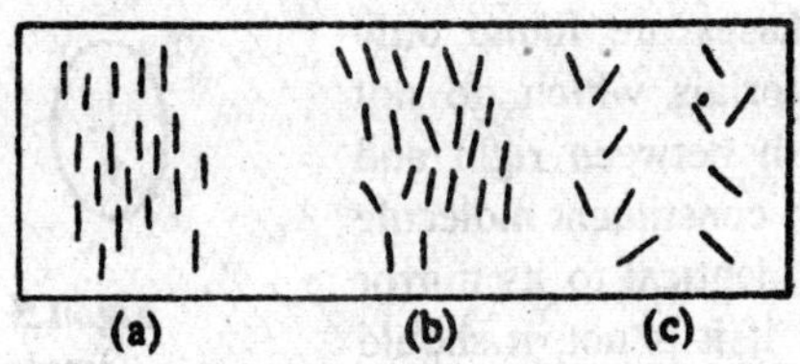

Fig. 1.4 : Nematic liquid crystals.

Properties : (i) Because of the absence of the translational order, nematic crystals flow like a normal liquid.

(ii) Because of the orientation of the molecules about a preferred axis (known as the director axis) nematic state gives rise to anisotropy in several properties, *e.g.*, optical, diamagnetic, viscosity and elasticity.

(iii) Degree of orientational order has been found to decrease with increase of temperature and nematic-isotropic transition occurs with a weak latent heat of transition ($\approx$ 1 kilo Joule per mole) and a small volume increase ($\approx$ 0.5%).

(iv) The centres of gravity of the molecules have no long range order, and therefore there is no Bragg peak in the X-ray diffraction pattern. The positions between the centre of gravity of neighbouring molecules are correlated similarly as those of conventional liquid.

(v) The molecules tend to be parallel to some common axis, labelled by a unit vector, a. This is supported by all macroscopic tensor properties. The difference between refractive index measured with polarizations parallel or normal to a is quite large.

(vi) The direction of a is arbitrary in space. In practice it is imposed by the guiding effect of walls of the container.

(vii) The direction of a and –a are indistinguishable. If the individual molecules carry a permanent dipole (Fig. 1.5), there are many dipoles up and many dipoles down.

(viii) These phases are found only with materials which do not distinguish between right and left. Each constituent molecule must be identical to its mirror image or if it is not, it should be a racemic mixture.

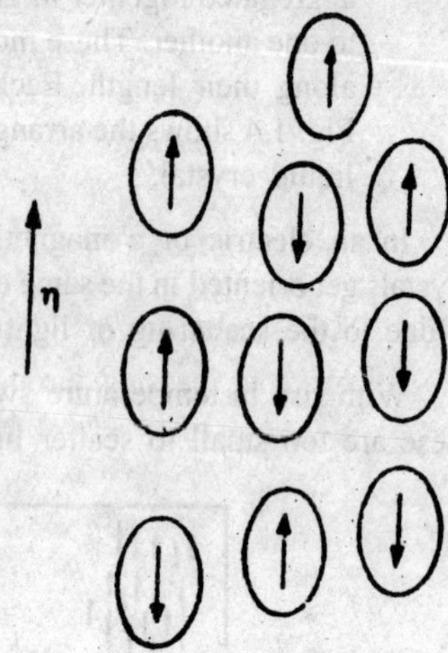

Fig. 1.5 : Nematic single crystal.

The substances which show nematic behaviour are given in Table 1.2.

Table 1.2 : Compounds yielding nematic type liquid crystals

Compounds	*Transition point (K)*	*Melting point (K)*
p-azoxy phenetole	410	440
Anisaldazine	438	453
p-Methoxy cinnamic acid	443	459
p-azoxyanisole	389	408
Dibenzal benzidine	507	533

2. *Cholesteryl Liquid Crystals* : Some liquid crystals, such as cholesteryl esters, show in addition to their nematic behaviour some colour effect under polarised light. These have been given the special name, cholesteryl phase.

Because of the optically active molecules, the cholesteric state has a spontaneous twist about an axis normal to the director. The helical arrangement is right or left handed. Cholesteryl phase possesses a very high optical rotatory power. The plane of polarisation of light is rotated by a few thousand degree per mm which is thousand times greater than the rotatory power of a solid crystal like quartz. The pitch of the helix decreases with increase of temperature and the wavelength of reflection follows out.

In this type of liquid crystals the molecular axes are aligned, and the molecules are arranged in layers in which the orientation of axis shifts in a regular way in going on layer to the next. These arc named as cholesteric crystals because the skeleton of the substance passing through this state is similar to that of the cholesterol. These liquid crystals are characterised by their very high optical rotation. The pitch of the spiral and the reflected colour depends sensitively on the temperature. The first substance in which mesomorphism was detected *i.e.*, cholesteryl benzoate, is of the cholesteric type; its transition temperature is 146°C and melting point 178 5°C.

3. *Smectic Liquid Crystals* : Smectic state is exhibited by common soaps at high temperatures (w 200°C) or id the presence of water and so the name given to this phase, *i.e.*, smectic meaning soap like.

The smectic liquid crystals do not flow as normal liquids but they flow in layers. Because of the weak forces between the layers, these can slide over one another.

The X-ray studies have revealed the existence of such parallel terrace-like planes in smectic liquids. But the spacing of the planes is changed from that in the solid state. It means that the lattice still exists in the liquid crystals though to some extent deformed. This may be depicted in Fig 1.6. Such smectic transformation has been observed in many cases such as ethyl-p-azoxy benzoate, ethyl-p *azoxy* cinnamate, ammonium oleate, etc.

Smectic is the name given by G. Friedel for certain mesophases with mechanical properties similar to soaps. All smectics have layered structure, with well defined interlayer spacing, which can be measured by X-ray diffraction. G. Friedel recognized only one type of smectics-the type which is now called smectic A.

Later on Vorlander showed that there are many different types of smectics, giving rise to different macroscopic textures readily recognized by optical observation. These are referred by the letters A, B and C.

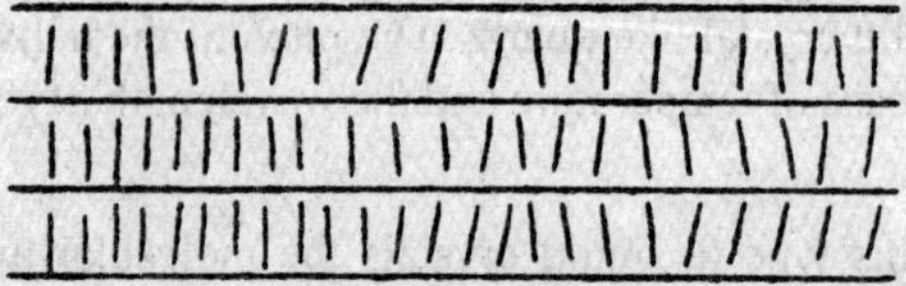

Fig. 1.6. Structure of Smectic A

Smectic A : A picture of the molecular arrangement for smectic A is shown in Fig. 1.6. The major characteristics are as follows:

(i) *A layer structure :* The thickness of the layer is very close to the full length of the constituent molecules.

(ii) Inside each layer. Each layer is a two dimensional liquid. The centres of gravity show no long range order.

Smectic B : The structure of a smectic B is defined as follows :

(i) Each layer is still a two dimensional liquid.

(ii) The material is optically biaxial.

These features are interpreted by assuming that in this type of smectic, the long molecular axis is tilted with respect to normal z of the layers (Fig. 1.7). This observation is supported by X-ray experiments, which give a layer thickness d = l cos ω, where *l* is the molecule, and of the tilt angle. The above two points (i) and (ii) are obtained only when the constituent molecules are optically inactive or with a recemic mixture. If optically active molecules are added to a smectic C, the structure distorts.

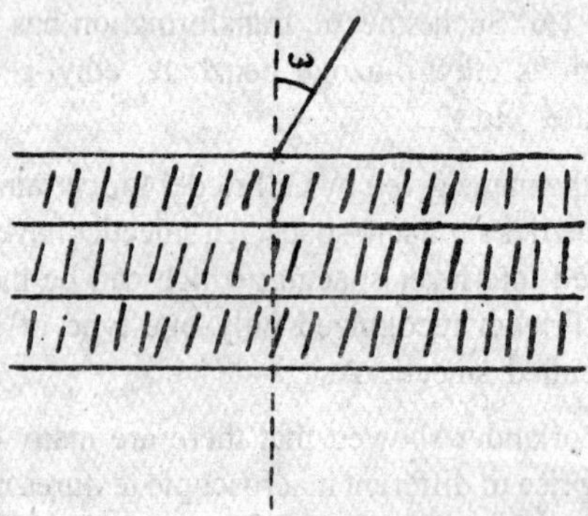

Fig 1.7 : Tilted arrangement of molecules

Smectic C : In smectic C, layers appear to have the periodicity and rigidity of a two dimensional solid. X-ray reflections, corresponding to an order inside each layer, are observed. The layers are not flexible, under the microscope the texture of the B phase (so called mosaic texture) shows domains inside which the layers are flat.

Thus, the B phase appears as the most ordered of the three major smectic phase A, B, C. If one substance is able to display all the three phases, the sequence in increasing temperature is always

$$\text{Solid} \rightleftharpoons \text{Smectic B} \rightleftharpoons \text{Smectic C} \rightleftharpoons \text{Smectic A.}$$

The smectic state is more ordered than the nematic state because it has additional one-dimensional translational order. In compounds where both the states occur, smectic phase occurs at a lower temperature.

Certain instances have been reported in which a crystalline solid passes through more than one smectic phase before being changed into the liquid. It is interesting to note that the temperature of the transition from one phase to another is also fixed. For example, cholesteryl myristate has two mesomorphic phases before it gets converted into the liquid.

$$\text{Crystal} \underset{}{\overset{72^{o}\,C}{\rightleftarrows}} \text{Mesophase I} \underset{}{\overset{78^{o}\,C}{\rightleftarrows}} \text{Mesophase II} \underset{}{\overset{83^{o}\,C}{\rightrightarrows}} \text{Liquid}$$

The above changes are exactly reversed on cooling.

A few examples of compounds yielding smectic mesophases are given in Table 1.3.

Table 1.3 : Compounds yielding smectic type liquid crystals

Compound	*Transition point*	*Melting point*
	°C	°C
Ethyl-p-azoxy benzoate	114	121
n-octyl-p-azoxy cinnamate	94	175
Ethyl-p-azoxy cinnamate	140	249

Compounds with Several Mesophases

Certain substances, notably the cholesteryl esters from the caprylate to the myristate show on heating three phase transformations instead of the usual two. They melt sharply at a certain temperature to give *smectic* type of liquid crystals in the first instance. Then they change sharply at a higher temperature to *Nematic* type which in turn melt sharply at a still higher temperature yielding a clear liquid. These three temperatures are 72, 78 and 83°C respectively. A few substances are known which exist in smectic and nematic forms, with a definite transition temperature, and some may have more than one smectic phase. For example, ethyl anisal-p-amino cinnamate, undergoes the following changes.

$$\text{Solid} \underset{83°C}{\rightleftharpoons} \text{Smectic B} \underset{91°C}{\rightleftharpoons} \text{Smectic A} \underset{118°C}{\rightleftharpoons} \text{Nematic} \underset{139°C}{\rightleftharpoons} \text{Liquid}$$

Similarly, terephthal-bis (–p–butylaniline) (TBBA), with formula

$$C_4H_9-C_6H_4-N=CH-C_6H_4-CH=N-C_6H_4-C_4H_9$$

gives the following set of transitions :

$$\text{Solid} \underset{113°C}{\rightleftharpoons} \text{Smectic B} \underset{144°C}{\rightleftharpoons} \text{Smectic C} \underset{172°C}{\rightleftharpoons} \text{Smectic A} \underset{200°C}{\rightleftharpoons} \text{Nematic} \underset{236°C}{\rightleftharpoons} \text{Liquid}$$

Theory of Liquid Crystals

The Swarm Theory

This theory was proposed by E. Bose (1909). According to this theory, since mesomorphic behaviour is shown by substances having long chains, the molecules will probably tend to arrange themselves parallel to each other. In this way a number of groups or *'swarms'* are obtained. The molecules in any one swarm are parallel to one another but may not be parallel to those present in another swarm. The swarms may be compared to the small crystals, although the molecules in the swarm are not arranged in rigid manner and also the size is not constant.

The turbidity of liquid crystals can be explained by *swarm theory.* Turbidity of liquid crystals is attributed to the scattering of light by the swarms. The effect is quite similar to that observed when a clear crystalline substance or glass is finely powdered. When the temperature is raised the size of the swarms diminishes because of increased thermal movement of the molecules. At the melting point when it becomes clear liquid, its size becomes too small to scatter the light. The double refraction, resulting from orientation within the swarms, will disappear at the same time.

It may be noted that the mesomorphic state is not very simple. This theory has not been finally accepted as there are still many facts which remain unexplained. It is not clearly understood why certain long chain molecules which might be expected to form liquid crystals do not do so.

Uses:

(i) Liquid crystals are becoming important substances because of their electrical and optical properties and use in electronic display instruments.

For example, these havc been used in gas liquid chromatography and in digital displays like pocket calculators, digital wrist watches, etc.

(ii) Liquid crystals are also used as commercial lubricants.

(iii) These are most suited to biological functions. Since the colour of liquid crystals depends sensitively on temperatures, these have been used to measure skin temperature.

(iv) These are used for detecting tumours in the body by employing method called thermography.

(v) These are employed as solvents for studying the structure of anisotropic molecules spectroscopically.

(vi) Due to their electrical and mechanical properties lying between crystalline solids and isotropic liquids these are used in gas liquid chromatography.

THE GLASSY STATE

If a liquid is cooled rapidly so that the atoms are not given an opportunity to arrange themselves properly, a crystalline packing is not obtained. Due to rapid cooling a solid material results in with very high viscosity. *The solid thus obtained which corresponds to a supercooled liquid in which the disordered arrangement of particles in the liquid state is frozen, is said to exist in a glassy state.* The substance existing in this state does not have a sharp melting point. An example is glass which is characterised by having several solid-like properties, viz., rigidity, toughness, hardness, elasticity, etc. but it does not possess a lattice structure of any form and does not possess a sharp melting point. The commonest example of the glass is SiO_2 (Quartz). The structure consists of three-dimensional network of SiO_4, as shown schematically in Fig. 1.8. The random arrangement of the chains facilitates entry of other ions into large openings of the solids. Such oxides as PbO_2 for example, modify SiO_2 glass in this way; the lead ion occupies a space between SiO_4 tetrahedron, thereby increasing the ratio of oxygen to silicon in the net works above the normal value of two. The regular structure of the net works in a glass leads to the variations in the inter atomic spacings and, therefore, to varying bond strengths within the solid. Because of this all the bonds do not break at a single temperature and there is no sharp melting point, as there is in crystalline solids.

Devitrification : If an old sample of glass is heated, it sometimes happens that small crystals grow out and the sample becomes unfit for glass blowing. This phenomenon of crystallisation is called devitrification. It is a slow process and is constantly going on in a sample of glass. This is clearly shown by glass panels of churches which are centuries old.

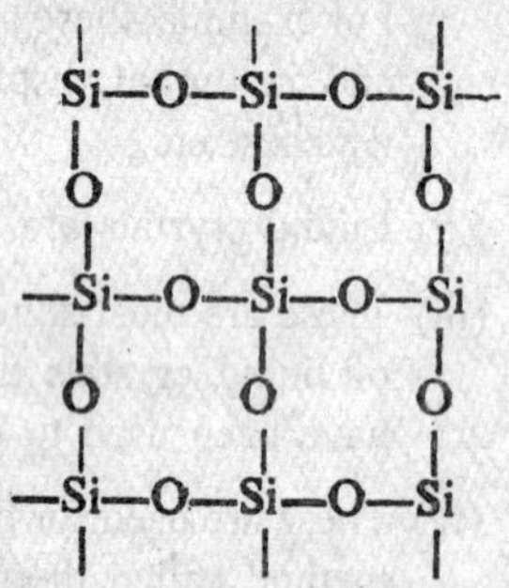

Fig. 1.8 : Quartz (Silica).

THE GROWTH OF CRYSTALS

1. *Introduction :* In general crystal growth involves a phase transformation, *i.e.*, when a substance changes from one state to another. The following are the basic conditions under which the growth of crystals takes place :

   (i) A change from the liquid phase to the solid occurs by crystallisation from a melt or a solution;

   (ii) A change from the gaseous phase to the solid crystalisation occurs by sublimation; and

   (iii) A change from one solid phase to another is accompanied by alteration in the shape of the crystal structure; this phenomenon is called *recrystallisation.*

The above three cases of formation of crystalline matter are not equally widespread; the first case is much more common than the second or the third. In crystal growth, nucleation and growth kinetics are both important. The nucleation of new crystals is an important aspect of crystal growing because one is always interested in getting a single crystal of maximum size and perfection, rather than a multiplicity of smaller crystals. One of the best ways to prevent the formation of more than one nucleus is to utilise a tiny crystal of a *"seed"* on which growth of the crystal occurs under such conditions which discourage the growth of many crystals.

The crystal may grow in a number of ways :

(i) Crystals may grow by layers.

(ii) Crystals may also grow dendritically.

(iii) Crystals may also grow to yield a cellular or honeycomb or interfacial structure.

The capacity of forming large crystals varies with substance and with the condition of crystallization. According to Hofer, the size (as measured by the mass M) of the crystals separating from in aqueous solution is related to the solubility, S by the equation :

$$M = \text{constant} \times S^2$$

Crystals can grow in opposition to considerable "forces. The disintegration of rocks by freezing of water to ice and the bursting of thick cast iron globes containing water on freezing are well known examples of crystalline forces. Ramberg calculated the force of crystallization per mm change of vapour pressure, P, caused by the hydrostatic pressure, p by means of Poynting's equation :

$$F = \frac{dp}{dp} = \frac{RT}{P} Ve^{PV/RT}$$

A substance usually has a definite crystal habit *e.g.*, its crystal may tend to grow more rapidly in one direction or have a prismatic habit, it may grow in two directions by a tabular habit forming plates or tablets etc. the habit often depends upon the speed of crystallization.

The presence of foreign bodies in solution often modifies the crystal form. A well known case is NaCl which grows from pure water solution as cubes but from 15% aqueous urea solution as octahedral. It is believed that urea is preferentially adsorbed on the octahedral faces preventing deposition of Na^+ and Cl^- ions and therefore, causing these faces to develop rapidly in urea.

2. *Theory :* The theory of growth can be explained in terms of entropy of melting. Materials which have a large entropy change are generally bound by large flat crystal faces during growth. Many organic compounds belong to this category.

   Materials with a small entropy change generally develop interfaces parallel to isotherms, yielding dendritic or cellular structures. Metals and certain spherically symmetric organic molecules belong to this category.

3. *Factors :* The following factors have the greatest effect on the shape of the growing crystal :

   (i) *Concentration currents :* The concentration currents strongly influence the shape of growing crystal; if the growing crystal is at the bottom of a vessel, the concentration currents tend to make it flat; if the growing crystal is suspended, the concentration currents tend to elongate it. The effect of the concentration currents can be eliminated by rotating the crystal about a horizontal axis or by shaking the crystalliser flask.

   (ii) *Impurities in the solution :* Impurities often strongly affect the crystal shape. A classical example is the introduction of carbamide into a solution of sodium chloride. With- out this addition, sodium chloride crystallises in cubes, with it in octahedra.

   (iii) *Effect of temperature :* The temperature variations often cause faces to appear which would never be formed under normal conditions of growth.

   (iv) *Viscosity of the solution :* If the viscosity is high enough, it will prevent the formation of concentration currents. Then, the crystal will be able to grow only through the diffusion of supersaturation which has a peculiar effect on the shape of the growing crystal.

4. *Growth of Crystals from the Melt :* Solidification of a melt provides a simple method for the growth of crystals with reasonably low melting points, and which melt congruently and without decomposition. The following methods of growing single crystals from a melt are generally employed :

(a) *The Bridgman method :* In this method, a heat-resistant glass or quartz tube with the bottom end drawn out into a capillary is filled with a melt and suspended in an electric tube furnace Fig. 1.9a.

The melt is allowed to cool by gradual lowering of the glass tube into another tube of lower temperature. Crystallisation begins in the capillary and construction of the cooler end of the crucible increases the chances that only a single nucleus and there- fore, a single crystal, will form. The Bridgman method (1925) is excellent for obtaining single crystals of zinc (M.P. 419°C).

(b) *The Czochralski (pulling) method :* The design of the apparatus required for this process is shown in Fig. 1.9b. In this method, a seed crystal is dipped into the melt and when an equilibrium has been established between the heat given to the melt and heat removed from the solid, the seed is slowly withdrawn by pulling up from the melt. As it is lifted, the material gets deposited on the seed. This method has been used for growing single crystals of germanium and silicon.

Czochralski's method has an advantage over the Bridgman method that the grown crystal is not contained by the crucible, so it is not subject to deformation or cracking by thermal stresses resulting from this contact.

(c) *The floating zone method :* This is shown in Fig. (1.9c). In this method, a short section of a vertically held rod is heated to melting by employing any heating method. If the length and diameter of the molten portion of the rod are not too big, the forces of surface tension are sufficient to hold the molten portion of the rod in the place.

A slow movement of the rod along the axis of the heater then leads to progressive melting of the solid rod on one side of the zone and a corresponding deposition on the other side. If a single crystal seed is placed on the deposition side, the whole rod can be converted to a single crystal. Single crystals of silicon (many inches in length and more than an inch in diameter) are grown by this method.

The floating zone method is better than the Bridgman's and Czochralski's methods because the liquid does not come into contact with the containing vessel.

(d) *Verneuil method:* This is shown in Fig. (1.9d). The substance to be converted into a single large crystal is placed in the form of a very fine powder into a vessel L with a wire net bottom. If a hammer is allowed to fall on the wall of a container B periodically, the powder gets moved into the tube A and then gets mixed up with hydrogen gas coming in at F. When the cone N of the heated powder attains a certain size and form, the conditions of combustion of the gas change as a result of which the apex of the cone becomes somewhat fused thus giving birth to a nucleus. After that the temperature of the flame is so adjusted that the melted particles of powder fall on the nucleus which, steadily growing upwards, produces a single crystal droplet. Large crystals of high melting point compounds such as ruby and corundum can be grown in this manner.

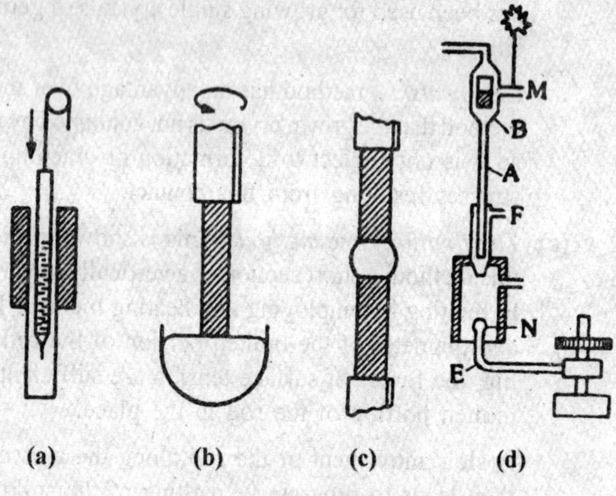

Fig. 1.9

Crystallisation from Solution

By this method, crystals can be formed only from supersaturated solutions. Supersaturation can be brought about in three ways :

(i) the temperature of the solutions may be gradually lowered, thereby progressively precipitating out the solute;

(ii) the solvent may be evaporated to maintain the necessary supersaturation;

(iii) solute may be continuously added, to compensate for the precipitated out.

All the above three principles are used practically to obtain crystals from the solution by the various methods which are as follows :

(a) *Hydrothermal method* : This method uses aqueous solutions under high temperatures and pressures, in a closed system, to dissolve and recrystallise such substances that are only slightly soluble under ordinary conditions. The main applications of this method is the growth of large synthetic crystals of quartz obtained from silica, SiO_2. The silica, SiO_2, is negligibly soluble under ordinary conditions, but at pressures of ~ 20,000 psi and at temperatures ~ 400°C it possesses reasonable solubility in 1M NaOH. The dissolving of silica is carried out in a sealed autoclave. After dissolving a temperature gradient is maintained within a sealed autoclave; this results in the deposition of a single crystal on a seed and the stocking of the solution at the other end. But by this method, single crystals of at least a kilogram in size have been grown.'

The hydrothermal method could not be applied to many other systems.

(b) *Flux method* : This method involves molten inorganic solvents as high temperature solvents which are used to dissolve refractory materials. For example, crystals of cobalt ferrite, $CoFe_2O_4$, can be grown satisfactorily from PbO solution, simply by heating the mixture of oxides of cobalt, iron and lead in a crucible to a high enough temperature to form a molten solution and then cooling slowly.

3. *Growth from the vapour* : During sublimation, crystals are formed directly from vapour, by passing the liquid phase. The specific feature of the crystals thus formed is usually their small size. Two general methods are used to grow crystals from the vapour phase :

(a) *Condensation method* : Crystal growth by the condensation of a vapour requires a temperature gradient with a high temperature source for the vapour to build up a supersaturation pressure condition in the colder deposition region to which the vapour is transported. On the colder side, a crystal will be formed. This method has been used to produce' crystals of a number of metals, inorganic compounds and organic compounds.

(b) *Epitaxial method* : In this method there is an .orientation of crystallisation on a foreign substrate. For example, the hydrogen reduction of $SiCl_4$ on the surface of a silicon single crystal yields single-crystalline epitaxial layers that preserve the orientation of the substrate, without a discontinuity at the interface. Similarly, crystals of CdS, CdSe and CdTe have been grown quite simply by reacting Cd vapour with H_2S, with Se vapour, and with Te vapour respectively. The two vapour streams are mixed and react with deposition on the cooler parts of the vessel.

Purification of Crystals : The crystals grown by the above methods can be purified by physical or chemical methods. Physical methods include distillation, sublimation, evaporation of volatile impurities, fractional electrolysis, ion exchange, chromatography and recrystallisation from the melt. Besides these methods, there is one more important physical method which is known as Pfann's "zone-refining" method. However, all the physical methods are useful in particular cases where zone-refining is not effective. In the *zone refining* process, a large number of successive crystallisations from the melt are to be carried out. For example, in a bar of impure solid, a short region is melted and the melted zone is allowed to move down slowly from one end to the other end of the bar where it is finally frozen. In order to get crystals of very high purity, the process is to be repeated many times.

In chemical methods for purifying crystals, the volatile intermediates such as halides, hydrides or carbonyls are used.

EXTERNAL FEATURES OF CRYSTALS

The external geometrical form of a crystal is obtained as a result of infernal regular arrangements of atoms of which it is built up. The regularity is that of a three dimensional pattern in which a certain unit of structure is repeated over and over again in space. It is apparent that internal constitution possesses an important influence on the crystalline forms. The different external features of crystals are briefly described below :

1. *Faces : Crystals are bounded by a number of surfaces which are generally perfect flat, These surfaces are known as faces. Faces are of two types : like and unlike.* There are some crystals which are bounded by faces which are all alike. For instance, fluorspar is generally obtained in cubes, alum in regular octahedron whereas galena crystal exhibits a combination of the cube and an octahedron (Fig. 1.10).

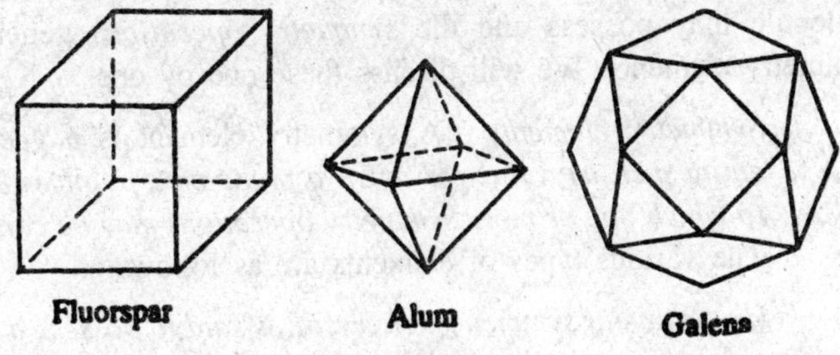

Fig. 1.10

2. *Form* : All the faces corresponding to a crystal are said to constitute a *'form'* A simple form of a crystal is entirely made up of like faces whereas a crystal which consists of two or more simple forms is known as a combination.

3. *Edges and interfacial angles :* An edge is formed by the intersection of two adjacent faces whereas the angle between any two faces of a crystal is known as the interfacial angle.

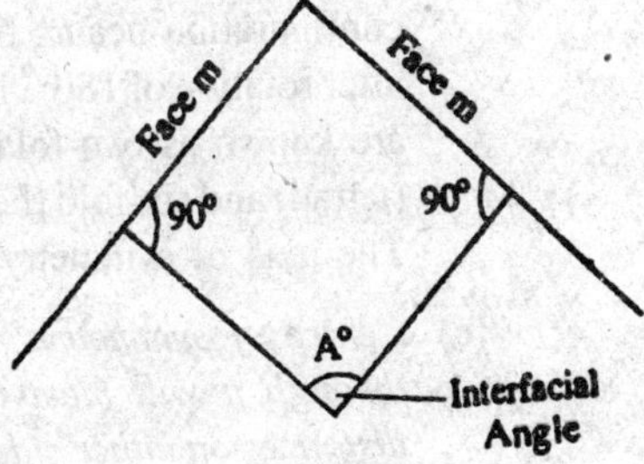

Fig. 1.11

The crystals are bounded by plane faces, straight edges and interfacial angles. The relation ship between these elements can be expressed by formula :

$$f + c = e + 2$$

where f, c and e denote the number of faces, angles and edges respectively.

4. *Zone and zone axis :* The faces of a crystal occur in sets, called zones, which meet in parallel edges, or would do so if the planes of the faces were extended. Each zone forms a complete belt around the crystal. A line drawn through the centre of a crystal in a direction parallel to the edge of a zone is called the zone axis.

SYMMETRY OF CRYSTALS

It is another important property of the crystals. In order to understand symmetry, we shall first consider the kinds of *symmetry elements* a molecule may possess and the *symmetry operations* generated by symmetry elements. We will discuss these one by one.

1. *Symmetry elements :* A symmetry element is a *geometrical entity such as a line (or axis), a plane or a point with respect to which one or more symmetry operations may be carried out.* The various types of elements are as follows :

   (a) *Plane of symmetry : A crystal is said to possess a plane of symmetry when an imaginary plane passing through the centre of crystal can divide it into two parts such that one is the effect mirror image of the other.* The standard rotation for a plane of symmetry is indicated

   (b) *Axis of symmetry : It is a line about which the crystal may be rotated so-that it represents the same appearance more than one during a complete revolution.* If the equivalent configuration occurs twice, thrice, four and six times, *i.e.,* after rotation of 180°. 1.20° 90° and 60°, the axes of rotation are known as two-fold (diad), three fold (triad), four-fold (tetrad) and six-fold (hexad), axes of symmetry respectively. The axis of symmetry is indicated by C

   (c) *Centre of symmetry : It is a point that any line drawn through it will meet the surface of the crystal at equal distances on either side.* It is important to mention here that a crystal may possess a number of planes or axis of symmetry but it can only have one centre of symmetry.

Now, we will illustrate the various elements of symmetry as shown in Fig. 1.12.

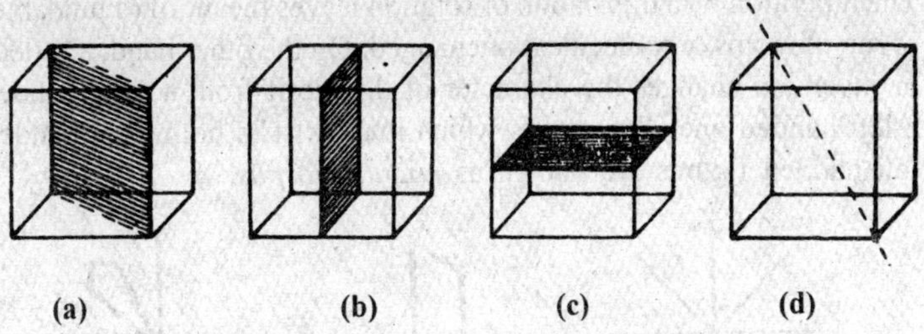

Figs. 1.12 : (a), (b) and (c) show the plane of symmetry. Figs. (d).

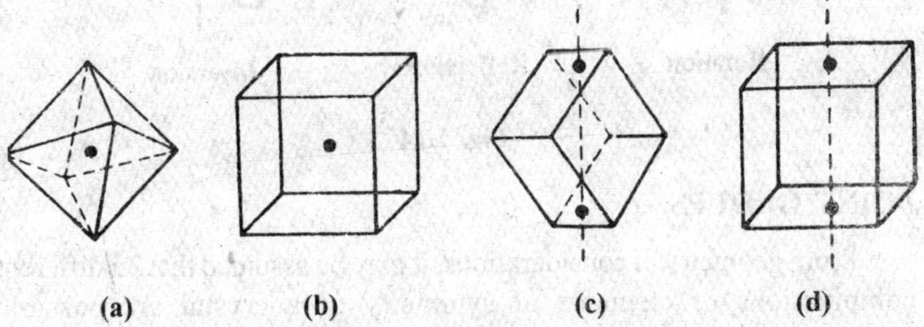

Fig. 1.13 : (a) and (b) represent the three-fold axis, four-fold axis and two-fold axis of symmetry respectively. Figs. (c) and (d) represent the centre of symmetry.

A cube has thirteen axis of symmetry (three four-fold, four threefold, six two-fold), nine planes of symmetry and one centre of symmetry, *i.e.*, 23 elements of symmetry altogether.

2. *Symmetry Operation :* A symmetry operation is a movement of a body, such that, after the movement has been carried out, every point of the body is coincident with an equivalent point (or perhaps the same point) of the body in its original orientation There are four principal operations for repeating figure :

   (a) Translation operation

   (b) Rotation operation

   (c) Reflection operation across a line in two dimensions or plane in three dimension, and

   (d) Inversion through a point Fig. 1.14.

The rotation and reflection operations are also known as *point operations*. These point symmetry operations carry the structure itself.

The repetition by a translation or rotation leaves the motif (a fundamental group of atoms or molecules) unchanged. On the other hand, a reflection or inversion changes the character of the motif from a right-handed to a left handed one. The motifs which may exist in both right-handed or left handed figures are known as *enantiomorphs*.

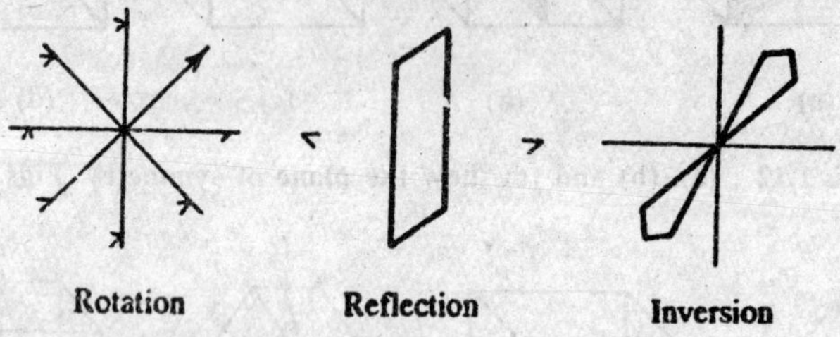

Fig. 1.14

POINT GROUPS

From geometrical considerations, it may be assumed that 32 different combinations of elements of symmetry of a crystal are possible theoretically. These are known as 32 point groups or 32 systems. But some of these could be grouped together. Therefore, these 32 systems could be grouped together in seven different categories which are known as the *seven basic crystal systems*. These seven systems are *cubic or regular, tetragonal, hexagonal, orthorhombic or rhombic, monoclinic, triclinic and rhombohedral or trigonal*. All these systems with the maximum number of planes and axes symmetry elements have been given in the following Table 1.4.

Table 1.4

System	*Symmetry*	*Examples*
1. Cubic or Regular	Nine planes Thirteen axes	NaCl, KCl, CaF_2 Cu_2O, ZnS Pb, Ag, Au, Hg, alums, diamond
2. Tetragonal	Five planes Five axes	KH_2PO_4, SnO_2, $ZrSiO_4$, TiO_2 Sn, $PbWO_3$
3. Hexagonal	Seven planes Seven axes	znO, CdS, HgS, Graphite, Ice, PbI_2, Beryl, Mg, Zn, Cd
4. Orthorhombic or rhombic	Three planes Three axes	$PbCO_3$, $BaSO_4$, K_2SO_4, KNO_3 Rhombic sulphur, Mg_2SiO_4

5. Monoclinic	One plane One axes	Na_2O_4. $10H_2O$, Na_2SO_4. $10H_2O$ Monoclinic sulphur, $CaSO_4$, $2H_2O$
6. Triclinic	No planes No axes	$CuSO_4$. $5H_2O$, H_3BO_3
7. Rhombohedral (or Trigonal)	Seven planes Seven axes	$NaNO_3$, ICl, Magnesite. As, Sb, Bi, Quartz

SPACE LATTICE OR LATTICE

Every solid substance possesses a definite geometrical shape which is characteristic and distinctive for that particular substance. In other words, a solid forms crystals.

A crystal is a homogeneous portion of a solid substance made up of regular pattern of structural units (atoms, molecules or ions) bounded by plane surfaces making definite angles with each other resulting in a definite and distinctive geometric form. *The regular pattern of points which describe the three dimensional arrangement of particles (atoms, molecules or ions) in a crystal structure, is called the space lattice or crystal lattice.*

Fig. 1.15 represents an array of points in two dimensions. It is important to note that environments about any two points is the same. Hence, Fig. 1.15 represents a lattice.

Fig. 1.16, represents an array of points in two-dimensions. In this case environments about any two points are not the same. Thus, Fig. 1.16 does not represent a lattice whereas Fig. 1.15, does so.

Similarly one can construct a three dimensional space lattice. *Thus a space lattice is one in which a three dimensional collection of points is present provided that the environment about any particular point is in every way same.*

The crystal structure basis and a, lattice : It is well known that a point is an imaginary *and infinitesimal spot in space and hence lattice of points is an imaginary concept.* The lattice can be distinguished from a crystal structure as described below :

A crystal structure is formed by associating with every lattice point a unit assembly of atoms or molecules identical in composition. This unit assembly is called the 'basis'. The crystal structure is not lattice, as a lattice is an imaginary concept, but can be said as a latticed array of atoms.

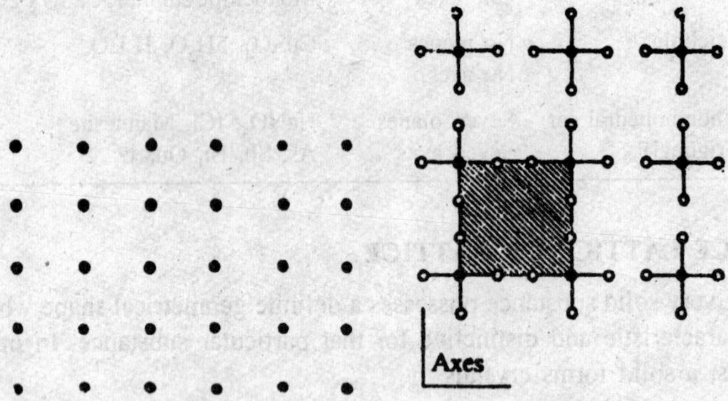

Fig. 1.15 : Two dimensional lattice.

Fig. 1.16 : Two dimensional collection of points but not a lattice.

Fig. 1.17 clearly illustrates the difference between the crystal lattice, the basis, and the crystal structure. In Fig. 1.17(a) the arrangement of points has been shown in which any point is completely equivalent to any other point in the structure (apart from the boundaries of course), thus it represents what we call the crystal lattice. Fig. 1.17(b) illustrates a particular arrangement of three ions denoted by a solid dot, smaller hallow circle and bigger circle, which represent the basis. Then this basis is attached to every point of the lattice of Fig. 1.17(a) which gives the structure shown in Fig. 1.17(c) to this which we identify as the crystal structure. All this has been shown in two dimension and is equally applicable to the three-dimensional crystals.

Lattice + basis = crystal structure.

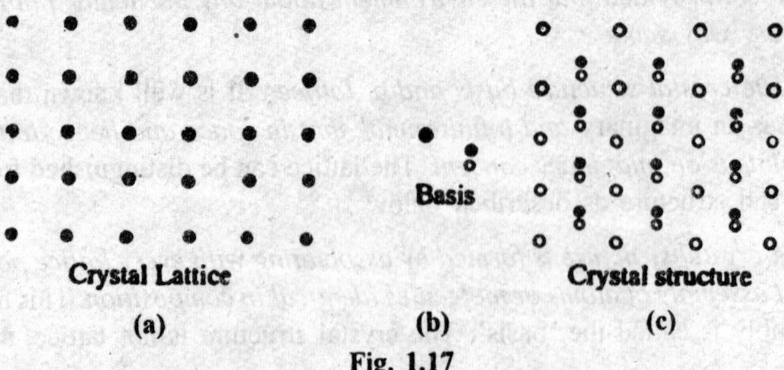

Fig. 1.17

THE UNIT CELL AND PRIMITIVE LATTICE CELL

The smallest portion of a space lattice which can generate the complete crystal by repeating its own dimensions in various directions is called unit cell.

Space lattice is a repetition of unit cell in three dimensions. The choice of a unit cell is by no means unique, There are many ways in which a cell can be drawn in a unique space lattice. It is usually convenient to choose a parallelogram whose edges are parallel with the crystallographic axes a, b, c, and with the shortest possible sides. In Fig. 1.18, region ABCD is called a unit cell and a, b, as the basic vectors. Choice of a unit cell is not unique and region A' B' C' D' or A" B" C" D" would serve the purpose equally well. (Fig. 1.18).

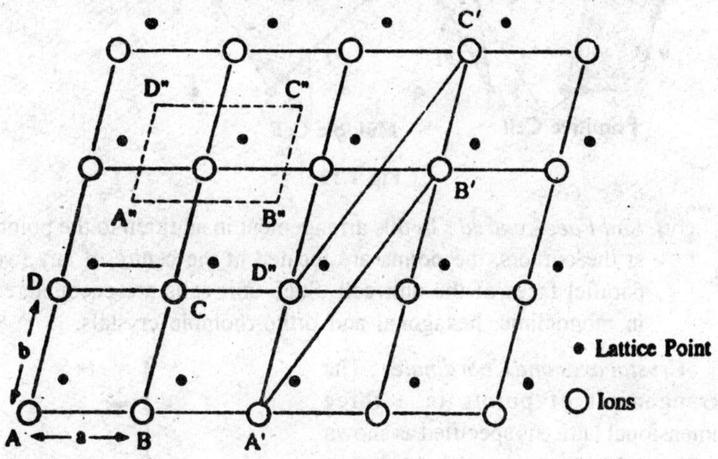

Fig 1.18

Parallelograms formed by lattice points only at their corners are known as primitive or simple cells. There is only one lattice point which is associated with each primitive cell. If we choose translations like a_2 b_2, etc., the unit cell will contain more than one lattice point and is, therefore, called *non-primitive multiple cell* (Fig. 1.19).

Broadly speaking, the three following types of unit cells are known to exist (Fig. 1.15) Art. 9.10) :

(i) *Simple* : It is that arrangement in which the atoms, ions or molecules are present only at the corners of the unit cell.

(ii) *Face-centred :* When in the unit cell, besides the points at the corners there is one point present in the centre of each face, it is called face-centred arrangement.

(iii) *Body centred :* In this arrangement in addition to the points at the corners, there is one point at the centre within the body of the unit cell.

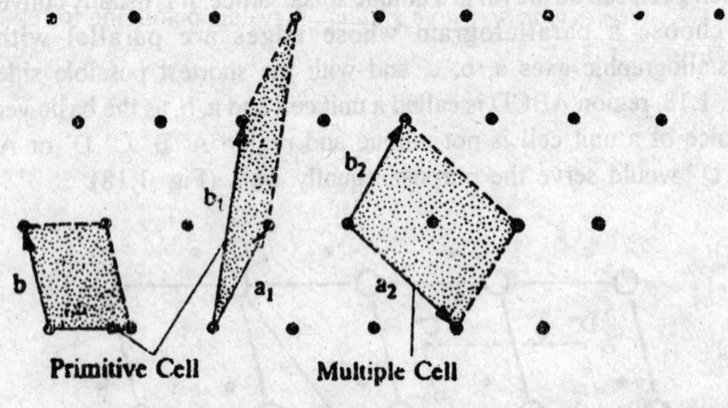

Fig 1.19

(iv) *End Face-centred :* In this arrangement in addition to the points at the corners, the points are located at the centre of any two parallel faces of the unit cell. Such unit cells are encountered in monoclinic, hexagonal and ortho-rhombic crystals.

Crystal axes and Coordinates : The arrangement of points in a three dimensional lattice is specified as shown in Fig. 1.20. The adjacent sides (*i.e.*, concurrent edges) or a unit cell are denoted by $\vec{a}, \vec{b}, \vec{c}$ respectively and the angles included between $\vec{a}, \vec{b}$; $\vec{b}, \vec{c}$ and $\vec{c}, \vec{a}$ are denoted by γ, α and β respectively. Generally, the magnitudes of the primitive vectors are mentioned.

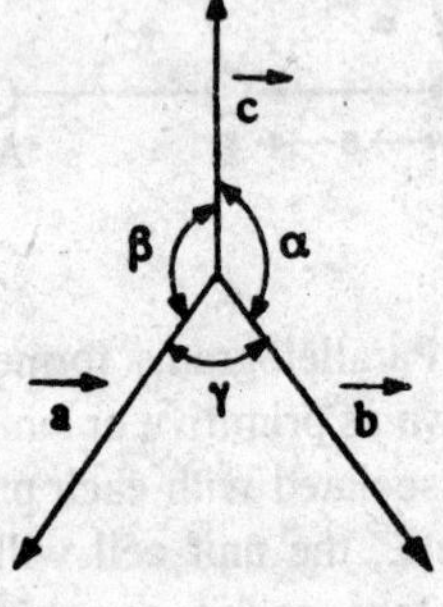

Fig. 1.20

CLASSIFICATION OF CRYSTALS

Although, there is no unique way of classifying all the crystalline solids found in nature, the division is mainly based upon :

(A) The classification of crystals by their shapes, or

(B) The nature of the chemical bonding of the constituent atoms and also on their thermal, electrical and magnetic characteristics. Let us discuss these one by one.

(A) *Classification of Crystals by Shapes :* There are 230 crystal forms possible, and practically all have been observed. On the basis of their symmetry, these 230 crystal forms may be grouped into 32 classes, and these in turn may be grouped into seven crystal systems. The seven systems are :

1. *Regular or Cubic system :* Crystals belonging to regular or cubic system are built up on three equal axes at right angles to one another. The crystals belonging to this class can have three types of lattices depending upon the shape of the unit cells.

   (a) Simple Cubic. In this there are particles only at the corners of the cube, (Fig. 1.21a).

   (b) Face Centred. In this, there are particles at the corners as well as at the centre of each of the six faces cube Fig. 1.21b.

   (c) Body Centred. In this, particles are located at the corners as well as at the centre of cubes. Fig, 1.21c. Among the crystals belonging to this type are diamond, zinc sulphide and anhydrous chlorides of alkali metals.

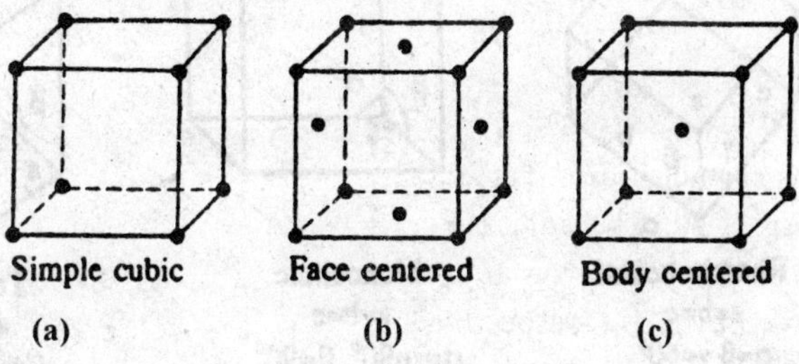

Fig. 1.21

2. *Hexagonal system :* In this three equilateral axes intersect at angles of 60° and with a vertical axis of variable length at right angle to the equilateral axes. This system includes five classes of crystals (Fig. 1.22a). Examples of this systems are beryl, apatite zincite and ice.

3. *Tetragonal system :* In this the three axes are at right angles to each other, but the two lateral axes are equal. This system includes seven classes of crystals and each class is characterized by a four fold or tetragonal axis of symmetry. Fig. (1.22b). Examples of this system are thorite (THO_2), anatase (TiO_2), white tin, indium and cassiterite (SnO_2),

4. *Rhombic system :* This system includes such crystals in which there are three rectangular but unequal crystallographic axes at right angles to each other. In these systems vertical axis is one of two fold or diagonal symmetry only (Fig. 1.22c).

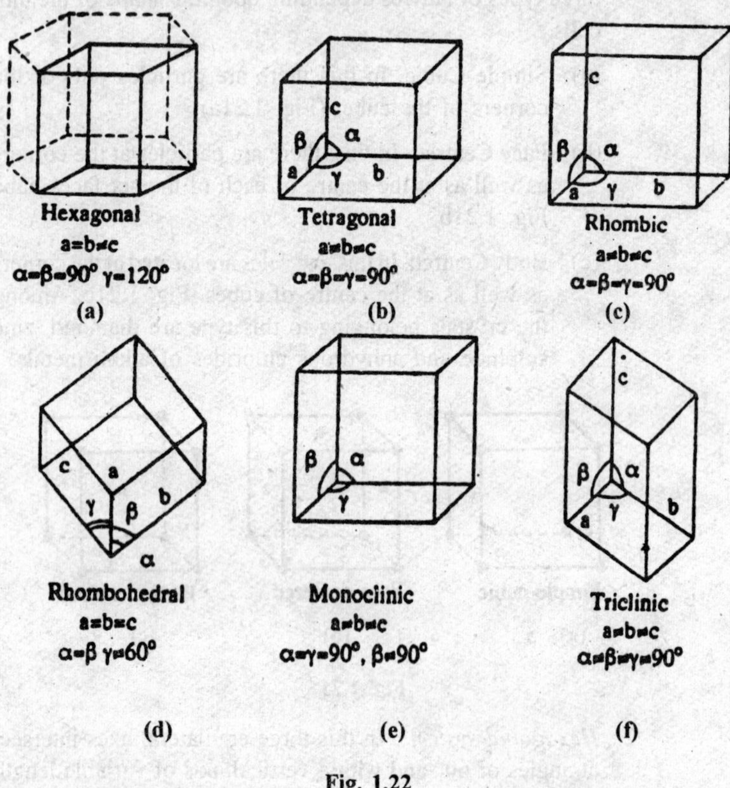

Fig. 1.22

This system includes three classes of crystals and examples of this are celestine ($SrSO_4$), epsomite ($MgSO_4.7H_2O$), olivine (Mg_2SiO_4) and carnallite ($KCl.HgC_2.6H_2O$).

5. *Trigonal or Rhombohedral system :* Some authors regard this type of the system as a part of hexagonal system Fig. (1.22d). However we are considering this as a separate system because in this the vertical axis is having only three fold symmetry where as in hexagonal system the vertical axis is six-fold symmetry.

   It comprises seven classes of crystals. Quartz and calcite are examples of this system.

6. *Monoclinic or Monosymmetric system :* This system includes such types of crystals, each of which possesses three unequal crystallographic axes of which one is perpendicular to the other two. Fig. (1.22e). Examples are Mohr's salt, orthoclase $KAlSi_3O_8$, cryolite (Na_3 AlF_6) and borax.

7. *Triclinic or Asymmetric system :* This system includes such crystals, all of which possess three unequal crystallographic axes all inclined to each other; none of the angles being right angles, there being no axes of symmetry, and the crystallographic axes are chosen parallel to three crystal edges, (Fig. 1.22f). Examples are blue vitriol ($CuSO_4 . 5H_2O$), albite ($NaAlSi_3O_8$) and potassium dichromate.

(B) Classification of Crystals Based on the Nature of the Forces. It is frequently more useful to classify crystals on the basis of nature of forces which hold the crystal lattices together. On this basis there are four types of crystals. These are :

(1) *Molecular Crystals :* In these crystals the units occupying the lattice points are molecules. The forces which hold covalent molecules together within a crystal are van der Waals forces and the type of lattice in which a given substance crystallizes depends upon the size and shape of its neutral molecules.

As the van der : Waals forces are much weaker, it means that molecular crystals are generally soft, easily compressible and can be easily distorted.

As molecular crystals contain no ions or charged particles, that is why the molecular crystals are bad conductors of electricity in solid, liquid as well as in the dissolved state (Fig. 1.23).

For example, iodine crystallizes with an orthorhombic lattice. At each lattice point there is situated a diatomic molecule of iodine. The interatomic distance in the iodine molecule (bond length) is 2.70Å, but the distance of closest approach of two iodine molecules (as measured by the distance between two nearest atoms from different molecules) is 3.54 Å.

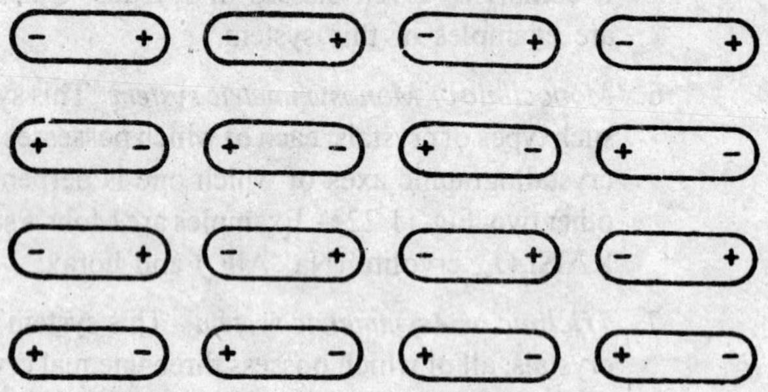

Fig. 1.23 : Molecular Crystal.

This indicates the weakness of the van der Waal's forces compared to the strength of the covalent bond itself.

(2) *Metallic Crystals* : The metallic crystal lattice consists of positive ions permeated by a cloud of valence electrons commonly referred to as the *electron gas* (Fig 1.24). The binding force is the attraction between the +ve ions of the metal and the electron cloud. The valency electrons may be considered to have been denoted by the atoms of the metal and belong to the crystal as a whole. These electrons are free to migrate throughout the crystal lattice giving rise to the high electric conductivity associated with metals. Sodium, iron, tungsten, copper and silver are typical examples of metallic crystals.

The force that binds a metal ion (kernal) to a number of electrons within its sphere of influence, is known as a *metallic bonding*. The properties of metals such as lustre, electrical and thermal conductance are due to the free motion of electrons in them. Metals are opaque to light as the light falling on them is either adsorbed or reflected back.

Depending on the number of electrons available in the valency shell of its atoms and the magnitude of forces binding them to the nuclei, the forces holding the crystal lattice together may be weak or strong. That is why it is not easy to predict the melting points of metallic crystals. Thus, mercury, cesium, and gallium, are low melting metals and the alkali metals, in general are soft while others, like iron, copper, silver etc., are hard and have high melting points.

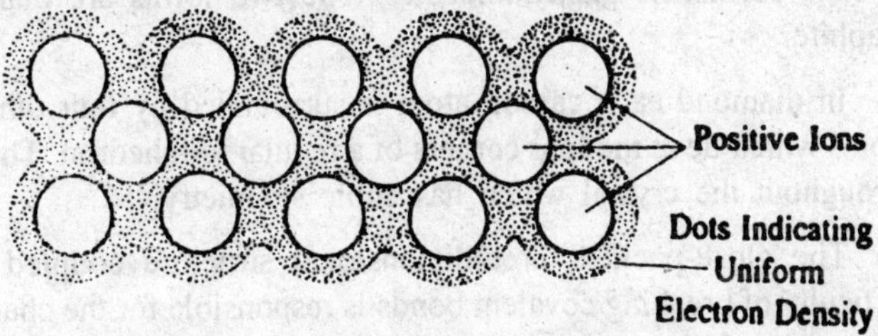

Fig. 1.24

A metallic bond diners from covalent bond in following respects :

(i) A covalent bond has a directional character whereas the metallic bond is non-directional in character.

(ii) Metallic bonds art weaker than the covalent bonds. For this reason the energy required to vapourise a mole of metal (Copper) to the free atoms is smaller than the energy required to vapourise a mole of crystal containing covalent bond.

(3) *The Covalent Crystals :* There is another class of solids in which atoms at their lattice points are bound together by sharing pairs of electrons between them as in the formation of a covalent bond.

This covalent bonding extends throughout the crystals and it has no small molecules in the conventional sense. The continuous bonding in the lattice makes a the whole crystal a single large molecule.

Some main properties of the covalent crystals are summarised below :

(i) These are poor conductors of electricity because ions or free electrons are not present in them.

(ii) These crystals have very high melting points and are quite hard and difficult to break because the shared pairs of electrons give rise to very strong forces between atoms.

(iii) The covalent solids may have fibrous or granular structures.

(iv) As the building units in the covalent crystals are atoms, they are sometimes called atomic crystals. Some examples of covalent crystals are :

Carbon : Carbon can exist in two crystalline forms in both of which crystals consist of giant molecules. The two forms are diamond and graphite.

In diamond each carbon atom is surrounded by four other carbon atoms which lie at the four corners of a regular tetrahedron. This extends throughout the crystal which has cubic symmetry.

The close-packed three-dimensional structure coupled with the difficulty of breaking covalent bonds is responsible for the characteristic properties of diamond :

(a) Diamond is the hardest substance known.

(b) Diamond possesses a high density (3.1 gm/cm^3) and refractive index.

Consequently, diamond has great value as a cutting tool and as a gem.

A crystal of graphite actually consists of a pile of these plates of giant flat molecules, superimposed one on the top of the other and held together loosely by van der Waals forces. The separation of the sheets is 3.40 A°, showing again the weakness of these forces as compared to the actual covalent bonds. The structure of graphite is shown in Fig. 1.25. This structure is responsible for the two characteristic properties of graphite.

(a) *Graphite is a good lubricant :* Owing to the weakness of the interplanar forces, the layers can be easily separated by sliding them one over the other, hence graphite is a soft substance, though in one plane only, and is used in so-called lead pencils and as a lubricant.

(b) *Graphite is a good conductor of electricity :* This is a consequence of the mobility of the bonding electrons, which arises from delocalized electron bondings throughout a vast planar system.

Silicon carbide : The crystal lattice of this molecule is the same as that of diamond in which alternate carbon atoms have been replaced by silcon atoms. Hence silicon carbide like diamond is extremely hard and is used as an abrasive.

Miscellaneous compounds : The valency shells in illenium and tellurium are completed by sharing only two electrons or two adjacent atoms. As a result of which chains of atoms extending through the crystal are formed. This is the case of one-dimensional covalent bonding. Same is the case with $BeCl_2$, or SiS_2.

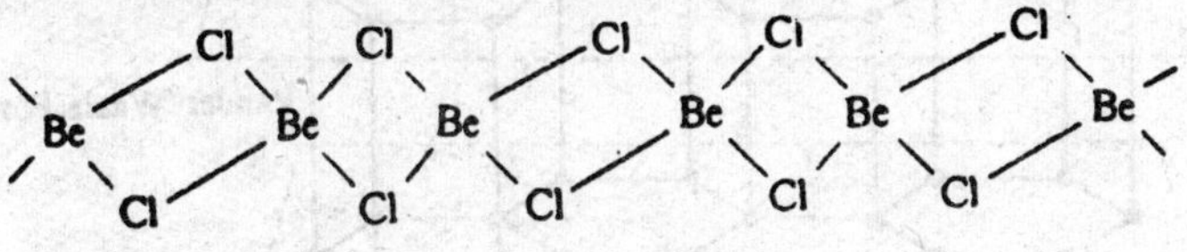

Fig. 1.25 : Structure of one dimensionally covalently bonded crystal $BeCl_2$.

The structure of mica consists of two dimensional covalent or ionic bonds with the layers so-formed being held together by relatively weak forces.

(4) *Ionic Crystals : The lattices in such crystals consist of alternate positive and negative ions in equivalent amounts.* The forces binding the constituent particles together are electrostatic in nature. Electrovalent compounds, in general, belong to this type. In sodium chloride the units are Na^+ and Cl^- ions.

Each ion of a negative sign is held by coulombic forces of attraction to all ions of positive sign. The ionic crystals have the following characteristics:

(i) Since the electrostatic attractions between oppositely charged ions, so closely situated, will be pretty strong, these crystals are usually hard. They have a close-packed structure and usually possess high melting points.

(ii) Inspite of being made up of ions, ionic solids do not conduct an electric current because the ions are immobilized.

(iii) In liquid state (molten form), ionsic solids are good conductors because the arrangement of ions in liquid state is quite disorderly and this helps in their movement under the influence of applied electric field.

(iv) Ionic solids are soluble in water and also in other polar solvents. They are insoluble or very slightly soluble in non-polar solvents such as benzene and carbon tetrachloride.

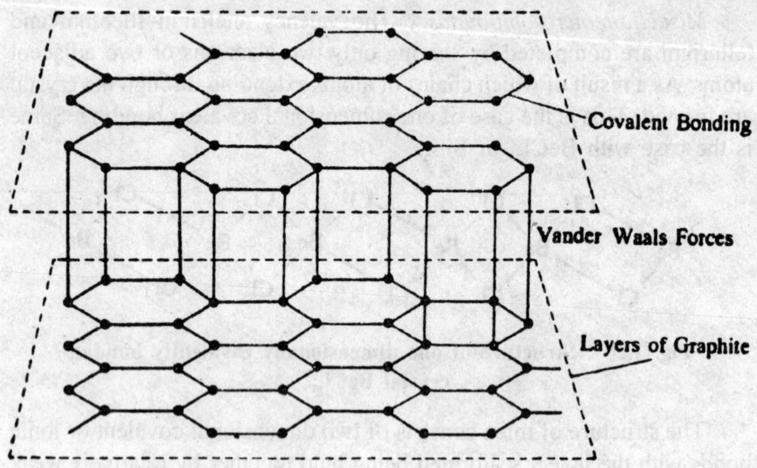

Fig. 1.26.

BRAVAIS LATTICES

In 1848 Bravais showed that the lattice points can be arranged in fourteen ways in space so that the environment looks like the same from each point. These fourteen arrangements are known as Baravais-lattices, named after the name of discoverer.

A combination of Bravais-lattices, with the elements of symmetry, produces 230 different arrangements which are known as space groups. However, in normal course seven systems of crystals are considered, which some modifications produce fourteen Bravais-lattices.

All the crystals in this system are made up of one or more of three kinds of lattices :

(i) the simple cubic or cubic P type lattice,

(ii) the body centered cubic (bcc) or cubic I lattice, and

(iii) the face centred cubic (fcc) or cubic F type lattice.

Table 1.5 lists the seven crystal systems.

Table 1.5

System	*Axes*	*Angles*	*Number of lattices in system*	*Lattice symbols*
Cubic	a = b = c	α = β = γ = 90°	3	P, I, F
Hexagonal	a = b ≠ c	α = β = 90° γ = 120°	1	P
Tetragonal	a = b ≠ c	α = β = γ = 90°	2	P, I,
Rhombic	a ≠ b ≠ c	α = β = γ ≠ 90°	4	P, I, C, F
Rhombohedral	a = b = c	α = β = γ ≠ 90°	1	P,
Monoclinic	a ≠ b ≠ c	α = γ = 90° β = 90° 2		P, C
Triclinic	a ≠ b ≠ c	α ≠ β ≠ γ ≠ 90°	1	P

CONDUCTION IN SOLIDS

From the point of electrical conductivity, solids can be broadly classified into three types :

(i) Conductors,

(ii) Insulators, and

(iii) Semiconductors.

The electrical conductivity of solids varies from 10^{-8} ohm^{-1} cm^{-1} in metals to 10^{-22} ohm^{-1} cm^{-1} insulators.

Conductivity of a solid substance is a characteristic property and reflects the internal structure and bonding in the solid.

The conductivity in a solid is due to the mobility of electrons or ions and imperfections which are charged. Conductivity inmates arises due to the movement of free electrons in metal crystals Although ionic solids are insulators, yet they conduct electricity to a very small extent. This arises due to the migration of ions or other charged species under an electric field to the vacancies or interstices in the ionic soloists. If an ion is moving from its lattice site to occupy a 'vacancy', it creates a new vacancy and in this way a vacancy may migrate across a crystal, which is effectively the same as moving a charge in the opposite direction.

Semi-conductors

These are solids which are perfect insulators at absolute zero but conduct electric current by passage of electrons at room temperature. Semi-conductors possess conductivities in the range 10^{-2}–10^{-9}ohm^{-1}. Silicon and germanium are two examples. In both these examples, there are four valency electrons which form four covalent bonds with other atoms and thus give rise to a tetrahedral structure.

SEPARATION BETWEEN LATTICE PLANES

The separation between successive lattice planes of a cubic, tetragonal and orthorhombic crystal system in which the edges are mutually perpendicular may be calculated as follows :

Suppose Fig. (1.27) represents a two-dimensional crystal lattice and O is origin of co-ordinate system at one of the lattice point. Suppose a plane AB is cutting intercepts 2a and 3b on X and Y respectively. The entire lattice is supposed to be made up of lattice planes which are parallel to AB. Suppose a perpendicular is drawn from O on AB. As the triangles A O D and AOB are similar, it means that

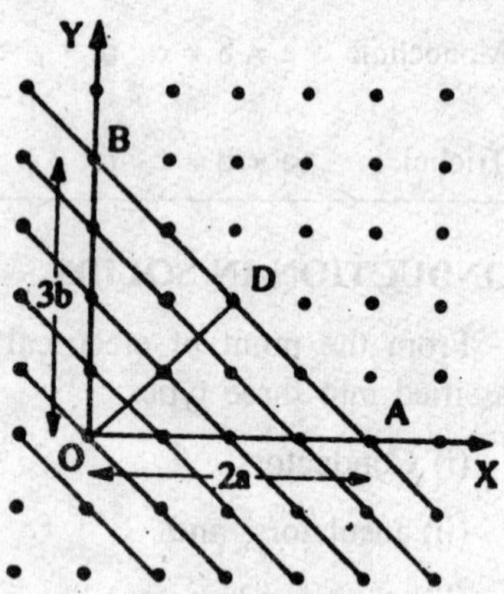

Fig. 1.27

$$\frac{OD}{OA} = \frac{OB}{AB} \text{ or } OD = \frac{OB}{AB} \times OA$$

$$OD = \frac{3b \times 2a}{\sqrt{\left[(2a)^2 + (3b)^2\right]}} = \left\{\frac{1}{\frac{1}{(2a)^2} + \frac{1}{(3b)^2}}\right\}^{-1/2}$$

From Fig. (1.27), it follows that there are five more planes between O and AB. It means that there will be six planes between O and AB. Thus, the separation between successive lattice planes is given by

$$d = \frac{OD}{6} = \frac{1}{6}\left[\frac{1}{(2a)^2} + \frac{1}{(3b)^2}\right]^{-1/2} = \left[\frac{36}{(2a)^2} + \frac{36}{(3b)^2}\right]^{-1/2}$$

$$= \left[\frac{1}{(a/3)^2} + \frac{1}{(b/2)^2}\right]^{-1/2}$$

As the separation ratio of the plane AB with OX and OY axes is 1.2 : 1/3, the Miller's indices h, k of this plane will be 3 : 2. Thus, the value of d becomes as

$$d_{h,k} = \left[\frac{1}{(a/h)^2} + \frac{1}{(b/k)^2}\right]^{-1/2} = \left(\frac{h^2}{a^2} + \frac{k^2}{b^2}\right)^{-1/2}$$

Similarly, one can consider the three-dimensional lattice, with Miller's indices h, k and l and may prove that the separation between successive planes is given by

$$d_{h,k,l} = \left(\frac{h^2}{a^2} + \frac{k^2}{b^2} + \frac{l^2}{c^2}\right)^{-1/2}$$

If we consider a cubic lattice, it means that a = b = c; the above expression becomes as

$$d_{h,k,l} = \frac{a}{\left(h^2 + k^2 + l^2\right)^{-1/2}} \qquad ...(1)$$

We will now determine the spacing for various cubic systems :

(A) Simple Cubic Lattice

In a simple cubic lattice one unit is situated only at each corner of the cubic cell. In simple cubic lattice the (100) planes cut the X-axis and the are parallel to the Y and Z-axis; (110) planes cut obliquely across the X- and Y-axis, but cut the Z-axis at infinity, *i.e.*, parallel to Z-axis and the (111) planes cut all the three axes obliquely and intercept each axis at the same distance from the origin (Fig. 1.28).

The perpendicular distance between successive (100) planes is denoted by d_{100} and is given by putting h = 1, k = 0 and l = 0 in eq. (1).

$$d_{100} = \frac{a}{\left[(1)^2 + (0)^2 + (0)^2 +\right]^{1/2}} = a$$

Similarly for (110) planes,

$$d_{110} = \frac{a}{\left[(1)^2 + (1)^2 + (0)^2 +\right]^{1/2}} = \frac{a}{\sqrt{2}}$$

Similarly for (111) planes,

$$d_{111} = \frac{a}{\left[(1)^2 + (1)^2 + (1)^2 +\right]^{1/2}} = \frac{a}{\sqrt{3}}$$

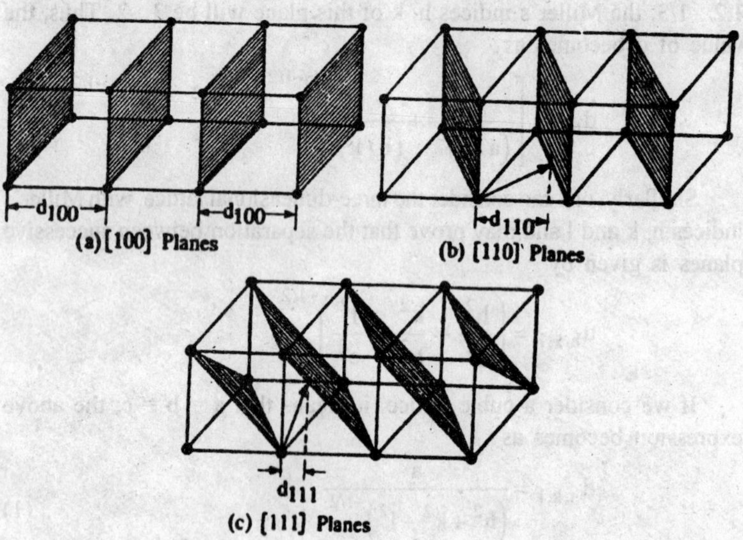

(a) [100] Planes

(b) [110] Planes

(c) [111] Planes

Fig. 1.28

Hence, the ratio of the separation between successive (100), (110) and (111) planes will be

$$d_{(100)} : d_{(110)} : d_{(111)} = a : \frac{a}{\sqrt{(2)}} : \frac{a}{\sqrt{(3)}}$$

or
$$\frac{1}{d_{(100)}} : \frac{1}{d_{(110)}} : \frac{1}{d_{(111)}} = 1 : \sqrt{(2)}; \sqrt{(3)}$$

(B) Face-centred Cubic Lattice

In face-centred cubic lattice, one unit is situated at each corner and one at the centre of each of the faces. In this case, additional (100) planes can be passed through the black dots. They pass through the centre of the front and back faces vertically (Fig. 1.29)

$d_{(100)}$, the distance between successive (100) planes, is a/2 assuming $d_{(100)}$ for simple cubic lattice to be a.

On the assumption that an additional set of (110) planes parallel to the first set can be passed through this type of lattice, these additional planes include the include the units in the centre of the front and the side faces and are midway between the planes of the first type of lattice.

Therefore, $d_{(110)}$ for f. c. l. = a/2 $d_{(100)}$ s.c.l.

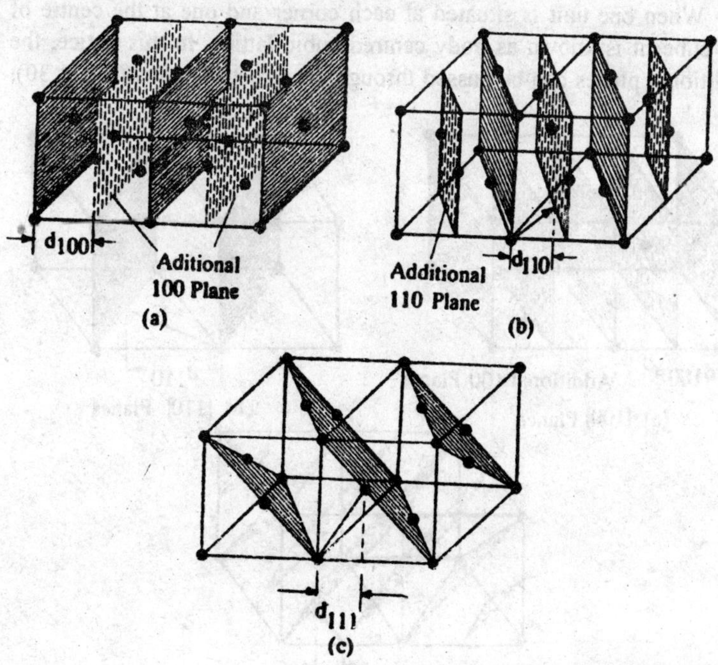

Fig. 1.29 : Showing the separation between successive planes in fcc.

$$= \frac{a}{2\sqrt{(2)}}$$

$d_{(111)}$ remains the same as that for a simple cubic lattice since (111) planes of the type already pass through the centre of all the faces of the face centred lattice.

$$\therefore \quad d_{(111)} = \frac{a}{\sqrt{(3)}}$$

The ratio between these distances is given by :

$$d_{(100)} : d_{(110)} : d_{(111)} : \; = \frac{a}{2} : \frac{a}{2\sqrt{(2)}} : \frac{a}{\sqrt{(3)}} = 1 : \frac{1}{\sqrt{(2)}} : \frac{2}{\sqrt{(3)}}$$

$$= 1 : 0.707 : 1.154.$$

or $$\frac{1}{d_{(100)}} : \frac{1}{d_{(110)}} : \frac{8}{d_{(111)}} = 1 : \sqrt{(2)} : \sqrt{(3)/2}$$

(C) Body Centred Cubic Lattice

When one unit is situated at each corner and one at the centre of the cube, it is known as body centred cubic lattice. In this lattice, the additional planes can be passed through the structural units (Fig. 1.30).

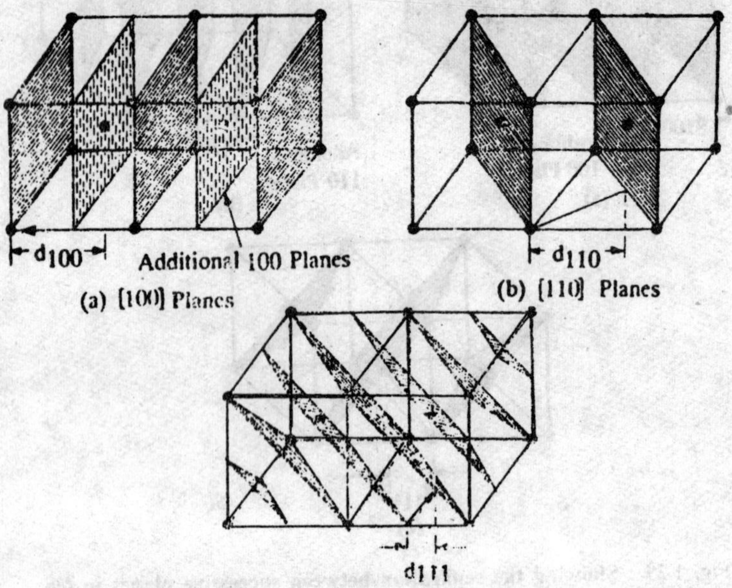

Fig. 1.30 : Showing the separation between successive planes in b c c.

$d_{(100)}$ for this lattice = 1/2a, since these new planes lie midway between the planes of the simple cubic lattice.

$d_{(110)}$ for this lattice $= \frac{a}{\sqrt{(2)}}$ since the simple cubic lattice as (110) planes already pass through the structural unit in centre of the cube.

But in case of (111) faces, the (111) planes do not pass through the structural units in the centre of the cube and so another set of planes can be inserted though these units which lie midway between (111) planes of the simple cubic lattice.

Hence $$d_{(111)} = \frac{a}{2\sqrt{(3)}}$$

The ratio between these distances is given by

$$d_{(100)} : d_{(110)} : d_{(111)} : \; = \frac{a}{2} : \frac{a}{\sqrt{(2)}} : \frac{a}{2\sqrt{(3)}}$$

$$d_{(100)} : d_{(110)} : d_{(111)} : = 1 : \sqrt{(2)} : \frac{1}{\sqrt{(3)}} = 1 : 1.404 : 0.577$$

$$\frac{1}{d_{(100)}} : \frac{1}{d_{(110)}} : \frac{1}{d_{(111)}} = 1 : \frac{1}{\sqrt{(2)}} : \sqrt{(3)}$$

It should be taken into consideration that these are the points and not the lines which constitute a space lattice. The lines are just put for convenience of representation of shape or dimension of the unit cell. They do not have actual existence.

COORDINATION NUMBER

The coordination number of any given sphere in a crystalline structure is the number of other spheres by which it is immediately surrounded. In the hexagonal and face centred cubical close packed structures the coordination number is 12, in the body centred cubical structure it is 8.

Greater is the coordination number, the more closely packed up will be the structure. The coordination number of the three cubic lattices may be determined as noted below :

1. *Simple cubic lattice :* In the simple cubic lattice, one atom is present at each of the eight corners of the unit cell. The geometry of the structure suggests that if we consider an atom at one corner as the centre, it will be surrounded by six equidistant neighbours and hence the coordination number of the simple cubic lattice is six. If ‘a’ represents the side of the unit cell, the distance between the nearest neighbours will be ‘a’.

   In most of the simple cubic lattices, the coordination number is six. However, an exception is found in the case of CsCl. In this case the coordination number is eight because each chlorine atom is surrounded by eight nearest neighbouring ions of opposite type. Another difference is found that the nearest distance between two neighbouring ions of opposite kind is $a\sqrt{3}/2$ but not ‘a’ as in other cases.

2. *Body centred cubic (b c c) lattice* : In this type of lattice the atoms are situated at the corners of unit cell and at its body centre. On drawing a sphere of this radius with any atom at centre then it is easier to see that eight atoms fall on it. Hence the coordination number of cubic I (bcc) lattice is 8. The nearest distance between two atoms is $a\sqrt{3}/2$.

3. *Face centred cubic (f c c) lattice* : In the face centred cubic lattice, the atoms are at the corners of the unit cell. The nearest neighbour to any atom is situated at the distance $a\sqrt{2}$ from it. From geometry of the arrangement it could be seen that there are twelve such atoms at a distance $a\sqrt{2}$ from any atom. Thus for the face centred cubic lattice the coordination number is 12.

CALCULATION OF NUMBER OF ATOMS PER UNIT CELL

There are two methods for calculating the number of atoms per unit cell.

Method I

This method is based upon the simple fact that one should be familiar with the shapes and structures of crystal lattices. Let us illustrate this method by considering the following examples.

(a) *Simple cubic or cubic P lattice* : An example of this type is cesium chloride. One cesium ion is present at each of the body centre of each unit cell. The simple geometrical consideration reveals that every lattice point is situated at the corner of the cell and is shared by eight unit cells and hence its contribution to each cell is 1/8 only.

Thus, the total contribution of eight lattice points of eight corners of a simple cubic will be $1/8 \times 8 = 1$. It means that there is one cesium atom per unit cell and one chlorine atom in the centre. In other words there is one molecule per unit cell.

(b) *Face centered cubic (f c c) or cubic F lattice* : A unit cell of this type is shown in Fig. 1. Art. 9.10. In the face centred cube, each cell corner represents one lattice point per unit cell. In addition each face centre has one lattice point. As one face is common to two unit cells, it means that there is a contribution of lattice point per unit cell per unit face. As there faces, it means that there are $6 \times 1/2$ lattice points per unit cell. Thus there are

8/8 + 6 × 1/2 atoms per unit cell. Hence in face centred cubic lattice there are four atoms per This represents closer packing of the atoms, molecules or ions and a higher symmetry than in the simple cubic lattice.

(c) *Face centred cubic (f c c) or cubic F lattice :* As there are eight atoms at the eight corners of the cell, the contribution of each atom to the cell is 1/8 *i.e.*, the contribution of these eight atoms is 1/8 × 8 = 1 atom per unit cell. Furthermore, there is one atom per unit cell at the body centre. In this way there are two atoms per unit cell.

Method II

Knowledge of the volume density and molecular weight of the constituent atoms of the cell also give information about the number of atoms or lattices per cell.

Consider a unit cell with volume V (in cm^3) which can be calculated from the unit cell dimensions.

Let ρ (gm/cm^2) be the density of the crystal. Then the weight of the matter in the unit cell =volume × density = V × ρ.

If 'n' represents the number of atoms or molecules per unit cell and M is the e atomic (molecular) weight of one atom or molecule, the weight of the matter in a unit cell = $n \times M \times 1.66 \times 10^{-24}$ gm, where 1.66×10^{-24} is the weight of an hydrogen atom in grams used in converting molecular weight in grams. Thus, we have

$$n \times M \times 1.66 \times 10^{-24} = V \times \rho$$

$$n = \frac{V \times \rho}{M \times 1.66 \times 10^{-24}} = \frac{V \times \rho}{M} \times 6.023 \times 10^{23} = \frac{V \times \rho \times N}{M}$$

where $N = 6.023 \times 10^{23}$ is the Avogadro's number.

The volume of a unit cell for different lattices is given below :

Cubic $V = a^3$; Hexagonal = abc sin 60°

Ortho-rhombic V = abc; Rhombohedral = $1/2\ a^3 \sin^2 a/\cos a/2$

The above results can also be utilized to calculate the lattice dimensions for cubic lattices if we know the density of the crystal and the number of molecules per unit cell.

RADIUS RATIO RULE

Goldschmidt found out the crystal structure of large number of inorganic compounds *i.e.*, NaCl, KCl, CsCl, ZnS, ZnO, BN, CaF_2, CO_2, SiO_2, TiO_2, CuO_2, FeS, CdI_2 etc. and observed that their structures are dependent on (i) ratio of radii (ii) ratio of lattice co-ordination number which is defined as the number of equidistant neighbours that an atom has in the given structure, and (iii) properties of polarisation of building units, atoms or ions. It is evident from the following data that NaCl, CsCl, FeS and ZnS are having face centered cubic lattice but lattice coordination number decreases from 8 in CsCl to 3 in BN.

Lattice type :	CsCl	NaCl	ZnO	ZnS	BN
Co-ordination no :	8	6	4	4	3

This difference in co-ordination number is attributed to the difference in the size of ion. For example, if a small positive ion gets surrounded by a large negative ion, the co-ordination number will be small. On the other hand, if large positive ion gets surrounded by a small negative ion a large co-ordination number is possible.

Lattice co-ordination number gets related to the ration of radii of positive and negative ions. For example, if a positive ion of radius r_+ gets surrounded by negative ions of radius r_-, the radii ratio, r^+/r^- is related to co-ordination number as follows .

Co-ordination No.	8	6	4	3
r^+/r^- ratio	> 0.73	0.41 – 0.73	0.22 – 0.41	0.15 – 0.22

This implies a positive ion gets surrounded by 8 negative ions if r^+/r^- ratio is greater than 0.73 and surrounded by 3 negative ions if r^+/r^- ratio is greater than 0.15 but smaller than 0.22.

Thus the higher r^+/r^- ratio the higher the lattice co-ordination number or in other words the larger the positive ion, the larger would be the number of positive ion surrounding it in a closed sphere.

The structures of compounds having general formula AX_2 type *i.e.*, CaF_2 also get related to r^+/r^- ratio as follows :

Lattice	**Co-ordination number**		**r^+/r^- ratio**
	A	X	
CaF_2	8	4	> 0.73
FeS	6	3	0.41-0.73

Cu_2O	4	2	0.22-0.41
CO_2	2	1	–

It is observed that number of X atoms surrounding each A atom must be twice that of A atom due to the AX^2 type. Further, the crystal structures of compounds of AX_3, AX_4 etc. type which are more complicated than those explained above can also be explained by radius-ratio rule. The explanation also involves polarisation of atoms or ions in them. Some representative ionic radii are given in Table 1.6.

Table 1.6 : Ionic radii (Entry × 10^{-2}) (m)

Ions	*Radii*	*Ions*	*Radii*	*Ions*	*Radii*
Li^+	0.60	Mg^{2+}	0.65	Se^{2-}	1.98
Na^+	0.95	Ca^{2+}	0.99	Te^{2-}	2.21
K^+	1.33	Sr^{2+}	1.12	F^-	1.36
Rb^+	1.48	Ba^{2+}	1.35	Cl^-	1.81
Cs^+	1.69	O^{2-}	1.40	Br^-	1.95
Be^{2+}	0.31	S^{2-}	1.84	I^-	2.16

Ion Size and CsCl Crystal Structure

The effect of relative sizes of the cations and anions on the crystal structures of CsCl crystal can be readily demonstrated.

We will now examine the cesium chloride structure shown in the Fig. 1. There are 8 negative ions around a positive ion. Let us assume that ions are spherical, incompressible hard spheres. Further, the anions are in contact with the cations and every anion is in contact with the neighbouring anions. The distance between $Cs^+ - Cl^-$ is therefore $r_+ + r_-$ respectively. Suppose d be the distance between $Cl^- - Cl^-$. KLMN is a vertical diagonal plane, KLO is a right angle triangle, hence

$$KL^2 = KO^2 + LO^2 = a^2 + a^2 = 2a^2 \quad ...(1)$$

Similarly.

$$KM^2 = KL^2 + LM^2 \quad ...(2)$$

or $$[2(r_+ + r_-)]^2 = 2a^2 + a^2$$

$\therefore$ $$KP = PM$$

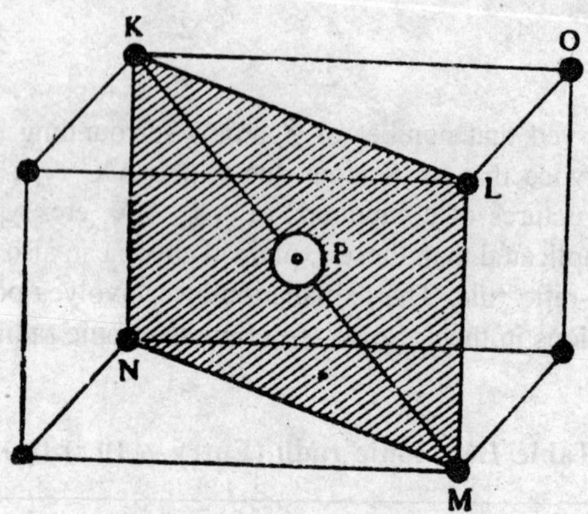

Fig. 1.31 : Cesium chloride crystal.

or $$4\,(r_+ + r_-)^2 = 3a^2$$

$$a^2 = \frac{4}{3}(r_+ + r_-)^2, \quad a = \frac{2}{\sqrt{3}}(r_+ + r_-)$$

The ions are just in contact, hence,

$$a = 2r_-$$

$$\because \quad 2r_- = \frac{2}{\sqrt{3}}(r_+ + r_-)$$

$$2r_- = \frac{2}{\sqrt{3}} r_- = \frac{2}{\sqrt{3}} r_+$$

$$r_-\left(2 - \frac{2}{\sqrt{3}}\right) = \frac{2}{\sqrt{3}} r_+$$

$$\frac{r_-}{r_+} = \frac{2}{\sqrt{3}} \times \frac{1}{\left(2 - \frac{2}{\sqrt{3}}\right)} = \frac{2}{\sqrt{3}} \times \frac{\sqrt{3}}{\sqrt{3 \times 2 - 2}}$$

$$= \frac{1}{\sqrt{3} - 1} = 1.366$$

The above ratio reveals the state when the ions are most closely packed. This is not feasible for ions of the similar charge. Hence the size of the anions would be smaller than that required for the anions to touch one another.

The ratio 1.366 reveals the upper allowed limit for this crystal.

ISOMORPHISM

Introduction : Substances are said to be isomorphous if :

(i) They crystallise with the same lattice;

(ii) They have the same chemical composition and arrangement of atoms or ions, *i.e.*, they are represented by the same type of chemical formula;

(iii) The atomic or ionic radii of the analogous atoms or ions (these which replace one another in the different formulae) are very similar, so that the unit cell dimensions are very nearly the same.

Examples : Some examples of isomorphous substances are :

(i) Sodium selenate and sodium sulphate, Na_2SeO_4 and Na_2SO_4.

(ii) Chromic oxide, ferric oxide and aluminium oxide, Cr_2O_4, Fe_2O_3 and Al_2O_3.

(iii) Sodium nitrate and calcium carbonate (calcite form). In this case both crystallise in rhombohedra of very similar dimensions, have similar arrangements of the component atoms and the corresponding ions are of closely similar size.

Criteria of Isomorphism : Various criteria adopted to decide about the isomorphism of substances are described below :

(i) *Similarity of crystalline forms* : This is the first criterion of isomorphism. It must, however, be noted that no system of crystal structure, except the regular cubic system, shows perfect identity. Generally, there are minute differences in the crystal angles.

(ii) *Formation of mixed crystals* : In many isomorphous substances ions of closely similar size can replace each other indiscriminately in the crystal lattice. A mixed solution of the isomorphous substances deposits crystals which may contain a wide range of proportions of the isomorphous substances. The cell dimensions are intermediate between those of the pure substances. These so-called mixed crystals are really solid solutions of one substance in another. For example, mixed crystals are formed by potassium sulphate and ammonium sulphate, K_2SO_4 and $(NH_4)_2SO_4$:

(iii) *Formation of isomorphous growths* : A crystal of one substance will frequently continue to grow when placed in a saturated solution of an isomorphous substance. This constitutes an overgrowth. For example, overgrowths are formed between the alums and also between the simple sulphates of some transition elements such as ferric sulphate, $Fe_2(SO_4)_3$ and chromic sulphate, $Cr_2(S_4)_3$. Isomorphous substances which form overgrowths are therefore able to induce crystallisation from a saturated solution of the other.

Discussion : It is important to remark that the fulfilment of the criteria (i) and (ii) is not realized in all cases. Several examples have been reported in which either criteria (ii) or (iii) or both may fail. It is generally accepted that if two of the above three conditions are satisfied, the substances concerned may be regarded as isomorphous.

MITSCHERLISCH'S LAW OF ISOMORPHISM

A generalization, called law of isomorphism, was put forward by Mitscherlisch in 1819, which may be restated in an amended form as follows :

"Isomorphous crystals have similar constitution, i.e., they contain the same total number of atoms arranged in similar manner in the crystal lattice."

This statement means that the molecules of truly isomorphous substances consist of the same number of atoms combined together in an identical pattern. Some examples are :

(i) Sodium nitrate and calcium carbonate (calcite form), $NaNO_3$ and $CaCO_3$. In this case, although, the compounds are chemically different, they both crystallize in rhombohedra of very similar dimensions, have similar arrangements of the component atoms and 'the corresponding ions are of closely similar size.

(ii) We may consider the minerals rhodochrosite, $MnCO_3$ and calcite $CaCO_3$. The crystals of these two substances resemble one another closely. Similarly, ferrous sulphate,

$FeSO_4$. H_2O and zinc sulphate, $ZnSO_4.7H_2O$ constitute another example.

Applications of the Law of Isomorphism : The law is of considerable importance as it has played very important role in the early development of chemistry. The following are some of the important applications in which the law of isomorphism was successfully applied.

(i) *Determination of Atomic Weight* : This law has been utilized in determining atomic weights of various elements in the various compounds. An interesting example is the determination of atomic weight of chromium when potassium chromate is isomorphous with potassium sulphate and is found to contain 26-78% of chromium.

Since the formula for potassium sulphate is K_2SO_4, that of potassium chromate, according to the law of isomorphism must be K_2CrO_4. If the atomic weight of chromium is x, the formula weight of potassium chromate must be

$$39.1 \times 2 + x + 64 \text{ or } 142.2 + x$$

This contains $\dfrac{100}{142.2 + x}$ per cent of chromium

But the percentage of chromium in potassium chromate is found from the analysis to be 26.78. Thus,

$$\frac{100x}{142.2 + x} = 26.78$$

Hence the atomic weight of chromium is 52.

(ii) *Correction of atomic weights* : The correction of the atomic weight of vanadium is the best example of the application of the law of isomorphism. In 1831, *Berzelius* assigned an atomic weight of 61 on the simple assumption that the element vanadium was hexavalent. This value was accepted upto 1868 when *Roscoe* applied the law of isomorphism and confirmed that the element must be pentavalent because its mineral vanadium was isomorphous with the following minerals.:

Fluoro Apatite, $3Ca_3(PO_4)_2$, CaF_2

Pyromorphite, $3Pb_3(PO_4)_2$, $PbCl_2$

Mimetite, $3Pb_3(AsO_4)_2$, $PbCl_2$

On the basis of above isomorphous substances vanadinite which is found to have vanadium, lead, oxygen and chlorine must be assigned the following formula;

$3Pb_3(VO_4)_2$, PbC_2

The above formula must contain vanadium as pentavalent like phosphorus and arsenic.

Therefore, its atomic weight was corrected to its present value.

(iii) *Determination of valency* : This law also utilized to determine the exact valency of beryllium, *Mendeleev* suggested that it resembled aluminium. Therefore, its valency should be three. Later *Mallard* showed that beryllium oxide is isomorphous with zinc oxide = ZnO) and must therefore possess the formula BeO. Hence, the valency of beryllium is two out not three.

Limitations of the law of isomorphism : Mitscherlisch's law has lost much of its dignity as a law because it has been found to be only approximately true. A number of exceptions have been reported. Some of these are summarised below :

(i) Some of the compounds which have the same number of atoms, possess similar chemical formula and yet they are not isomorphous. Thus, the law of isomorphism is only an approximate generalisation and, at best may be termed as a rule of convenience.

(ii) A large number of substances have been reported which are found to possess different structures and contain unequal number of atoms and yet these are isomorphous. An interesting example is that Ag_2S and PbS are isomorpho'is, even though they contain unequal number of atoms.

(iii) There are some substances which do not have exactly the same crystalline forms but still we call them isomorphous substances. For example, we call $FeSO_4.7H_2O$ (green vitriol) and $ZnSO_4.7H_2O$ (white vitriol) as isomorphous substances. Their crystals are same but not exactly identical in the sense that they slightly differ in their interfacial angles.

LAWS OF CRYSTALLOGRAPHY

There are three fundamental laws of crystallography.

A. Steno's Law : The first formal statement of law describing the geometric properties of crystals was made by Stensen (sometimes referred to be the Latin form, Steno) in 1669. *Stensen's law*, confirmed by many measurements during the century following its pronouncement, may be stated as follows;

"The crystals of same substance can nave different shapes depending upon the number and size of faces but the angles between the

corresponding faces remain constant."

In Fig. (1) two crystals are represented in two dimensions. The shapes are different but the angles remain the same. The constancy of interfacial angles was demonstrated by Stensen for two substances : quartz (SiO_2) and haematite (Fe_2O_3). The validity of the law of constancy of angles for crystals of substances was confirmed by *Rome Deliste* much later, in 1783.

Let us consider another example to understand Steno's law. A crystal of quartz has hexagonal symmetry. If a section is cut perpendicular to the hexagonal axis, it should have a regular hexagonal shape (Fig. 132b) in which each interior angle being 120°. In actual practice, the growth of certain faces in the crystal is inhibited by the crowding of neighbouring crystals and other influences, thereby causing the faces of a real crystal to develop unequally and hence, the section of real quartz crystals may have shapes such as those in Fig. 1.32c and 1.32d. However, in all possible distortions, the angle remains same at 120°.

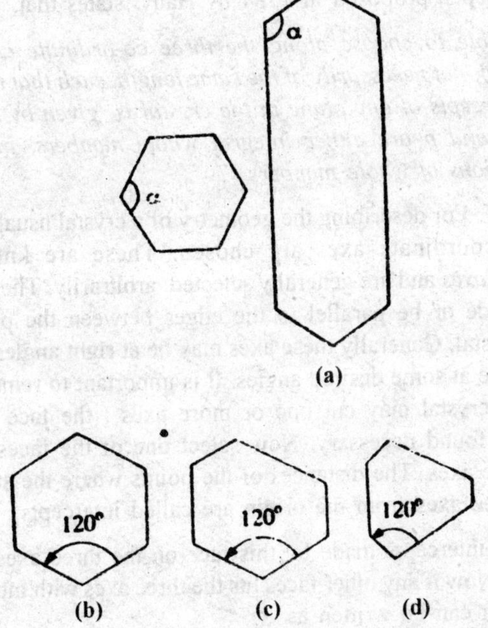

Fig. 1.32 : The constancy of interfacial angles : (b) Section of an ideal quartz crystal; (c) and (d) possible shapes of the quartz crystal system.

Significance : It follows from this law that a substance might be identified by the shape of the crystal, *i.e.*, by the exact values of the angles that its faces and edges form. Measurements of angles are ordinarily made only between faces because of the experimental difficulties encountered in measurement involving the edges. The instruments used for this purpose are known as *goniometers*, and they operate either by contact or through reflection of light beams from the face.

B. Law of Constancy of Symmetry : According to this law,

"All crystals of one and the same substance have the same symmetry."

This law of crystallography may/also be stated as :

"All crystals of the same substance possess the same elements of symmetry."

The total number of planes, lines and centres of symmetries possessed by a crystal is termed as the elements of symmetry of the crystal.

C. Hauy's Law of rational intercepts : This law of rationality of indices or intercepts, proposed in 1784 by Hauy, states that,

"It is possible to choose along the three co-ordinafe axes unit distances (a, b, c), not necessarily of the same length, such that the ratio of the three intercepts of any plane in the crystal is given by ma: nb; pc, when m, n and p are either integral whole numbers, including infinity, or fractions of whole numbers."

Explanation : For describing the geometry of a crystal usually three non-co-planar coordinate axes are chosen. These are known as *crystallographic axes* and are generally selected arbitrarily. These three axes may coincide or be parallel to the edges between the principal phases of the crystal. Generally these axes may be at right angles to one another or may be at some desired angles. It is important to remark that any face of the crystal may cut one or more axes ; the face can be extended if it is found necessary. Now select one of the faces which intersect the three axes. The distances of the points where the standard face cuts the three axes from the origin are called intercepts.

Suppose the intercepts made by this face on the three axes are a, b and c, (Fig. 3). Now if any other face cuts the three axes with intercepts x, y and z, that it can be written as

$$x : y : z = ha : kb : lc \quad ...(1)$$

where h, k and *l* are simple small integers (1, 2, 3, ...). From Eq (1) it follows that all faces cut a given axis at distances from-origin which bear a simple ratio to one another. This is the *law of rational intercepts.*

D. Weiss Indices : Let us consider another face which makes intercepts x', y' and z' on the three axes where the standard intercepts are a, b and c such that

$$x' : y' : z' = 2a : b : 2c \qquad ...(2)$$

From Eq. (2) it follows that the ratio of intercepts in terms of the standard is 2 : 1 : 2. These ratios characterize and represent any plane of the crystal, *These co-efficients are generally known as Weiss indices of the plane.* In some other words, Weiss indices may be defined as follows :

"The ratio of the distances from the origin at which a face intersects the crystallographic axes are called the Weiss's parameters of the crystal face."

E. Miller's Indices : Miller indices of a plane are the reciprocals of the distances from the origin at which a given face intersects the three axes. Let us consider a plane which is parallel to Z-axis. Then the plane will cut the Z-axis at infinity. Immediately one would suggest that the ratio of the intercepts would be

$$x : y : z = 3a : b : \infty c \qquad ...(3)$$

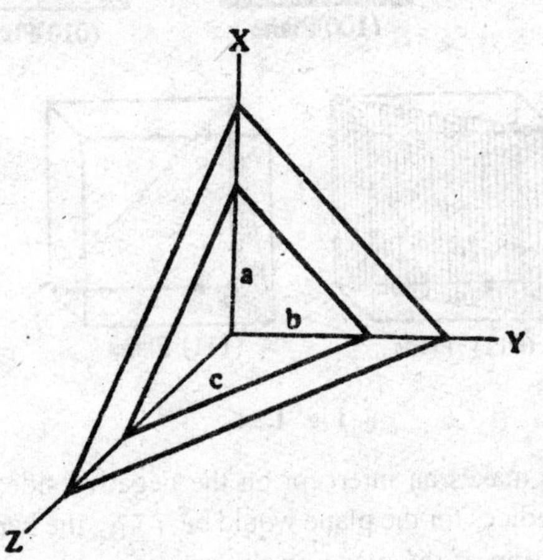

Fig. 1.33.

In the above case the use of Weiss indices is award and has been universally replaced by Miller's indices. Miller's indices are generally obtained by taking the reciprocals of the coefficients of a, b and c and when it is found necessary, the ratio is multiplied by the least common multiple to get integral values. This Miller indices of Eq. (3) will be $1/2 : 1 : 1/\infty$, *i.e.*, $1 : 2 : 0$. This plane or face is simply indicated as (120).

Let us consider another plane which is perpendicular to one axis and parallel to the other axis, then the Weiss indices will be

$$1 : \infty : \infty$$

and the Miller's indices will be

$$\frac{1}{1} : \frac{1}{\infty} ; \frac{1}{\infty} \text{ or } 1 : 0 : 0$$

Chemical Crystallography and liquid Crystal

This face is indicated by (100) plane.

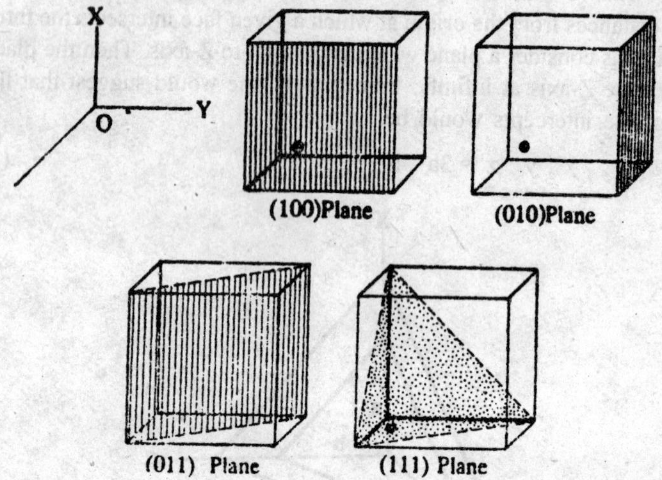

Fig. 1.34.

If the plane makes an intercept on the negative side, say, a :–b : ∞ c, the Miller indices for the plane would be $\bar{1}\,\bar{1}\,\bar{0}$; the bar above would indicate intersection of the plane on the negative side of the axis. The symbol $\bar{1}\,\bar{1}\,\bar{0}$ means minus unity.

SOLID STATE DEFECTS

An ideal crystal is one which has the same unit cell containing the same lattice points across the whole of the crystal. At absolute zero, crystals tend to have perfectly ordered arrangement of ions and there are no defects. As the temperature increases, the chance that a lattice site may be unoccupied by an ion increases. *This constitutes a defect.* Since the number of defects depends on the temperature, they are sometimes called thermodynamic defects. The number of defects formed per cm^3, n, is given by $n = Ne^{-W/2RT}$ where N is the number of sites per cm^3 which could be left vacant, W is the work necessary to form a defect, R is the gas constant and T, the absolute temperature.

Types of Defects. It is customary to divide the defects under two sub-heads:

1. *Stoichiometric defects :* In stoichiometric compounds, irregularity in the arrangement of the ions in a lattice can occur due to a vacancy at a cation and an anion site or by the migration of an ion to some other interstitial site. Two types of such defects are common. These are :

   (a) *Schottky defect :* This defect arises if some of the lattice points are not occupied. The points which are unoccupied are known as *lattice vacancies* or *holes* (Fig. 1.34a).

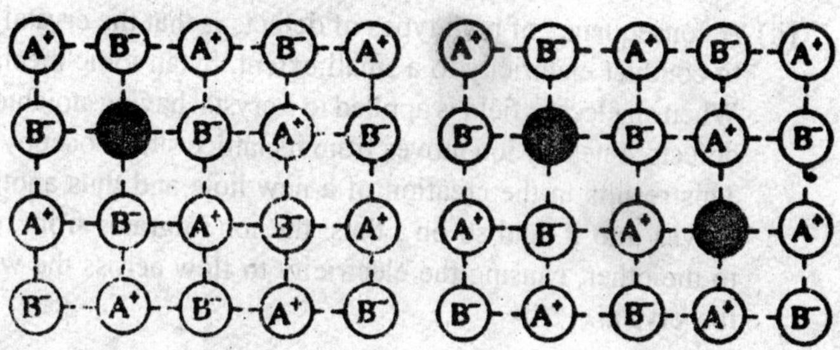

Fig. 1.34(a) : Schottky Defect. **Fig. 1.34(b) : Frenkel Defect.**

This sort of defect tends to be formed in compounds with high coordination numbers, and where the positive and negative ions are of similar size, *e.g.*, NaCl and CsCl furnish good examples of ionic solids in which Schottky defect appears.

(b) *Frenkel defect :* This defect arises when *an ion occupies an interstitial position between the lattice points*. A hole may exist in the lattice because an ion occupies an interstitial position rather than its correct lattice site (Fig. 1.34b). Examples of this defect are ZnS and AgBr. In silver bromide. AgBr, some of the Ag^+ ions are generally missing from their regular positions and occupy positions in between the other ions in the lattice. Zinc sulphide, ZnS, provides another example in which Frenkel defect arises. Zn^{2+} ions get entrapped in the interstitial space leaving holes in this lattice. The Frenkel defect is generally favoured by the following type of compounds:

(i) *Which have low coordination number* : In compounds with low coordination number the attractive forces are lesser which are easy to overcome so that the cation can easily move into the interstitial site.

(ii) Which have ions of different sizes.

(iii) Which have highly polarising cation and an easily polarisable anion.

Consequence of stoichiometric defects : Schottky and Frenkel defects in crystals lead to the following results :

(i) The closeness of similar charges brought about by the Frenkel defect leads to an increase in the dielectric constant of the crystals. The density of the medium however remains unchanged.

(ii) A consequence of both types of defects is that the crystal is able to conduct electricity to a small extent, by an ionic mechanism. When an electric field is applied to a crystal having stoichiometric defects, a nearby ion moves from its lattice site to occupy a hole. This results in the creation of a new hole and thus another ion moves into it, and so on ; thus, the ion migrates from one end to the other, causing the electricity to flow across the whole of the crystal.

(iii) The presence of 'holes' *lowers the density of the crystal*, as expected.

(iv) The presence of 'holes' also lowers the *lattice or the stability of the crystal*.

II. Non-Stoichiometric Defects : Non-stoichiometric compounds are those where the ratio of the number of atoms of A^+ to the number of atoms of B^- does not correspond to a simple whole number as suggested by the formula. Non-stoichiometric compounds do not obey the law of constant composition. This makes the structure irregular in some way, *i.e.*, it contains defects, which are in addition to the normal thermodynamic (Schottky and Frenkel) defects.

Non-stoichiometric defects, evidently, are of two types depending upon whether *positive charge is in excess or negative charge is in excess*. These are known as metal excess defect and *metal deficiency defects*, respectively.

Metal excess defects : In these defects, positive charge is in excess. These defects may arise in two ways :

(a) *A negative ion may be missing from its lattice site leaving a 'hole' which is occupied by an extra electron to maintain the electrical balance* Fig. (1.35a). There is evidently an excess of positive (metal) ions, although the crystal, as a whole is neutral. This defect is somewhat similar to Schottky defect but differs in having only one hole and not a pair as in the latter case. This type of defect is not very common. For example, when NaCl is treated with sodium vapour, a yellow non-stoichiometric form of NaCl is obtained in which there is excess of sodium ions. In such a case, the extra positive charge due to extra sodium ion is balanced by the presence of free electron in the lattice.

(b) The metal excess defect may also appear in another way. An extra positive ion may also occupy an interstitial position in the lot f ice and to maintain electrical neutrality, an electron is also present in the interstitial space Fig. (135b).

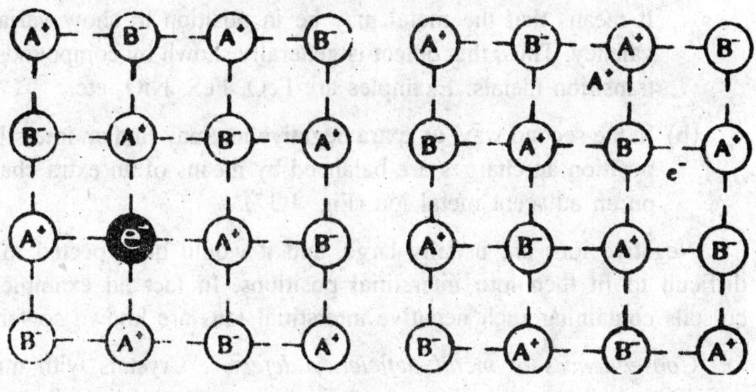

Fig. 135(a). Fig. 135(b).

Although, this type of defect is somewhat similar to Frenkel defect yet it differs from that in having no holes and in having interstitial electrons. This defect is much more common than the first type of metal excess defect. This type of defect is shown by such crystals which are likely to develop Frenkel defect. An interesting example is zinc oxide crystal.

Consequences of Metal Excess Defects

(i) As the crystals associated with metal excess defects of the first or second kind contain free electrons, they can conduct electricity to some extent because the number of defects and hence the number of electrons are very small. Such crystals are generally termed as semiconductors.

(ii) The crystals showing metal excess defects ate generally coloured. This is due to the presence of free electrons. When these electrons arc excited to higher energy levels by absorption of certain wavelengths from the visible white light, these compounds appear coloured. For example, zinc oxide is white in cold but appears yellow when hot.

Metal deficiency defects : Theoretically metal deficiency can occur in *two ways*. Both require variable valency of the metal, and might therefore be expected with the transition metals.

(a) In the first way, a positive ion may be missing from its lattice site, and the extra negative charge is balanced by an adjacent metal ion having two charges instead of one (Fig. 136).

It means that the metal may be in position to show variable valency. Thus, this defect is generally shown by compounds of transition metals. Examples are FeO, FeS, NiO, etc.

(b) In the second way, an extra negative ion may find an interstitial position an charges are balanced by means of an extra charge on an adjacent metal ion (Fig. 1.37).

Negative ions are usually large, and it would be expected to be difficult to fit then into interstitial positions. In fact no example of crystals containing such negative interstitial ions are known so far.

Consequences of metal deficiency defects : Crystals with metal deficiency defect are semi-conductors. This property arises from the movement of an electron from one to another ion. The substances permitting this type of movement are known as o-type semiconductors.

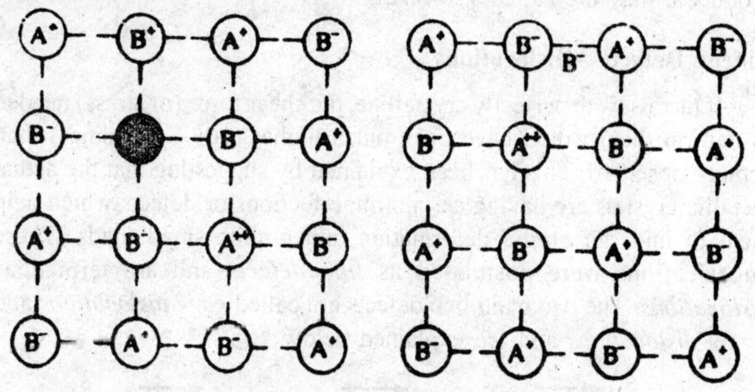

Fig. 1.36. Fig. 1.37.

Colour Centres

Crystals of pure alkali metal halides such as NaCl, KCl, etc. are white. However, it is observed that if NaCl crystal is heated in presence of sodium vapour, it becomes yellow. Similarly, if a KCl crystal is heated in the presence of potassium vapour, it acquires magenta colour. These colours are produced due to some imperfections introduced in the crystal as explained below for the case of NaCl:

When NaCl is heated in an atmosphere of Na vapour, the excess Na atoms deposit on the surface of the NaCl crystal. Cl^- ions then diffuse to the surface where they combine with the Na atoms which become ionized by losing electrons. These electrons diffuse back into the crystal and occupy the vacant sites created by the Cl^- ions (Fig. 1.35a, $A^+ Na^+$, $B^- = Cl^-$). This electron is shared by all the Na^+ ions present around it and is thus considered as a delocalized electron. When light falls on it, it absorbs some energy for excitation from ground state to the excited state. This gives rise to colour. Such points are called *F-centres* : the name comes from the German word for colour, *Farbe*. Actually, a range of energies can excite this electron so that an absorption band called the *F-band* is observed.

It is possible to produce halide crystals containing excess helogen ions. Such excess halogen ions are accompanied by positive ion vacancies which serve to trap holes quite similar to the way the anion vacancies trapped the electrons. In order to distinguish the colour centres thus produced, they are called *V-centres.*

Linear Defects—Dislocations

If metals were perfectly crystalline, the shear force (or stress) needed to deform them would have been much higher (10^2 – 10^4 times) than actuary observed. This has been explained by suggesting that the actual metallic crystals are having certain imperfections or defects which help them to undergo elastic deformation under quite small loads. These imperfections were postulated as *line defects* and are termed as *dislocations.* The two main line defects are called *edge dislocations* and *screw dislocations* and are explained below :

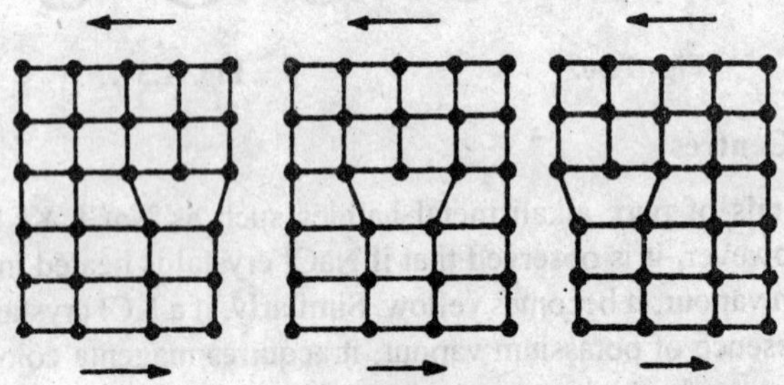

Fig. 138 : Motion of an edge dislocation under shear.

(1) *Edge dislocations :* This defect has seen so named because it arises due to the termination of the edge of an atomic plane within the crystal instead of passing all the way through. The way in which an edge dislocation moves under the shear and helps in the deformation of a crystal is shown in Fig. 1.38.

Thus because of the shear, the dislocation moves across the crystal with the net result that the top half of the crystal gets displaced one lattice distance with respect to the bottom half of the crystal.

(2) *Screw dislocations :* To understand screw dislocations, imagine a cut in a crystal, with the particles to the left of the cut pushed up through the distance of one unit cell (Fig 1.39). The unit cells now form a continuous spiral around the end of cut, the screw axis. In other words, the atomic planes around the screw axis (dislocation line) from a spiral or helical ramp. Such a displacement of atoms is termed as screw dislocation.

The existence of screw dislocation would help in the further growth of the crystal because atoms can easily settle down on the step. However when the lower terrace of the crystal is completely covered, further growth ceases. Just like the edge dislocations, the screw dislocations can also move under the influence of appropriate shear force and thus help in the deformation.

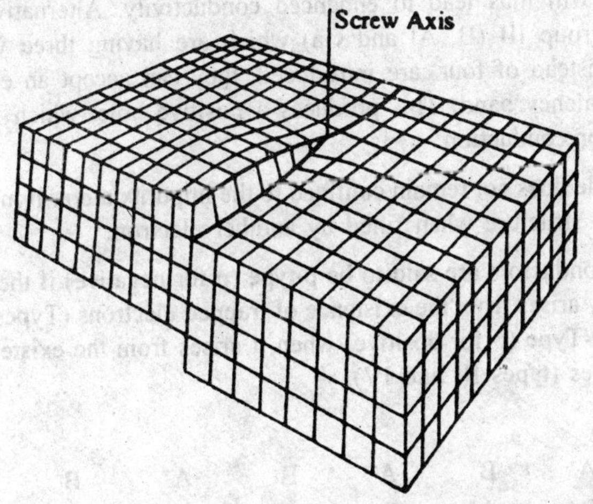

Fig. 1.39 : Screw dislocation at the surface of a crystal.

Semiconductor

A semi conductor may be defined *as an electronic conductor with electrical conductivity in the range* 10^{-7} *to* 10^{4} *ohm*$^{-1}$ *m*$^{-1}$. The electrical conductivity of a pure semi conductor crystal is termed as intrinsic conductivity. It is ascribed to the existence of a vacant conduction band that gets separated by an energy gap from the filled valency band. The smaller the energy gap, the higher the concentration of electrons in the conductor band at a given temperature.

The addition of impurities to a semi-conductor is termed as doping. Impurities have a very large effect on electrical conductivity. Semi-conductors exhibit conductivities which are considerably lower than those of metals (say $10^{-5} \sim 10^{-9}$ ohm^{-1} cm^{-1} compared with 10^3 to 10^5 ohm cm^{-1}) Silicon and germanium in the pure state are very poor conductors of electricity. Impurity doped germanium and silicon, however, exhibit semi-conductivity.

Impurities are termed as donors if they can give up an electron to the valency band and as *acceptors* if they can accept electrons from the valency band, leaving holes in the band.

Atoms of group V (P, As. Sb) arc donor impurities in silicon or germanium. Four electrons in this atom will be used in the formation of covalent bonds and the remaining excess electrons will be free to move and will thus lead to enhanced conductivity. Alternatively, if atoms of group III (B, Al and Ga) which are having three valency electrons instead of four, are introduced, they can accept an electron from the valency band. This produces a positive hole, which is then available for conduction.

The hole does not remain confined to the impurity atom. It migrates through the structure when filled by another electron.

Semi conductors are said to be p-type (p for negative) if the semi-conductivity arises from the existence of trapped electrons (Types I and II) and of p-Type (p for positive) when it arises from the existence of positive holes (types III and IV).

Type I

A^+ B^- A^+ B^- A^+ B^-

B^- A^+ B^- A^+ [e] A^+

A^+ B^- A^+e B^- M B^-

B^- A^+ B^- A^+ B^- A^+

A^+ [e] A^+ B^- A^+ B^-

B^- A^+ B^- A^+ B^- A^+

Excess metal (A^+) with anion (B^-) vacancies.

Type II

A^+ B^- A^+ B^- A^+ B^-

B^- A^+ B^- A^+ B^- A^+

e A^+

A^+ B^- A^+ B^- A^+ B^-

B^- A^+ B^- A^+ B^- A^+

A^+ B^- A^+ B^- A^+ B^-

e A^+

B^- A^+ B^- A^+ B^- A^+

B^- A^+ B^- A^+

Excess metal with interstial cations (A^+)

Type III

A^+	B^-	A^+	B^-	A^+	B^-
	B^-				
B^-	A^{++}	B^-	A^+	B^-	A^-
A^+	B^-	A^+	B^-	A^+	B^-
B^-	A^+	B^-	A^+	B^-	A^+
			B^-		
A^+	B^-	A^+	B^-	A^{++}	B^-
B^-	A^+	B^-	A^+	B^-	A^+

Non-metal with interstitial anions.

Type IV

A^+	B^-	A^{++}	B^-	A^+	B^-
B^-	[]	B^-	A^+	B^-	A^+
A^+	B^-	A^+	B^-	A^+	B^-
B^-	A^+	B^-	A^{++}	B^-	A^+
A^+	B^-	A^+	B^-	[]	B^-
B^-	A^+	B^-	A^+	B^-	A^+

Excess non-metal with cation vacancies.

It is not always easy to decide whether a given substance refers to an n-type or p-type semi-conductor. If the solid is an oxide, and the variation of conductivity with the pressure of oxygen gas in contact with the solid is noted, p-type shows an increase in conductivity with increasing oxygen pressure because the number of positive holes is increased by oxidation and the n-type shows a corresponding decrease in conductivity because the excess of metal ions is decreased by the addition of electronegative constituent.

If a p-type semi conductor is brought into contact with an n-type semi conductor, a double layer of charge would be set up at the junction. These p-n junctions are useful because they can be used as rectifiers, solar cells, and other economic devices.

When an electron and a hole recombine, light may be emitted, and this refers to the mechanism for luminescent p-n diodes.

PACKING OF UNIFORM SPHERES

There are two ways in which spheres can be packed efficiently; the resulting structures are called closest-packed structures. In one of these, called cubical close packing, (ccp), a first layer of spheres is packed together as tightly as possible. A second layer of sphere, is placed on the first with each sphere in the second layer resting in a depression, or hole, between spheres in the first layer. The third layer lies above holes in the first layer.

In the second way of arranging the spheres, the first and second layers are arranged as above but the third layer is directly placed above there in the first layer. This arrangement is called hexagonal-closed packing (h c p) (Fig. 1.40).

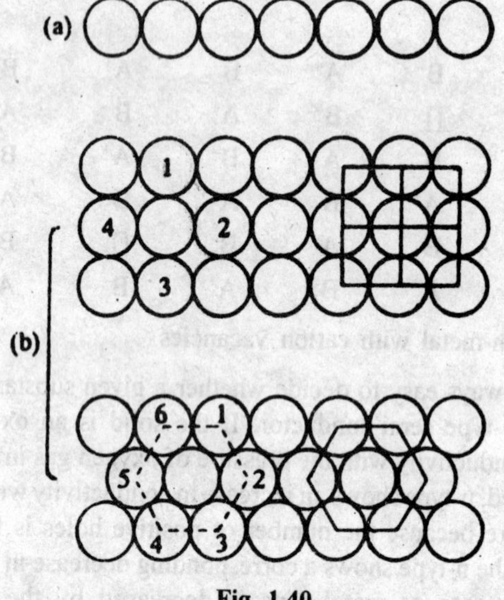

Fig. 1.40

In the c c p structure there are two types of vacant sites, tetrahedral and octahedral. A tetrahedral site is an empty space surrounded by four spheres and an octahedral site is an empty space surrounded by six spheres (Fig. 1.41).

The holes are also called *interstices*. In the hydrides, borides, carbides and nitrides of transition metals the non-metals hydrogen, boron, carbon and nitrogen occupy intersticial positions. The sizes of crystals acquired importance on account of interstices.

In the preceding articles we have considered the occupants of lattice sites to be spheres of unspecified dimensions. From this, one may get the impression that they were small compared with the distances between them. Actually, this is not the case. It becomes very useful to consider a lattice which is composed of rigid spheres of definite dimensions packed tightly enough to touch each other. On this basis one may understand the packing differences involved in the various types of lattices and also arrive at the radii of the spheres involved.

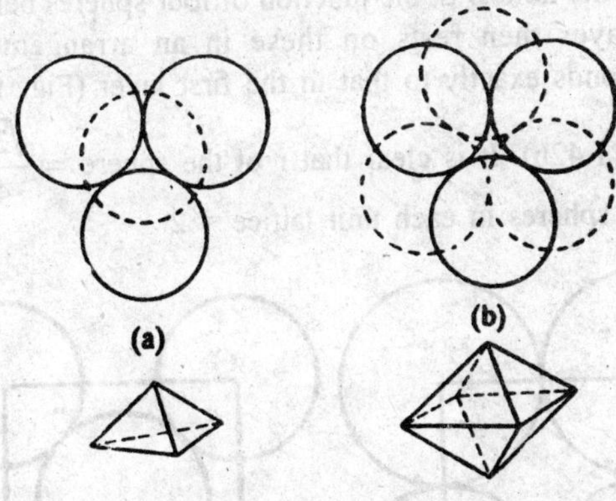

Fig. 1.41

For sake of simplicity we will consider a cubic unit lattice of edge length 'a' composed of uniform spheres of radius r. Suppose the spheres are of the same size. We will now consider the cases one by one.

1. *In a simple cubic lattice* : Spheres are packed by laying down a base of spheres, and then piling upon the base other layers in such a way that each sphere is immediately above the one below it. From the figure (1.42a) it may be seen that the diameter of the spheres is a, the radius of the sphere is a/2, *i.e.*,

   $r = a/2$

Volume of the sphere's $= \frac{4\pi^{r2}}{3} = \frac{4\pi\left(\frac{a}{2}\right)^3}{3}$

In simple cubic lattice, one sphere is present in the unit lattice. Thus,

Volume of the lattice $= a^3$.

$\therefore$ Fraction of the total volume occupied by the sphere is

$$\frac{\left(\frac{4\pi}{4}\right)\left(\frac{a}{2}\right)^3}{a^3} = \frac{\pi}{6} = 0.5236$$

Hence, void space = 1 – 0.5236 = 0.4764 of the total volume = 47.6%.

2. *In the body centred cable lattice* : Here the packing consists of a base of spheres, followed by a second layer where each sphere rests in the hollow at the junction of four spheres below it. The third layer then rests on these in an arrangement which corresponds exactly to that in the first layer (Fig. 1.42b).

From Fig. (1.42b). It is clear that r of the sphere $= \frac{\sqrt{3}a}{4}$

Number of spheres in each unit lattice = 2.

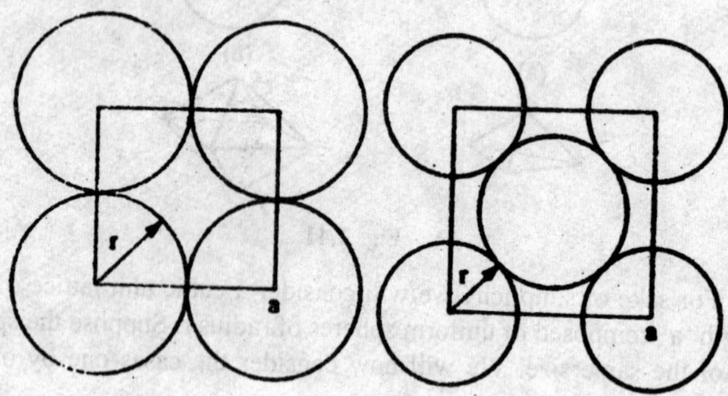

Fig. 142(a) **Fig, 1.42(b)**

Volume occupied by these 2 spheres $= 2 \times \left[\frac{4\pi}{3}\left(\frac{\sqrt{3}a}{4}\right)^3\right]$

Volume of the lattice = a^3.

Fraction of the total volume occupied by the spheres

$$= \frac{2 \times \left[\frac{4\pi}{3}\left(\frac{\sqrt{3}a}{4}\right)^3\right]}{a^3}$$

Void space = 1 = 0.68 = 0.32 = 32%.

3. *In the face centred cubic lattice* : A front view of the packing in a face-centred cubic unit lattice is shown in Fig. (1.43). Considerations similar to those used above show that the sphere radius in such packing is given by √(2a)/4. As a face centred cubic unit lattice consists of four spheres, the volume fraction occupied by these is

$$\frac{4\left[\frac{4\pi}{3}\left(\frac{\sqrt{2}a}{4}\right)^3\right]}{a^3} = \frac{\sqrt{2}\pi}{6} = 0.7404$$

Thus, the void space

$$= 1 - 0.7404 = 0.2596$$

From the above calculations it is evident that the packing is least compact for simple cubic arrangement, followed by the body-centred cubic and then by the free-centred cubic packing.

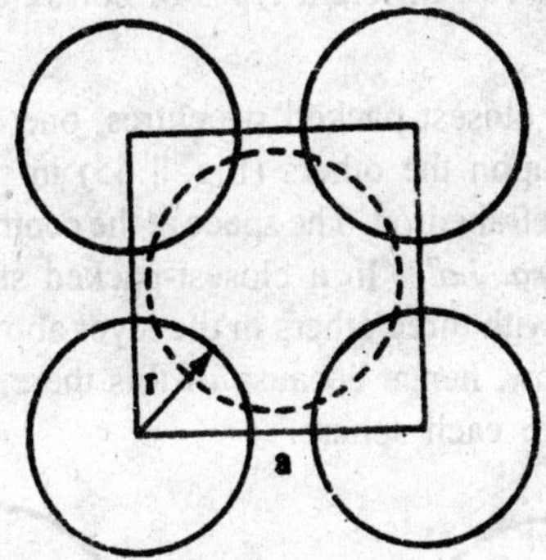

Fig. 1.43

Packing of Non-Uniform Spheres

When a system has spheres of two different sizes, the large-sized spheres will generally take up their definite positions whereas the smaller spheres will have to accommodate themselves in the voids between the large spheres. An example of this is sodium chloride which is shown in Fig. 1 in which smaller sodium spheres have to be accommodated in the voids between the larger chlorine spheres. The face-centred cubic arrangement of the sodium chloride crystal is apparent. If a is the edge of unit lattice, it is related to r_+ and r_-, the radii of two ions, by the equation

$$2r_+ + 2r_- = a$$

or $$r_+ + r_- = \frac{a}{2}$$

This sum is obviously the interionic distances between the sodium and chloride ions. If we consider a body-centred lattice of the CsCl type. The inter-ionic distance is given by

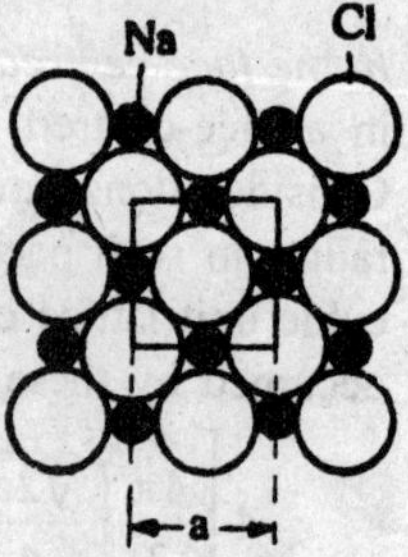

Fig 1.44

THE CLOSED PACKED STRUCTURES AND THE NATURE OF VOIDS

It is known that even in the closest packed structures only 74% of the available space is utilized. How this 26% of the space gets distributed ? Quite obvious, this *unutilized* space gets distributed throughout the structure in the form of voids or holes. A closer view of these structures reveals that there are two different types of *holes* : the *tetrahedral*, and the *octahedral*.

In either of the closest packed structures, one of the basic unit is, one sphere resting upon the others (Fig. 1.45) in such a way that the four spheres form a tetrahedron. The space at the centre of this tetrahedron is called a *tetrahedral hole*. In a closest-packed structure, since each sphere is in contact with three others in the layer above it, and with three more in a layer below, hence because of this there are two tetrahedral holes associated with each sphere.

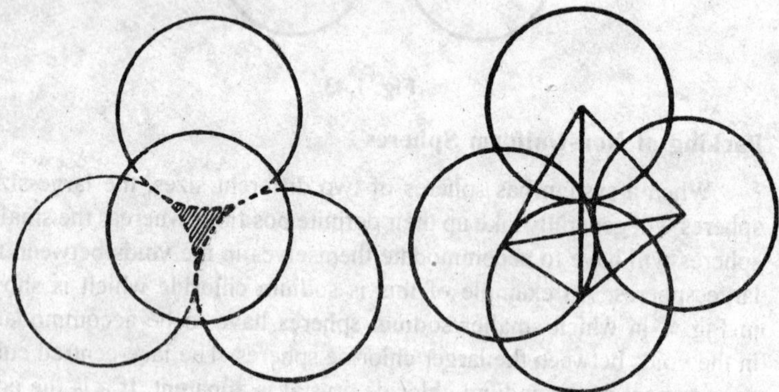

Fig. 1 : A tetrahedral hole in the closest packed structure.

The other type of hole in the closest-packed structures is formed when six spheres form a regular octahedron (Fig. 146).

This is termed as an octahedral hole. Another way of locating the octahedral holes is to regard a face-centered cubic lattice. The centre of any regular octahedron falls on an equatorial plane formed by the centres of four spheres ; these locations are marked X in Fig. 146. It is evident from Fig. 146 that there is only one octahedral hole for every one sphere in the structure. This conclusion holds true for the hexagonal closest-packed structure as well. Hence there are half as many octahedral holes in a closest-packed structure as there are tetrahedral holes.

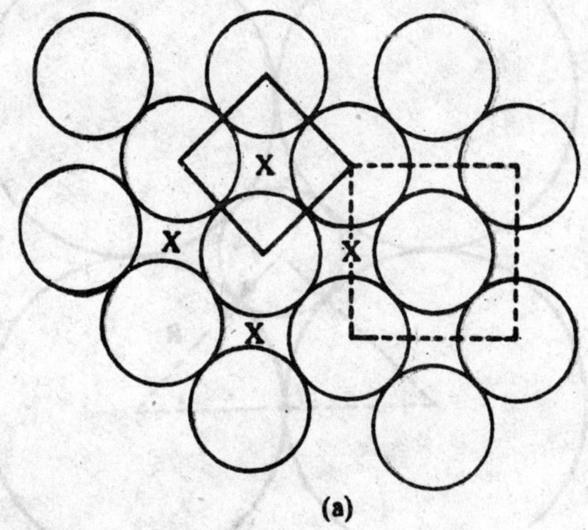

(a)

Fig. 146(a) : The octahedral hole in a closest-packed structure.
(b) Location of the octahedral hole in a face-centred cubic lattice.

DIMENSIONS OF THE HOLES

The dimensions of the holes in the closest-packed structures are related to the size of the spheres forming the structure. A mathematical analysis is given below.

The octahedral hole. Fig. 1.47 shows a section of the structure through an octahedral hole.

Suppose r be the radius of the octahedral hole formed within the structure formed by the spheres of radius R. Then,

$$2(r + R)^2 = (2R)^2$$

or $$R + r = 2R$$

or $r/R = 0.414$

i.e., the radius of the sphere which can be accommodated in an octahedral hole, without disturbing the structure should not exceed 0-414 times that of the large sphere.

The tetrahedral hole : A tetrahedral hole in a closest-packed structure can be produced by placing four spheres of radius R at alternate corners of a cube. Since the spheres are in contact, the face-diagonal is equal to 2R.

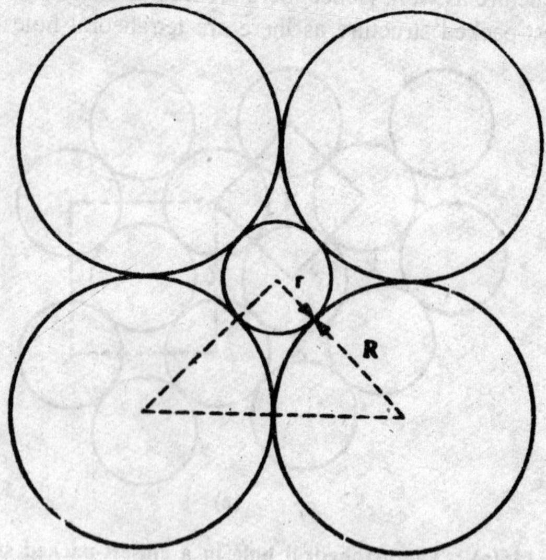

Fig. 1.47 : The geometry of an octahedral hole.

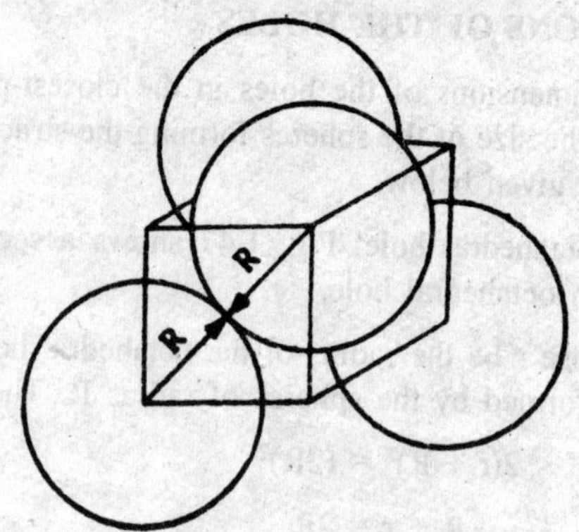

Fig. 1.48 : A tetrahedral hole in a closest packed structure.

Then,

The edge length of the cube = $2R/\sqrt{2} = \sqrt{2}R$

Then,

The body diagonal of the cube = $\sqrt{3} \times \sqrt{2}R$

From geometrical considerations,

$$R + r = 1/2 \text{ (body diagonal)} = 1/3 (\sqrt{3}\ \sqrt{2}R) = \sqrt{\frac{3}{2}}\ R$$

$$\text{or } r = \left(\frac{1.732}{1.414} - 1\right) R$$

$$\text{or } \frac{r}{R} = \frac{0.318}{1.414} = 0.225$$

i.e., the sphere which can be placed in a tetrahedral hole without disturbing the closest-packed structure should not possess a radius larger than 0.225 times the radius of the sphere forming the structure.

CLOSE PACKING OF SPHERES

The study of crystals reveals that there are a number of ways in which identical solid spheres can be packed together. An essential feature in each case is that the spheres prefer closest packing possible so that a maximum possible density is reached. Another reason of closest packing is on account of their mutual attractive interactions. Hence, closer they lie, the greater the stability of the packed system. As the constituent particles are of various shapes, the mode of closest packing of particles will vary according to their shapes.

We will now discuss here the packing modes of simple spherical particles of almost equal size to which the common constituent particles of crystals belong. First of all we will consider the packing of a number of spheres in one plane. The arrangement is shown in Fig. 149. As this is the closest packing, it is, therefore, the most stable arrangement possible. From calculations it can be shown that in this arrangement only 60.4 per cent of the space is occupied by the species whereas the remaining 39-6 per cent of the space is empty and is known as void volume.

In this arrangement, any of the spheres has six nearest neighbours. Therefore, the coordination number of each sphere when packed in this manner is six. Around each sphere, a sort of a hexagonal pattern is formed which is shown in Fig. 1 as shaded spheres surrounding spheres A or B.

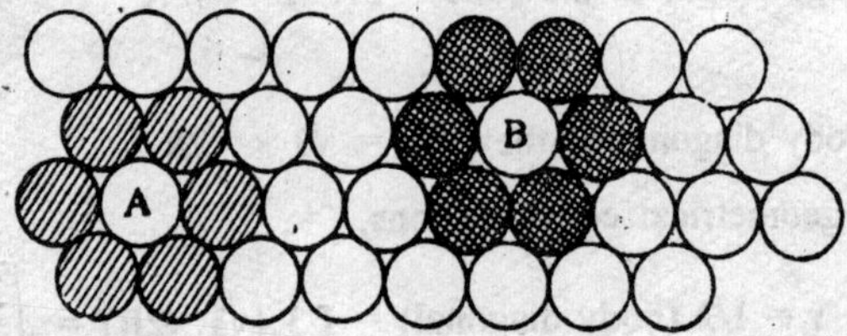

Fig. 1.49

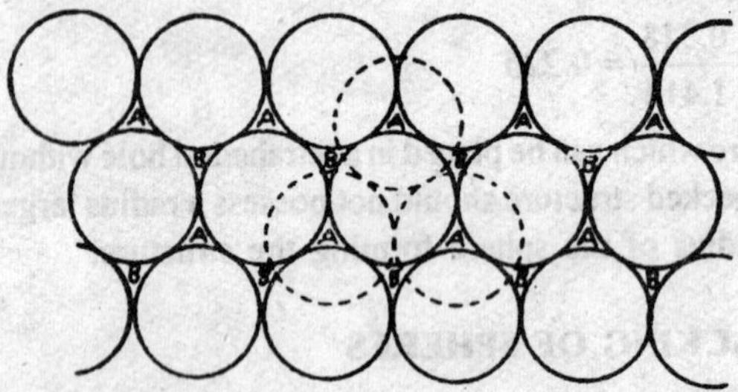

Fig. 1.50

Suppose we want to build up another layer on the lower layer. One method is to fit a sphere above a hollow at the centre of three touching spheres in the first layer. Now there will be four spheres, three of first layer and one of second layer, which will be situated at the four corners of a regular tetrahedron. This type of arrangement is shown in Fig. 1.50, in which the full-line circles are representing the spheres in the first layer whereas the dotted circles are representing those in the second layer. In each layer, the packing closest is possible.

Now for building the third layer, we have a choice of spheres in two different ways. These are :

(i) *Hexagonal Close Packing :* One way is to put the spheres vertically above those in the first layer so that each sphere of the third layer lies strictly above a sphere of the first layer. This arrangement, if continued indefinitely in the same sequence, is represented as ABABABA. This is illustrated in Fig. 151. When this arrangement is examined carefully, it is found to represent *hexagonal close packing (hcp) symmetry which means that* the whole structure possesses one-six fold axis of symmetry. The same appearance is also obtained by rotating the crystal through an angle of 60" (Fig. 1.51). It may be noted that the symmetry axis is perpendicular to the layers of close packed spheres.

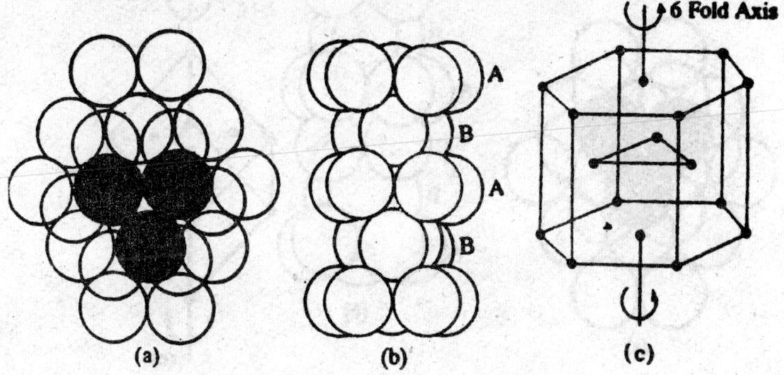

Fig. 1.51

(ii) *Cubic Close Packing* : The second way is to put the spheres on the other set of hollows marked B. In this way that the spheres in the fourth layer will correspond with those in the first layer. This arrangement, if continued indefinitely in the same sequence, is represented as ABC ABC ABC A... This arrangement is shown in Fig. 1.52. This arrangement possesses cubic close packing (ccp) symmetry. This structure has four 3-fold axes of symmetry which are passing through the diagonals of the cube. One of these axes is shown in Fig. 1.52.

If we carefully examine the ABC ABC A...... system of close packing, it is observed that there is a sphere at the centre of each face of the unit cube and hence this structure is also known as face centred cubic structure (Fig. 1.52).

In this arrangement, each sphere is in contact with the six nearest spheres. In addition to this, each sphere is also in contact with three spheres, in layer above and three spheres in the layer below. Thus, in both hcp and ccp arrangements each sphere is surrounded by twelve equidistant spheres. Therefore, the coordination number is 12 in each case.

It can be shown by calculations that in the arrangement shown in Fig. 1.52, only 74 per cent of the total space is occupied by the spheres whereas the remaining 26 percent of the space is empty and is called *void volume.*

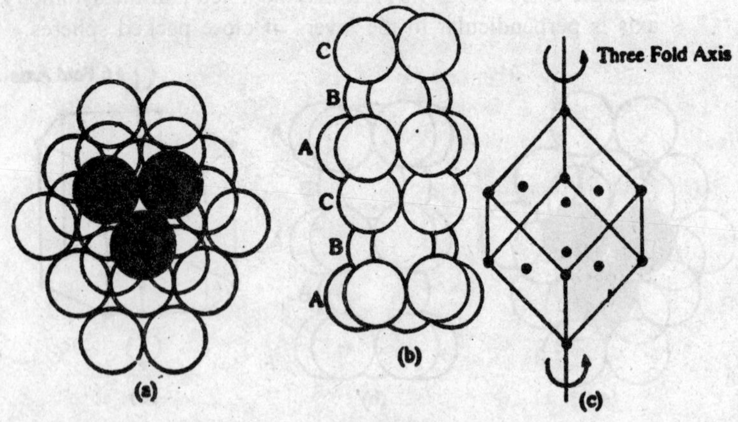

Fig, 1.52

Body Centred Cubic Arrangement : Another possible arrangement of packing of spheres is body centred cubic (bcc) arrangement. This structure will be obtained if the spheres in the first layer (marked A) of cubic closest packing are slightly opened up. Due to this, none of these are in contact with each other. Such an arrangement is shown in Fig. 1.53(a).

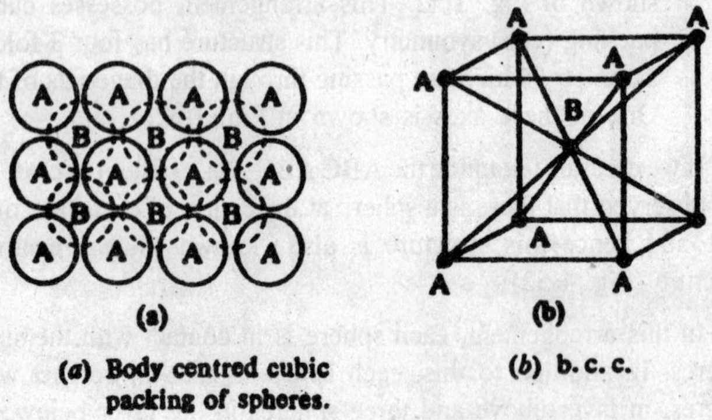

(*a*) Body centred cubic packing of spheres.

(*b*) b. c. c.

Fig. 1.53.

One can build the second layer of the spheres (marked B) on top of the second layer of the spheres (marked B) on top of the first layer in such a way that each sphere of the second layer will be in contact with four spheres of the layer below it. If we do the successive building of the other layers, the third layer will be exactly like the first layer, *i.e.*, on the top of A. If this pattern of building layers is repeated infinitely, an arrangement as shown in Fig. 1.53 (b) is obtained; in this arrangement, there is a sphere at each corner and at the centre of each unit cube. Thus, in this arrangement, each sphere is in contact with eight other spheres (four spheres) in the layer just above and four spheres in the layer just below) so that the coordination number in this type of arrangement is eight.

Most of the metallic elements possess crystal lattice of one or the other of the above three types. Some examples of these with their coordination numbers are given in Table 1.7.

Table 1.7 : Crystal Lattices and Coordination Numbers of the Various Types of Metallic Crystals

Metallic Elements	*Lattice Type*	*Coordination Number*
Be, Mg, Cd, Zn, Ti	Hexagonal close packing (hcp)	12
Cu, Ca, Sr, Ag, Au	Cubic close packing (ccp)	12
Li. Na, K, Rb, Cs	Body centred cubic arrangement	8

INTERNAL STRUCTURE OF CRYSTALS

X-rays can be used to investigate the internal structure of crystals. It was *Laue* (1912) who first suggested the use of X-rays in determining the structure of crystals. His experiment was further developed by *W.L. Bragg* and others. However, the following methods have been developed for the study of X-ray diffraction of crystals.

1. *Laue's Photographic Method :* Laue (1912) showed that if X-rays are allowed to pass through a crystal, a diffraction pattern is obtained which gives the positions of molecules or ions in a crystal. The experimental equipment employed by Laue is relatively simple and shown in Fig. (1.54).

'A' is a source of X-rays which is usually obtained from a tungsten target at about 60,000 volts. This source gives white X-rays, *i.e.*. X-rays beam of continuous range of wavelength. Before passing through the crystal, the X-rays are passed through a slit system (B) to obtain a fine pencil. Then, the X-rays are permitted to pass through a crystal, (C) which is set on a holder to adjust its orientation. The emergent beam from the crystal is made to fall on a photographic plate (S) which is kept at a few centimetres from the crystal. On doing this a Laue photograph is obtained on the photographic plate as shown in Fig. (1.55). The photograph consists of a central-spot which arises due to undeviated beam. The central spot is surrounded by spots which arise from different diffracted beams. From the arrangement of spots on a diffraction pattern, one can get some idea about the symmetry and the position of molecules in the crystal.

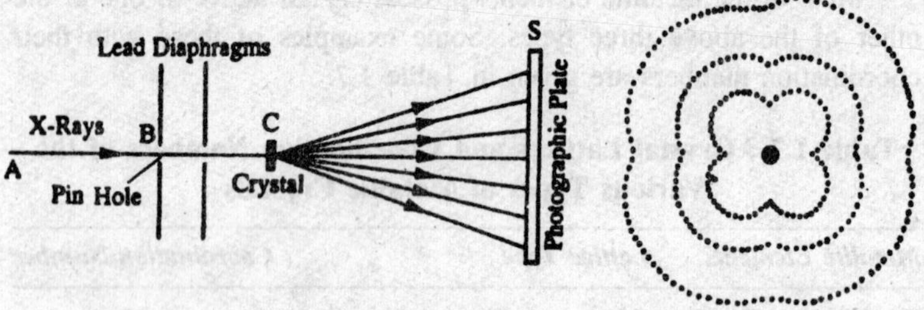

Fig. 1.54 : Diffraction of white X-rays by a single Stationary crystal.

Fig. 1.55.

From Laue's diffraction pattern, one can deduce the geometry of the structural units of the crystals, but the mathematics involved is very complicated. At the same time, there is another disadvantage that this method requires a big single-crystal which is a difficult task to get it.

2. *X-Ray Spectrophotometer Method :* A simple interpretation of Lane diffraction pattern was given by *W.L. Bragg.* In 1915 Bragg devised a simple method for the study of crystals by X-rays. He pointed out that when a beam of X-rays is passed through a crystal, each atom in the path acts as scattering centre and, thus emits secondary radiations. However, reflection of X-rays can take place at certain angles when the path difference of emitted ray from the successive planes is a whole number multiple of the wavelength.

Derivation of Bragg's Equation

It is now easy to derive a condition which should be satisfied so that the rays scattered by the various atoms in the crystal may undergo constructive interference to produce an intense spot on a photographic plate. Consider a set of parallel atomic planes whose spacing is d and let a narrow monochromatic beam of X-rays fall upon these parallel planes at a glancing angle 8, Fig. (1.56). Each atom in a given layer, according to Huygen's principle, becomes the centre of expanding wavelets whose envelope gives rise to reflected wave front. As in the case of optical reflection, the reflected wave front will have maximum intensity at an angle 6 to this plane which is equal to the glancing angle. Each parallel layer in the given set gives rise to the reflected wave front. The condition for constructive interference between the reflected wavefronts is that the path difference between the reflected wavefront from one layer and that from the next must be equal to an exact wavelength or an integral multiple of it.

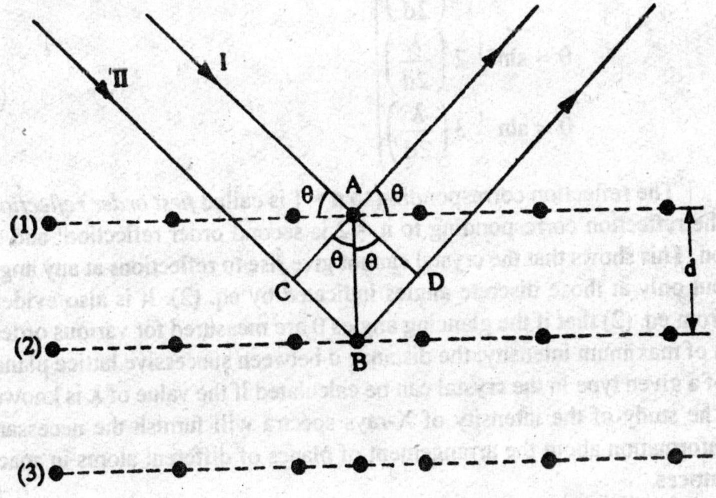

Fig. 1.56 : Reflection of X-rays from lattice planes of crystal.

In Fig. (1.56) two parallel rays I and II are reflected by two atoms A and B in two adjacent layers; B being vertically below A. The ray reflected from B travels a longer distance than that reflected from A. Draw AC and AD perpendiculars to the direction of the incident and the reflected rays. Each of these lines makes an angle 6 with AB whose length is equal to the spacing d. Further, CB = BD = AB sin θ = θ sin θ. The additional path travelled by the ray reflected from the second layer is equal to (CB + BD). Thus, the condition for constructive interference between the two rays is

$$CB + AD = n\lambda$$

or $$2d \sin \theta = n\lambda \qquad ...(1)$$

where n is an integer. This is known as *Bragg's equation* and gives the condition which must be satisfied for the reflection of X-rays from a set of atomic planes.

For a given set of lattice planes, d is a fixed value. If homogeneous X-rays of definite wavelength X are used, then the possibility of getting maximum reflection, depends on θ. If, θ, the angle which the rays make with the plane of the crystal, is regularly increased, a number of positions corresponding ton =1, 2, 3, etc., will be found at which the reflections will be maximum.

$$\left.\begin{aligned} \theta &= \sin^{-1} 1\left(\frac{\lambda}{2d}\right) \\ \theta &= \sin^{-1} 2\left(\frac{\lambda}{2d}\right) \\ \theta &= \sin^{-1} 3\left(\frac{\lambda}{2d}\right) \end{aligned}\right\} \quad ...(2)$$

The reflection corresponding to n = 1 is called *first order reflection*; the reflection corresponding to n = 2 is second order reflection; and so on. This shows that the crystal cannot give rise to reflections at any angle but only at those discrete angles indicated by eq. (2). It is also evident from eq. (2) that if the glancing angles θ are measured for various orders n of maximum intensity, the distance d between successive lattice planes of a given type in the crystal can be calculated if the value of λ is known. The study of the intensity of X-rays spectra will furnish the necessary information about the arrangement of planes of different atoms in space lattices.

Bragg was successful in obtaining the positions of maximum reflection intensity by means of an X-rays spectrometer which is discussed as below :

Construction of Bragg's Spectrometer : In order to measure the glancing angle θ, an X-ray spectrometer, similar in construction to an optical spectrometer, is used. The essential parts of Bragg's X-ray spectrometer are depicted in Fig. 1.57.

(i) X-rays from the anticathode A, limited to a narrow pencil by two adjustable slits B and B', are allowed to fall upon the crystal C which is mounted in wax on the table of spectrometer and its position can be noted from the vernier V which is capable of moving along the circular scale S.

(ii) The reflected rays after passing through the slit F enter the ionisation chamber E through a narrow aluminium window W.

(iii) The chamber is mounted on an arm and its position is noted from the vernier V.

(iv) One of the plates of the ionisation chamber is connected to the positive of a H.T. battery whereas the negative of which is connected to a quadrant electrometer. The other plate has been earthed.

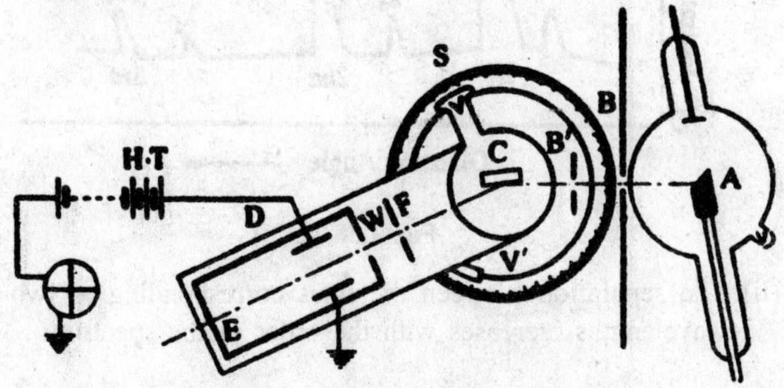

Fig. 1.57.

(v) The deflection in the electrometer gives the intensity of reflected X-rays. To increase the deflection, the chamber may be filled with sulphur dioxide or methyl iodide.

Working

(i) In using the apparatus, the crystal is kept in such a position that $\theta = 0$ and the position of ionisation chamber is adjusted to receive the X-rays.

(ii) The crystal and the ionisation chamber are now allowed to move in small steps so that the angle, through which the chamber has moved, becomes twice the angle, through which the crystal has rotated. The ionisation at first decreases as θ increases but for certain values of θ, it increases sharply. This corresponds to the direction of the X-ray spectrum.

The X-ray spectrum for two wave lengths λ_1 and λ_2 in the first, second and the third order is depicted in Fig. (1.58).

From the above it is evident that :

(i) The intensity of a line corresponding to a wavelength decreases with the order of spectrum, and

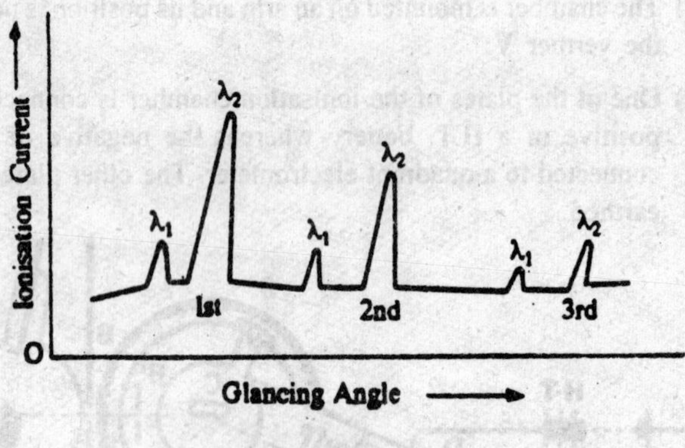

Fig. 1.58

(ii) The separation between the lines corresponding to two near wavelengths increases with the order of the spectrum.

Measurement of Wave Length

The wavelength of X-rays may be determined by employing the equation

$$2d \sin \theta = \lambda$$

The value of θ for various order spectra produced by reflection from a crystal, say of rock salt is measured and the mean value of λ/d is determined. The value of λ/d is termed as lattice constant, *i.e.*,

Lattice constant = λ/d.

Knowing d, the wavelength λ can be calculated. We shall now calculate 'd'

Calculation of 'd'

Let us calculate 'd' in the case of a rock salt (NaCl) crystal in which the atoms are arranged in a cubic lattice. The spacing 'd' is the distance between two adjacent atoms. Each elementary cube of rock salt crystal consists of eight atoms of sodium and four of chlorine. Thus each elementary cube consists of molecules of sodium chloride. But each atom lies at the junction of eight of elementary cubes and is thus shared by each one of them. Hence each cube contains only half a molecule of sodium chloride.

Now molecular weight of NaCl = 23 + 35.5 = 58.5

∴ Mass of one molecule of NaCl = 58.5 g

Since each mole of any substance contains = 6.023×10^{23} molecules

∴ The mass of each molecule of NaCl $= \frac{58.5}{6.023 \times 10^{23}} = 9.71 \times 10^{-23}$ gm/molecule or mass of half a molecule of that an elementary cube of NaCl

$= (9.71 \times 10^{-23})/2 = 4.855 \times 10^{23}$g m.

But the density of NaCl = 2.16 gm/cm^3.

∴ Volume of the elementary cube $\frac{\text{Mass}}{\text{Density}} = \frac{4.885 \times 10^{-23}}{2.16} \text{cm}^3$.

Let d be the spacing between two atoms. The volume of the elementary cube is then also d^3. Hence,

$$d^3 = \frac{4.885 \times 10^{-23}}{2.16} \quad \therefore d = \left(\frac{4.885 \times 10^{-23}}{2.16}\right)^{1/3}$$

or $\quad d = 1.82 \times 10^{-8}$ cm = 2.84 A°.

Thus, knowing θ and d, λ can be calculated.

Determination of Crystal Structure by Bragg's Law

First of all, allow the X-rays to fall on the crystal surface and then rotate the crystal. X-rays are reflected from various lattice planes. The intense reflections are noted with the help of Bragg's X-ray spectrometer. Record the glance angle for each intense reflection. By using the Bragg's equation, ratio of lattice spacing for various groups of planes can be obtained. This ratio comes to be different for different crystals. The experimentally observed ratio is then compared with the calculated ratio, and thus a particular structure may be identified. Some standard ratios are as follows :

(i) $d_{(100)} : d_{(110)} : d_{(111)} : = 1 : \frac{1}{\sqrt{2}} : \frac{1}{\sqrt{3}}$ (simple cubic lattice)

(ii) $d_{(100)} : d_{(110)} : d_{(111)} : = 1 : \frac{1}{\sqrt{2}} : \frac{2}{\sqrt{3}}$ (Face centred cubic lattice)

(iii) $d_{(100)} : d_{(110)} : d_{(111)} : = 1 : \sqrt{2} : \frac{1}{\sqrt{3}}$ (Body centred cubic lattice)

The above method is illustrated in the case of sodium chloride. In this the maximum intensity of reflections occurs at the glancing angles of 5.9°, 8.4°, and 5.20°, for (100), (110) and (111) faces respectively for first order reflections. Thus, in the case of NaCl

$$d_{(100)} : d_{(110)} : d_{(111)} := \frac{1}{\sin 5.9^0} : \frac{1}{\sin 8.4^0} : \frac{1}{\sin 5.2^0}$$

$$d_{(100)} : d_{(110)} : d_{(111)} :=$$

$$\frac{1}{0.103} : \frac{1}{0.146} : \frac{1}{0.091} = 1 : 0.704 : 1.155 = 1 : \frac{1}{\sqrt{2}} : \frac{2}{\sqrt{3}}$$

The above values are in full agreement with the values predicted theoretically. Thus, we can safely say that the crystal of NaCl has a face-centred cubic lattice.

3. Rotating Crystal Method

This is the most widely used technique for elucidating the precise structure of crystal. This method was developed by *Schiebold* in 1919. This arrangement of this method is outlined below :

(a) The X-rays are generated in the X-ray tune and then the beam is made monochromatic by a filter.

(b) The beam is then allowed to pass through a collimating system which permits a fine pencil or parallel X-rays.

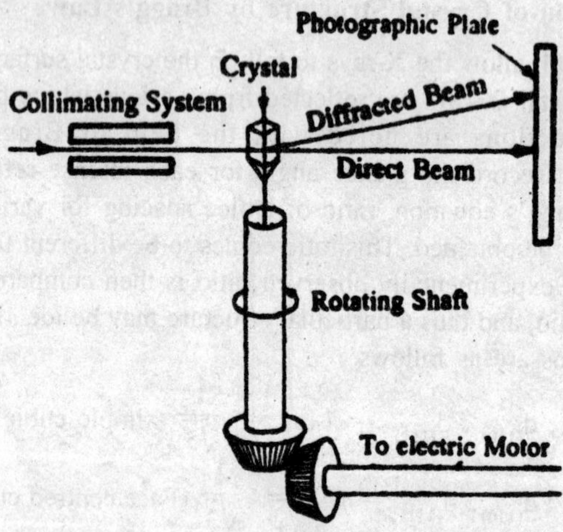

Fig. 1.59.

(c) Allow the X-ray beam to fall on a crystal mounted on a haft which can be rotated at a uniform angular rate by a small motor.

(d) Allow the shaft to move. This sets the crystal into slow rotating about an axis. This causes the set of planes coming successively into their reflecting positions, *i.e.*, the value of θ satisfies the Bragg's equation. Each plane will produce a spot on the photographic plate.

(e) The two types of photographs may be obtained :

(i) Complete rotation method : In this method there occurs a series of complete revolutions. It is observed that each set of planes in the crystal diffracts four times during the rotation. These four diffracted beams are distributed into a rectangular pattern about the centre point of the photograph.

(ii) Oscillation photograph method : In this method the crystal is oscillated through an angle or 15° or 20°. The photographic plate is also moved back and forth with the same period as that of the rotation of the crystal. The position of a spot on the plate indicates the orientation of the crystal at which the spot is formed. It is observed that the oscillation photograph does not show the symmetry whereas it is shown by the complete rotation photograph. The limited range in oscillation photograph reduces the possibility of overlapping reflections.

Theory

Let us now discuss the theory of the rotation photograph. Suppose the crystal to be rotated in the X-ray beams is mounted in such a way that one of the crystallographic axes gets coincided with the axis of rotation. Suppose this axis is z-axis.

Along the z-axis consider two neighbouring atoms A and B at lattice points. The distance between these atoms will be the promotive translation c (Fig. 1.60). In such a situation, the condition of diffraction is represented by

$$c\ (\cos\delta + \cos\phi)\ n\lambda \qquad ...(1)$$

where n denotes an integer, indicating the order of diffraction by the line grating. If the crystallographic axis is perpendicular to the rotation axis, $\phi = 90°$. Thus, the equation (1) becomes as

$$c \cos \delta = c \sin \mu = \lambda \quad ...(2)$$

For various orders of diffraction, n takes values such as 0, 1, 2, 3, 4, ... n, yielding the following series of equations :

$$\left.\begin{aligned} \cos \delta_o &= \sin \mu_o = 0 \\ \cos \delta_1 &= \sin \mu_1 = \lambda / c \\ \cos \delta_2 &= \sin \mu_2 = \frac{2\lambda}{c} \\ &\dots\dots\dots\dots\dots\dots \\ &\dots\dots\dots\dots\dots\dots \\ \cos \delta_n &= \sin \mu_n = \frac{n\lambda}{c} \end{aligned}\right] \quad ...(3)$$

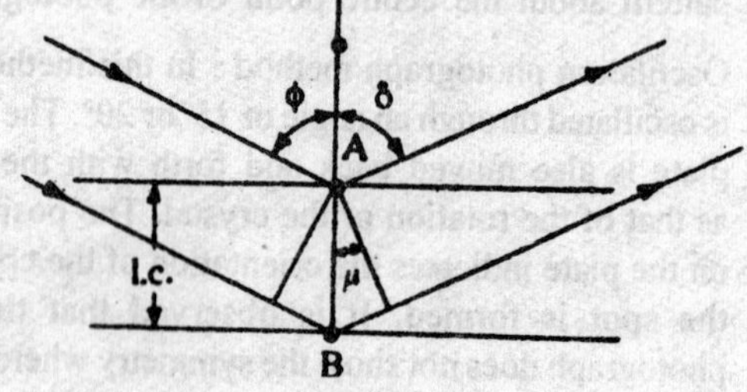

Fig. 1.60.

If λ is constant, equation (3) gives the loci of all possible diffracted rays as the crystal rotates, These loci are elements of a series of cones which are shown in Fig. 1.61. In this figure, the half apex angles are denoted by δ_o, δ_1, δ_2, ... etc. Any element of each of the cones makes angles μ_o, μ_o, μ_o, etc. respectively with the horizontal plane. The central horizontal plane includes such diffracted beams which are of order of zero. The miller indices of all planes yielding diffracted beams in horizontal plane should be represented by (h, k, o). Similarly, X-rays diffracted by planes of indices (h, k, l) lie on the first order cone which will be defined by

$$\mu_o = \sin^{-1} \lambda/c \quad ...(4)$$

It means that all the beams diffracted from planes of indices (h, k, n) will lie on the nth order cone which may be defined by

$$\mu_o = \sin^{-1} n\lambda/c \quad ...(5)$$

In Fig. (1.61), the cones are intersecting the cylindrical film in a series of circles lying in planes perpendicular to the axis of rotation. These will become a series of parallel straight line, when the film is unrolled. These lines are known ;as layer lines. The cone for n = 0 is a plane which is perpendicular to the axis and is having the direction of incidence, and the intersection of this with the film is known at the first second, third, etc. layer lines. In order to determine the dimension of unit cell, the following procedure is opted :

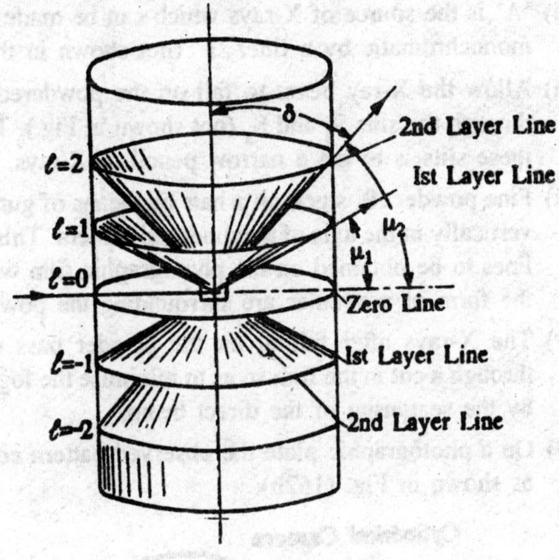

Fig. 8.

The distance between layer lines gives lattice translation. If one knows the distance of the film from the crystal, the distance of the layer lines from the equatorial lines will give the values of δ and as λ is also known, the value of c can be calculated by using equations (3) and (5). Similar to the determination of c, the other translations a and b can be calculated. The values of a, b and c give the dimensions of unit cell of the structure.

From above, it follows that the rotating crystal method is a very powerful method which gives the size of unit cell.

(III) Powder Crystal Method

In all the above described methods, a single crystal is required whose size is ;much larger than microscopic dimensions. However, in the powder method, the crystal sample need not be taken in large quantity but as little as 1 mg of the material is sufficient for the study. The powder method was devised independently by Debye and *Scherrer* in Germany and by *Hull* in America about the same time e (1916).

Arrangement

The experimental arrangement of this method is shown in the Fig. (1.62a). Its main features are outlined below :

(i) 'A' is the source of X-rays which can be made approximately monochromatic by a filter 'F' (not shown in the Fig.).

(ii) Allow the X-ray beam to fall on the powdered specimen 'P' through the slits S_1 and S_2 (not shown in Fig.). The function of these slits is to get a narrow pencil of X-rays.

(iii) Fine powder 'P' stuck on a hair by means of gum is suspended vertically in the axis of a cylindrical camera. This enables sharp lines to be obtained on the photographic film which is bent in the form of a circular are surrounding the powder crystal.

(iv) The X-rays after falling on the powder pass out of camera through a cut in the film so as to minimise the fogging produced by the scattering of the direct beam.

(v) On a photographic plate the observed pattern consist of traces as shown in Fig. (162b).

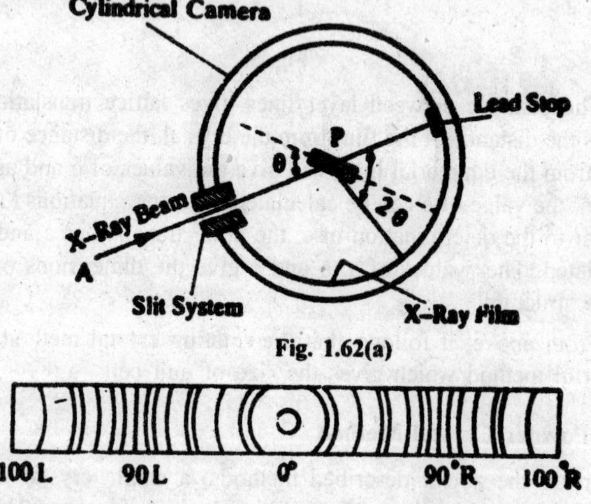

Fig. 1.62(a)

Fig. 1.62(b). Arrangement of lines in a powder photograph.

Theory and Calculations

When a monochromatic beam of X-rays is allowed to fall on the powder of a crystal, then

(i) There will be some particles out of the random orientation of small crystals in the fine powder, which lie with a given set of lattice planes (making the correct angle with the incident beam) fro reflection to occur.

(ii) While another fraction of the grains will have another set of planes in the correct position for the reflection to occur and so on.

(iii) Also, reflections are possible in the different orders for each set.

All the like orientations of the grains due to reflection fro each set of planes and for each order will constitute a diffraction cone whose intersection with a photographic plate gives rise to a trace as shown in the Fig. (1.62b).

The crystal structure can be obtained from the arrangement of the traces and their relative intensities. We will now proceed to do this.

If the angle of incidence is θ, then the angle of reflection will be 2θ as shown in Fig. (1.63).

If the film radius is 'r', the circumference 2πr corresponds to a scattering angle of 360°.

Then, we can write

$$\frac{1}{2\pi r} = \frac{2\theta}{360^\circ} \qquad \text{or} \qquad \theta = 360^\circ \times \frac{1}{4\pi r}$$

From the above equation, the value of θ can be calculated and substituting this value in the Bragg's equation, the value of d can be calculated.

$$n\lambda = 2d \sin \theta$$

It is important to remark here that in order to index reflections, one must known the crystal system to which the specimen belongs. This is usually done by microscopic examination.

Applications

(i) This method is most useful for cubic crystals.

(ii) It is also used for determining the complex structure of metals and alloys. However, their structures could not be revealed by the earlier studies.

(iii) The method also helps to distinguish between the allotropic modification of the same substance

(iv) The structure of rubber was fully revealed by this method. Rubber has been seen to crystallise on stretching and decrystallise on slakening.

STRUCTURE OF ROCK SALT, NACl

The structure of sodium chloride (Fig. 1.64) was studied by the Bragg's spectrometer. The intensity of the ionisation current was measured for different angles of rotation of the crystal. Then, a curve was plotted between the current intensity and glancing angles as shown in Fig. 1 for (100) and (110) and (111) faces.

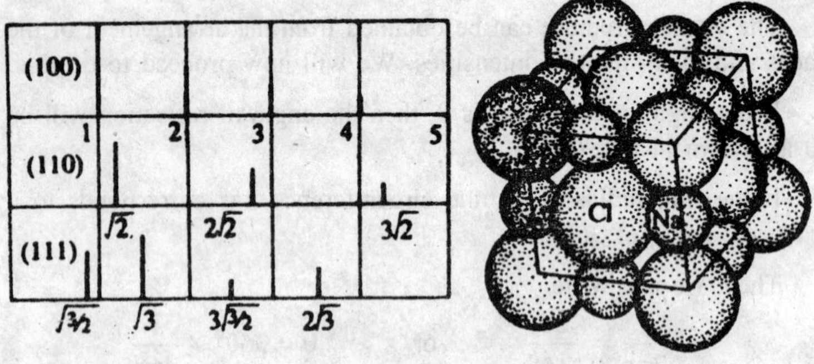

Fig. 1.63 Fig. 1.64

From the graph it is evident that,

(i) The intensities of current fall of regularly for the faces (100) and (110), and

(ii) There is a difference for the (111) face. In case of (111) face, the first order spectrum is abnormally weak, the second as abnormally strong, the third as abnormally weak and fourth as abnormally strong.

It was proved that the (100) face of a sodium chloride crystal yielded reflection maxima at glancing angles of 5.9°, 11.85°, and 18.15°. If we say that these are representing first, second, and third order reflections, respectively, then their sines must be in the ratio of 1 : 2 : 3. These are in good agreement with our expectations. Let us now consider the Bragg's equation,

$$2d \sin \theta = n\lambda$$

or
$$d = n\lambda/2) \sin\theta \qquad ...(1)$$

Therefore, for a particular order or reflection,

$$d \,\alpha\, \frac{1}{\sin \theta} \qquad ...(2)$$

The first order reflections are found to have the glancing angles as 5.9°,

8.4°, and 5.2° for the (100), (110) and (111) planes of sodium chloride. Applying equation 2), we have

$$d_{(100)} : d_{(110)} : d_{(111)} : = \frac{1}{\sin 5.9^o} : \frac{1}{\sin 8.4^o} : \frac{1}{\sin 5.2^o}$$

$$= \frac{1}{0.103} : \frac{1}{0.146} : \frac{1}{0.091} = 1 : 0.704 : 1.155$$

As we know that for a face centred cubic lattice

$$d_{(100)} : d_{(110)} : d_{(111)} = 1 : 0.704 : 1.154$$

These ratios are same as obtained for the unit cell of a face centred cubic lattice. Thus, the structure of NaCl is face centred lattice.

Now the problem is to decide about the structural units of the crystals. These units may be molecules, atoms of ions. This problem is solved by considering the plane (111).

In planes (111), the first order spectrum is abnormally weak, the second abnormally strong, the third abnormally weak and fourth abnormally strong. The alternation of these intensities are due to alternate layers or planes of chlorine atoms separated by layers of sodium atoms. Also, it is evident from the alternations in the intensities that the real unit of matter in crystal lattice is the atoms and not the ions.

Bragg gave the structure of NaCl as shown in Fig. (1.64). From the structure, it follows that :

(i) The rock salt crystals have face-centred cubic lattices with sodium and chlorine atoms arranged alternately in all directions parallel to the three rectangular axes, and

(ii) each unit cell of sodium chloride consists of 14 sodium atoms and 13 chlorine atoms, and

(iii) each chlorine atoms is surrounded by six sodium atoms and each sodium atoms is surrounded by six chlorine atoms.

STRUCTURE OF SYLVINE (KCl)

The study of structure of Sylvine was done by Bragg's X-ray spectrometer. The intensities of ionisation current were determined for glancing angles. By plotting the curve between the current intensity and glancing angles, two types of the curves are observed as shown in Fig. (1.1) for (100), (110) and (111) faces. The first order spectrum from the (100), (110) and (111) planes of KCl was observed at the glancing angles 5.38°, 7.61° and 9.38° respectively. Therefore,

$$d_{100} : d_{110} : d_{111} = \frac{1}{\sin 5.38^\circ} : \frac{1}{\sin 7.61^\circ} : \frac{1}{\sin 9.38^\circ}$$

$$= \frac{1}{0.0938} : \frac{1}{0.1326} : \frac{1}{0.1630} = 1 : 0.704 : 0.575$$

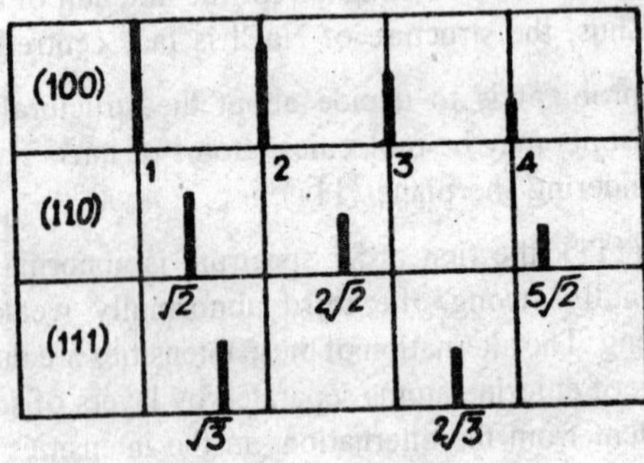

Fig. 1.65

But we know for a simple cubic lattice,

$$d_{100} : d_{110} : d_{111} : 0.707\ .0.575.$$

It, therefore, follows that KCl crystal has a simple cubic lattice whereas NaCl has a face-centred cubic lattice.

Expected similarity between NaCl and KCl: In almost every property, KCl resembles

NaCl : suggesting that the fundamental lattice should be the same, *i.e.*, the face centred lattice.

The expected behaviour is supported by the following observations :

The first order reflection from (100) planes of the rock salt and KCl occurred at 5.9° and 5.3° respectively. As we know that the value of d is inversely proportional to sin 6, it means,

$$\frac{d_{100}\ \text{KCl}}{d_{110\ \text{NaCl}}} = \frac{\sin 5.9^0}{\sin 5.3^0} = 1.11$$

Since the volume, of a small cubic lattice is $(d_{111})^3$. it follows that if the potassium and sodium chlorides have the same crystal lattice structure, the quantity $(1.11)^3 = 1.37$ should give the ratio of molecular volumes of the two salts but the experimental value is 1.38 which suggests that the two must have same space lattice, *i.e.*, face-centred cubic lattice. But the Bragg's study indicated that KCl should have a simple cubic lattice. Thus, there exists some anomaly.

How to solve the above anomaly?

The above anomaly can be explained on the basis that,

"the X-ray scattering factor for an atom is equal to the number of extraplanetary electrons. (viz., atomic number) at small grancing angles."

The atomic number of potassium and chlorine are 19 and 17 respectively. As these numbers are not very much different, the X-rays are unable to detect any difference between the two kinds of atoms. If we assume that all the atoms are identical, it means that the face- centred arrangement of NaCl (Fig. 1.65) becomes a simple cubic arrangement for KCl.

In (111) face of KCl, the atomic numbers of K and Cl are so identical-that the reflected intensities are nearly same and the odd orders are completely cut off.

Due to the above mentioned facts, KCl appears to be a simple cubic lattice. On the other hand, in the case of NaCl there is difference between the atomic numbers of sodium (11) and chlorine (17). It means that their scattering factors are different and hence, give rise to a face-centred cubic lattice. From the above discussion, we can safely claim that

"Both the rock salt (NaCl) and (KCl) have the same structure."

Although, the structures of NaCl and KCl are same, there is an important difference which may be observed by applying the expression (1) to both the crystals. Therefore,

$$\frac{d_{100}\ \text{KCl}}{d_{110\ \text{NaCl}}} = \frac{\sin 5.3^{o}}{\sin 5.9^{0}} = \frac{1}{1.11}$$

From the expression, it is evident that the fundamental units are more widely spaced in the crystal of KCl than in the crystal of NaCl.

STRUCTURE OF ZINC BLENDE

This belongs to the cubic system. The crystal of zinc blende is built up of the face-centred lattices as shown by the relative positions of the first order spectra from the (100), (110) and (111) faces, shown in Fig. (1.66); the angles 26 and intensities are indicated by the positions and heights of the vertical lines. From this spectra, the following important conclusions could be drawn.

(i) The (100) planes yield a very weak first order spectrum. This indicates equidistant alternating (100) planes. This is similar to the like (111) planes of sodium chloride.

(ii) For the (111) planes, the second order is abnormally weak. This indicates that the distance between the successive zinc layers is four times that between a zinc layer and the next sulphur layer.

From observations (i) and (ii), it is evident that the structure of zinc blende will have alternate planes of zinc alone and of sulphur alone.

(iii) The intensities of spectra due to (110) planes are normal; in that the intensity of spectrum decreases regularly as the order increases. It means that zinc blende structure must contain equal number of zinc and sulphur atoms; the sulphur atoms being placed centrally inside alternate cubic spaces.

The structure showing these requirements is shown in Fig. (1.67). From the structure it follows that

(i) Both zinc and sulphur atoms lie on the face-centred cubic, lattice, [for zinc it is shown in Fig. (1.67) while that of sulphur it is not shown in Fig. (1.67)].

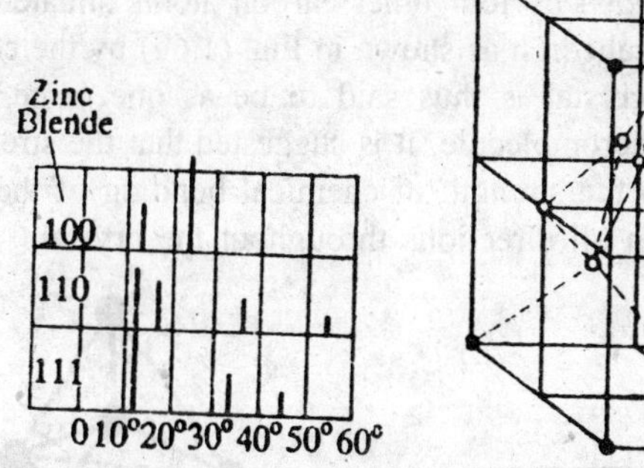

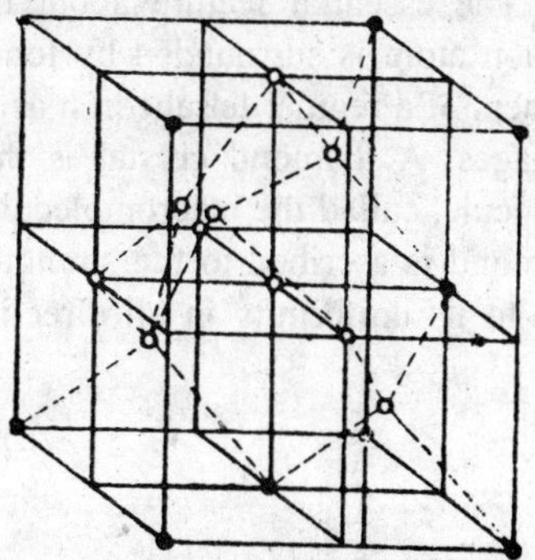

Fig. 1.66 : (Zinc Blende). **Fig. 1.67.**

(ii) The sulphur atoms are lying at the centres of alternate corners of a regular

(iii) Each zinc atom is surrounded tetrahedrally by four sulphur atoms.

(iv) From Fig. (1.67), it is also evident that alternate planes (100) consist of zinc or sulphur points only and also the planes of the latter lie midway between two of the former.

(v) Successive planes (110) contain both zinc and sulphur atoms.

(vi) The planes (111) show alternation; the separation between successive zinc and sulphur atoms being one quarter of the separation between two zinc planes.

STRUCTURE OF DIAMOND

It was one of the earliest crystals to be investigated by X-ray by W.H. Bragg and W.L. Bragg in 1913. It has the same space lattice as that of zinc sulphide, as shown in Fig. 1.68 with some differences given as below :

(i) In diamond carbon atoms take the place of zinc and sulphur atoms in zinc sulphide. This has been confirmed from the fact that from (100) face no first or third order spectra are obtained at all where as no second order spectrum is obtained from (111) face.

(ii) Diamond is considered to be built up of two intersecting or interpenetrating face centred lattices related to one another, in the same manner as zinc and sulphur lattices in zinc sulphide.

The essential feature about the diamond structure is that every carbon atom is surrounded by four other carbon atoms situated at the corners of a regular tetrahedron as shown in Fig. (1.69) by the covalent linkages. A diamond crystal is thus said to be as one large carbon molecule, called the macromolecule. It is suggested that the strength of diamond is ascribed to the strength of chemical bonding of the atoms and to its uniformity in all directions throughout the crystal.

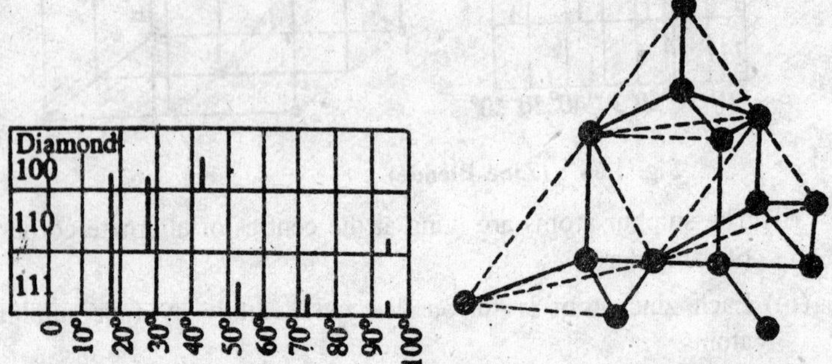

Fig. 1.68 : Diamond. **Fig. 1.69 : Structure of Diamond.**

The determination of the structure of diamond confirmed the correctness of the chemical theorem that the normal valencies of carbon atom are directed from the centre to the corners of a regular tetrahedron.

The procedure in case of diamond was very ingenious. X-rays from rhodium was used.

This gives A = 0.607×10^{-8} cm. The sine of the angle from (111) face of diamond was 0.1456 for first order; From Bragg's equation,

$$\lambda = 2d \sin \theta,$$

$$0.67 \times 10^{-8} = 2d \times 0.1495$$

or $$d = 2.032 \times 10^{-8}$$

If a face centered lattice is assumed, the number of carbon atoms in an elementary cube of side la was calculated by putting 4 C atom in the cube and density of diamond as 3.51.

$$4 \times \underbrace{12 \times 1.64 \times 10^{-24}}_{\text{mass of H atom}} = 3.51\lambda\ (2a)^3$$

$$2a = \sqrt[3]{(22.4 \times 10^{-24})} = 2.82 \times 10^{-8} \text{ cm.}$$

and the distance between consecutive (111) planes is $2a/\sqrt{3} = 1.63 \times 10^{-8}$ instead of 2.03×10^{-8} found. These two numbers are very nearly in the ratio of $1 : \sqrt[3]{2}$. It is clear that we must put 8, not 4 carbon atoms in the elementary cube. Then

$$\frac{2a}{\sqrt{3}} = 2.05 \times 10^{-8}$$

or $$2a = 3.55 \times 10^{-8}$$

We have, therefore, 4 C atoms which are to assign to the elementary cube in such a way as not to interfere with the characteristic of the face centred lattice.

The absence of a second order spectrum suggests that in addition to plane spaced 2.03×10^{-8} cm apart, there are other planes dividing the spacings of the first set in the ratio 1 : 3. The four extra C atoms should go into places to be found in the cubes in multiples of 4 and the simplest planes to put them to the centres of the eight similar cubes into which the main cube can be divided This gives the arrangement shown in Fig. 1.70.

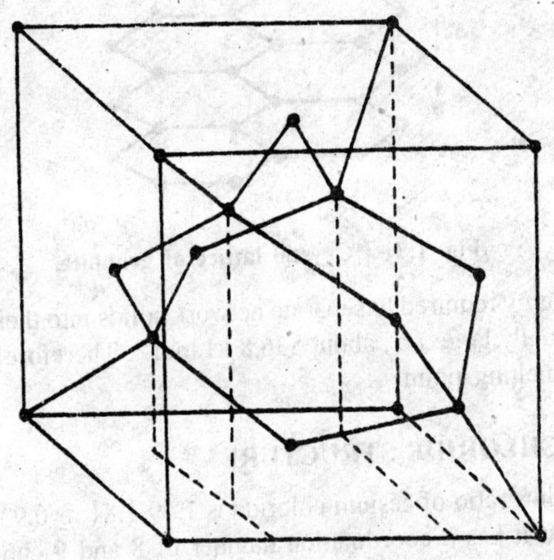

Fig. 1.70 : The diamond structure.

The crystal is therefore a three dimensional molecule without any limit on its extension giving rise to a giant molecule, a polymer of carbon. The crystal is a fee one where alternate four out of eight tetrahedral holes are occupied by carbon atoms. The C-C bond distance is 0.154nm.

THE GRAPHITE CRYSTAL

It is an entirely different type of crystal with atoms held in position by van der Waals forces of attraction. The structure is shown in the Fig. 1.71. In this structure carbon atoms lie in a series of parallel sheets in which they are arranged at the corners of plane regular hexagons. The distance between the atoms is 0.142 nm. The distance between the adjacent layers is 0.335 nm. Weak binding between the layers results the graphite crystals slippery and flaky confering on graphite its valuable lubricating property.

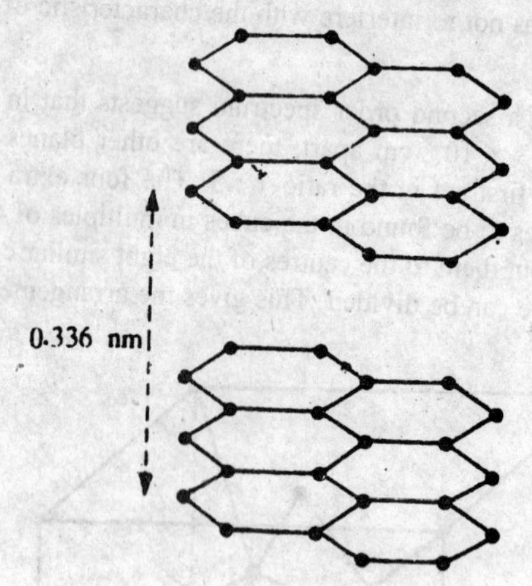

Fig. 1.71 : Crystal lattice of graphite.

The energy required to separate network solids into their constituent atoms is usually large *i.e.*, about 836.8 kJ mol^{-1}. Therefore graphite has very high melting point.

CESIUM CHLORIDE STRUCTURE

The radius ratio of cesium chloride is 1.69/1.81 or 0.93. This value suggests that it has a coordination number of 8 and 9 cubic structure. The Cl^- ions form the simple cubic arrangement while cesium ions occupy the cubic interstitial sites, *i.e.*, each cesium ion is having eight chloride ions as its nearest neighbours. From Fig. 1.72, it can be observed that each chloride ion is also surrounded by eight cesium ions which are also disposing towards the corners of, a cube. Thus, both types of ions in cesium chloride are in equivalent positions and the stoichiometry is 1 : I and the coordination number is 8 : 8.

As the coordination number of CsCl is higher than that of NaCl, it is expected that cesium chloride is more stable than sodium chloride because each cesium ion is having more ions of opposite charge as its neighbours. Actually, cesium chloride lattice has been found to be one per cent more stable than sodium chloride lattice.

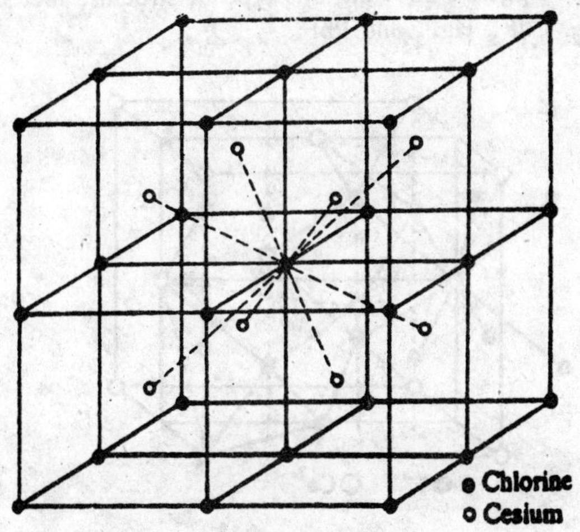

Fig. 1.72 : Body centred cubic structure of Old.

Bromides and iodides of cesium are having the same coordination number and the same structure as that of cesium chloride.

STRUCTURES OF IONIC SOLIDS OF GENERAL FORMULA, AX_2

Examples of these are calcium fluoride, titanium dioxide, calcium carbide and titanium dioxide. We shall discuss these one by one.

Calcium Fluoride Structure

This structure is shown in Fig. 1.73. The fluoride structure is found when the radius ratio is 0.73 or above.

In the fluoride structure, each Ca^{2+} ion is surrounded by eight F^- ions at the corners of a body centred cubic arrangement while each

F^- ion is surrounded by four Ca^{2+} ions arranged tetrahedrally. Thus, the coordination numbers of Ca^{2+} and F^- are 8 and 4. Such an arrangement is known as 8 : 4 arrangement. Calcium ions in fluoride structure are arranged in face-centred cubic close packed (ccp) type of arrangement. As there are two tetrahedral sites available for every calcium ion, the fluoride ions occupy all the tetrahedral sites Thus, the stoichiometry of the compound is 1 : 2.

Others compounds showing this type of structure are SrF_2. BaF_2. $BaCl_2$. $SrCl_2$. CdF_2, HgF_2 and PbF_2.

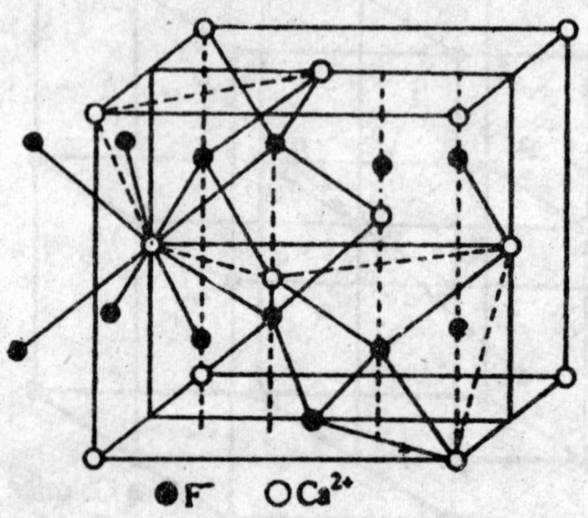

Fig. 1.73 : Body-centred cubic structure of calcium fluoride structure.

Structure of titanium dioxide, TiO_2 (Rutile structure). The rutile structure is shown in Fig. 1.74. In this structure, each Ti^{4+} ion is surrounded by six O^{2-} ions arranged octahedrally and each O^{2-} ion is surrounded by three Ti^{4+} ions arranged in a plane triangular manner. Thus, the coordination numbers of Ti^{4+} and O^{2-} are 6 and 3.

The structure of titanium dioxide cannot be considered to be cubic as one of the axes if shorter than the other by 30 per cent.

Structure of Calcium Carbide

This structure is shown in Fig. 1.75. This structure is similar to that of sodium chloride in which sodium ions are replaced by calcium ions while chloride ions by carbide ions. The carbon atoms in carbide are associated in pairs which are aligned in parallels.

Unlike sodium chloride, the cubic symmetry is distorted in calcium carbide.

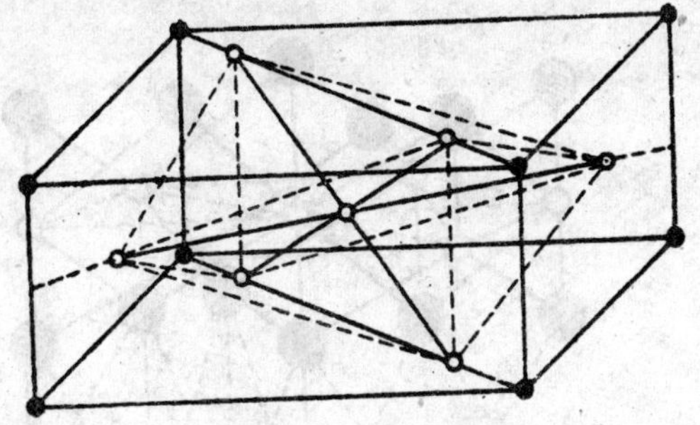

Fig. 1.74 : Structure of titanium dioxide (Rutile).

The structure of FeS_2 has been found to be similar to that of CaC_2. However, unlike CaC_2, the S_2 units in FeS_2, do not align in parallels.

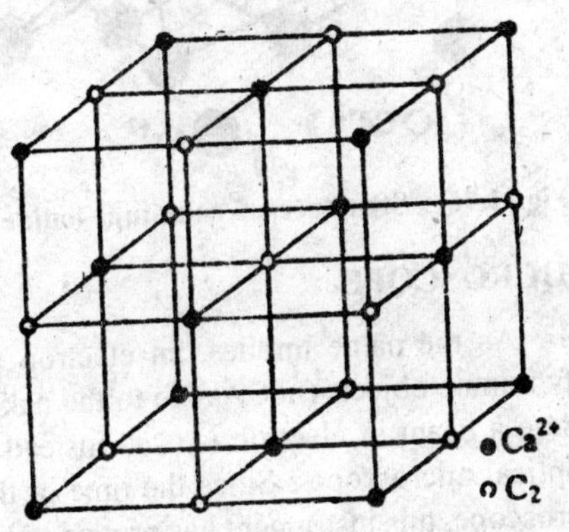

Fig. 1.75 : Structure of calcium carbide.

Structure of Cadmium Iodide

This structure crystallises in so called layer structure shown in Fig. 1.76. In this structure each cadmium atom is surrounded octahedrally by six iodine atoms while three cadmium atoms nearest to each iodine atoms are at the corners of a pyramid of which the iodine atom is the apex. The atoms thus form layers, and each sandwich consisting of a layer of cadmium atoms with layers on either side of it, is electrically neutral. As there are only van der Waal's forces between adjacent composite layers, crystals of this kind show a pronounced cleavage parallel to the layers. The unsymmetrical environment of the iodine atoms reveals that this structure is not a purely ionic one.

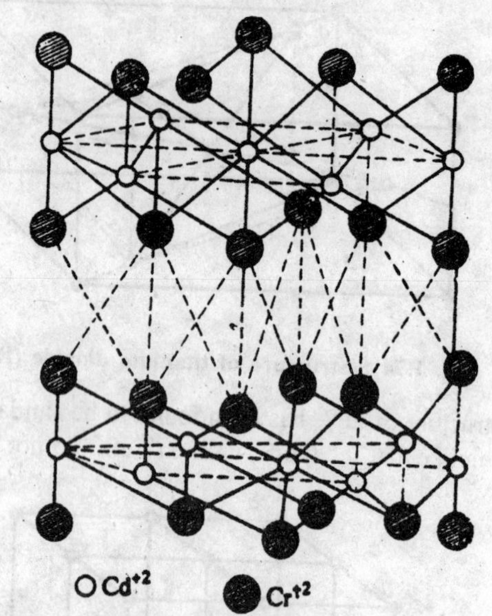

Fig. 1.76 : Structure of cadmium iodide.

ELECTRON MICROSCOPE

Introduction : As the name implies, an electron microscope is a device to magnify minute objects (not visible to the naked eye and light microscopes) where a beam of electron is used instead of light rays as in the ordinary optical microscopes. Since the time of the design of the first electron microscope, this instrument has proved to be of great value both in science and industry.

Principle : The working and construction of an electron microscope are based on the following two principles:

(i) Particles such as electrons possess a wave nature similar to light but of much shorter wavelength.

(ii) Similar to light rays, electrons can be focussed by suitable electric and magnetic fields.

In order for an electron microscope to have useful magnification two conditions must be satisfied, *i.e.*, high resolution for the details and perceptibility of the resolved details. It means that higher the resolving power of the instrument with a simultaneous greater perceptibility, the greater will be its magnifying power. If the perceptibility is decreasing with increase in resolution, the image starts blurring as its size increases. Thus, no useful purpose is served.

Types of Electron Microscopes

The various types of electron microscopes are as follows :

(i) magnetic electron microscope,

(ii) electrostatic electron microscope,

(iii) point projection electron microscope,

(iv) X-ray shadow electron microscope,

(v) emission electron microscope, and

(vi) proton electron microscope. We will discuss the first two types :

1. *Magnetic Electron Microscope* : The first magnetic electron microscope was devised by two German Scientists Knoll and Ruska (1931). It is shown in Fig. 1.77. Its various parts and their functions are summarised below :

(i) G is an electronic gun maintained at a potential difference of 50,000 volts. It emits a beam of electrons.

(ii) G is the first magnetic lens which is formed by a coil enclosed in an iron shield with a centred ring gap to regulate the convergence of electrons on the object O.

(iii) O is an object which is to be magnified. The object is generally of the order of 10 milli-microns thickness.

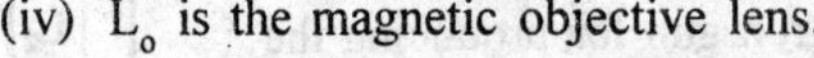

(iv) L_o is the magnetic objective lens.

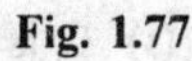

Fig. 1.77

(v) L_p is a projector lens where a second enlarged image of the portion of the first image is produced.

(vi) S is a fluorescent screen on which the image is made visible by scintillation for direct view or on a photographic plate for permanent screen.

(vii) The whole system is enclosed in a metallic case which is evacuated by high speed pumps.

Working : The electrons obtained from an electron gun G are allowed to pass through magnetic lens Lc which acts as a condenser to concentrate the electron beam. The converged beam from Lc is permitted to fall on the object O to be magnified. Then the scattered electrons pass through the magnetic objective lens Lo which forms the first magnified image which is arranged to be focussed just above the focal plane of the projection lens. The beam from the first image traverses a projector lens S_p where a second enlarged image of the portion of the first image is produced. The final image I_2 is to be projected on to a fluorescent screen S, where it is made visible by scintillations for direct view or on a photographic plate for permanent record.

Uses : Knoll and Ruska obtained a magnification as high as 10,000 with a resolving power of 500Å when the working voltage was 50,000 volts. Later Messers Metropolitan Vickers (1936–1937) constructed a modified version of magnetic electron microscope, entitled EM2 type. This microscope has a magnification of 1,00,000 with a resolving power of 100°. The photographic image I_2 can be magnified upto 5 times so that an overall magnification of 500,000 is possible. Later various versions were put in the market which have much higher magnification powers and are easy to handle.

2. *Electrostatic Electron Microscope :* The principle and construction of this type of electron microscope is same as that of magnetic electron microscope. The only difference is that magnetic lenses have been replaced completely or partly by electrostatic electron lenses. An electrostatic electron beam is schematically shown in Fig. (1.78). Its various constructional parts and their functions are summarised below :

(i) C is an electron gun for producing the electron beam. It is same as in magnetic electron microscope. It contains a hot wire tungsten filament (cathode), anode and grid. The grid is operated at a potential of 200 volts, negative with respect to the cathode

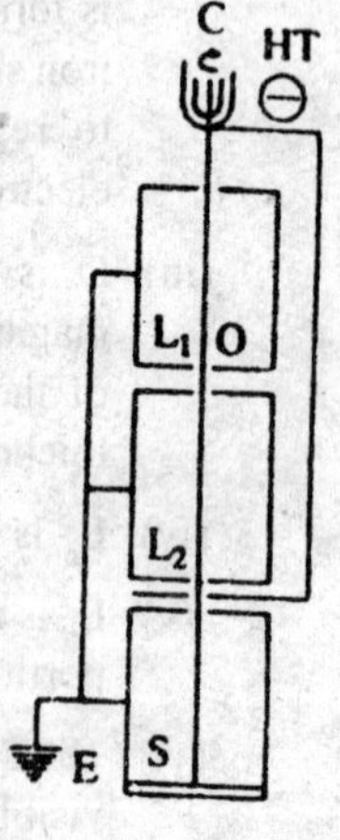

Fig. 1.78

(ii) L_1 and L_2 are objective and projection lenses respectively. These lenses are of disc type with minute aperture, instead of tube type. In order to prevent electrical discharges, the edges of the lenses are rounded and polished. The central elements of both lenses are connected to the cathode and to the same source of high negative voltage. The central elements of both lenses are connected to the cathode and hence to the same source of high negative voltage.

(iii) The anode of the electron gun and the outer elements of the lenses are all connected to the body of instrument and earthed.

(iv) S is a fluorescent screen or a photographic plate.

Working : An electron beam from the electron gun is made to pass through the lens L, which is refracted by the object O, so as to form the intermediate image. The central portion of this image after traversing through the lens is selected by an aperture and magnified once more, so that a final image is formed on a fluorescent screen or a photographic plate S by a projector lens Fig. (1.78).

Advcntages : The electrostatic electron microscope is better than the magnetic electron microscope because it can be operated very easily.

Disadvantages : Although, an electrostatic microscope is more simple than the magnetic electron microscope, it has the following disadvantages.

(i) The chromatic aberration and spherical aberration of the lenses employed in electrostatic microscopic are larger than those of the magnetic lenses.

(ii) For the same voltage, the focal lengths of electrostatic lenses can be made as short as of magnetic lenses.

(iii) In electrostatic type, there is a necessity of introducing high voltages at several points on the microscope.

From the above defects it may be concluded that the magnetic electron microscope remains as a standard instrument for working procedure due to its case of focussing and ready variation on its magnification.

Applications of Electron Microscope

This instrument can be used to make object visible which are too small to be observed by any other available instrument. For this reason, it is used for the following observations :

(i) It is used for the studying of the structures of bacteria, viruses, organic tissues, etc., because these are beyond the range of an ordinary microscope.

(ii) It is used to study the structures of textile fibres synthetic chemicals, composition of paper etc.

(iii) It is used to determine the shape and size of sub-microscopic particles which are generally present in paints, in smokes, in colloidal solutions and in the atmosphere. We need their sizes for controlling the quality of a product in the industries which involve submicroscopic particles.

(iv) It has been used to elucidate the velocity distribution of the transmitted electrons which depends upon the nature of the toms in the object.

(v) It is used for determining the size of particles in colloidal solutions and insecticides which are too small to the visible by an ordinary microscope.

(vi) The photographs taken by electron microscope yield great information about the size and shape of very small particles.

Although an electron microscope has wide applications, yet it has the following disadvantages :

(i) In some case, the electron beam destroys the specimen under study.

(ii) As all the specimens have to be kept in high vacuum they may undergo changes caused by drying and evaporation.

SOLVED EXAMPLES

Example 1:

A crystal plane intercepts the three crystallographic axes of the following multiples of unit distances, 3/2, 2 and 1. What are the Miller indices of the plane?

Solution:

Suppose the unit length is a. Then, the intercepts made on X, Y, and Z-axes are 3a/2, 2a and a respectively. Now the ratios of intercepts are

$$\frac{a}{3a/2} : \frac{a}{2a} : \frac{a}{a} : \text{or } \frac{2}{3} : \frac{1}{2} : 1$$

The miller indices (h, k, l) are

$$h : k : l : : \frac{2}{3} : \frac{1}{2} : 1$$

$$h : k : l : : 4 : 3 : 6$$

The plane under question is a (436) plane.

Example 2:

What will be wavelength of X-rays which gives a diffraction angle, 2θ equal to 16.8^o for a crystal, if the interplanar distance in the crystal is 0.400 nm and any second order diffraction is observed ?

Solution:

We know,

$$n\lambda = 2d \sin \theta$$

Here $n = 2$, $d = 0.400 \text{ nm} = 0.4 \times 10^{-9}$ m and

$$2\theta = 16.8o \text{ or } \theta = 8.4^o$$

Therefore, $2\lambda = 2 \times 0.4 \times 10^{-9} \times \sin 8.4^o$ m

or $$\lambda = \frac{2 \times 0.4 \times 10^{-9} \times \sin 8.4^o}{2} = 0.4 \times 10^{-9} \times 0.146 \text{ m}$$

$$= 0.0584 \times 10^{-9} \text{ m or } 0.584 \text{ Å}$$

Example 3:

Calculate the distance between two lattice planes which give first order diffraction at an angle of 26.42^o with molybdenum X-rays of wavelength 0.710 Å.

Solution:

We know, $n\lambda = 2d \sin \theta$

here $n = 1$,

$\lambda = 0.710$ Å, $\theta = 26.42^o$

$\therefore$ $1 \times 0.710 = 2 \times d \sin 26.42^o = 2 \times d \times 0.445$

or $d = 0.799$ Å.

Example 4:

Tungsten has a body centred cubic lattice, and each lattice point is occupied by one atom. Calculate the metallic radius of the tungsten atom given the density of tungsten is 19.30 g/cm^{-3} and its at. wt. is 183.9. Avogadro's number is 6.02 × 10^{23}.

Solution:

In a body-centred cubic structure, with one atom at each lattice point, unit cell is occupied by two atoms.

Let the unit cell edge length of tungsten by a Å.

Hence the volume of unit cell is $a^3 \times 10^{-24}$ cm^{-3}.

[∵ 1 Å= 10^{-8} cm.]

This volume is occupied by two tungsten atoms since in b.c.c. lattice the unit cell is occupied by two atoms,

∴ The volume occupied by one mole of tungsten atoms

$$\frac{a^3 \times 10^{-24} \times 6.023 \times 10^{23}}{2} \text{ cm.}^3$$

The mass of one mole of tungsten atoms = 183.9 g.

Therefore, the density of tungsten

$$\frac{\text{Mass of one mole}}{\text{Volume of one mole}} = 19.30 \text{ g cm}^{-3}.$$

$$\therefore\ 19.30 = \frac{183.9 \times 2 \times 10}{a^3 \times 6.023} \text{ or } a^3 = \frac{183.9 \times 2 \times 10}{19.30 \times 6.023}$$

$$= 31.6\text{Å or } a = 3.163\text{Å}.$$

In the body-centred cubic arrangement of the tungsten atoms, the atoms at the corners of the unit cell are in contact with the atom at the centre of the cell.

Thus, the metallic radius of the tungsten atom will be equal to one quarter of the length of a body diagonal of cell, *i.e.*,

$$r = \frac{\sqrt{3a^2}}{4} = \frac{\sqrt{3 \times (3.163)^2}}{4} = 1.37 \text{ Å}$$

Thus the metallic radius of the tungsten atom is 1.37Å.

Example 5:

The density of potassium chloride at 18° C is 1.9893 g/cm³ and the length of a side of the unit cell is 6.29082Å, as determined by X-ray diffraction. Calculate Avogadro's number [At. wt. of k = 39.102, Cl = 35.453].

Solution:

Density of KCl = 1.983 g./c.c.

a = 6.29082Å = 6.29082 × 10^{-8} cm.

Mol. wt. of KCl = 74.555

Avogadro's number = N = ?

Since KCl crystallizes in the NaCl type structure it has face centred cubic lattice. Hence there are 4 atoms per unit cell.

Volume of unit cell = $[6.29082 \times 10^{-8}]^3$.

Volume occupied by one mole of KCl

$$= \frac{\left[6.29082 \times 10^{-8}\right]^3 \times N}{4} \qquad ...(1)$$

$$\text{Density} = \frac{\text{Mass of one mol}}{\text{Volume of one mole}} = 1.9893 \text{ g/cm}^{-3} \qquad ...(2)$$

From equations (1) and (2), we get

$$\frac{74.555 \times 4}{10^{-24} \times (6.29082)^3 \times N} = 1.9893.$$

$$N = \frac{74.555 \times 4}{(6.29082)^3 \times 10^{-24} \times 1.9893}$$

$$= \frac{74.5554 \times 10^{24}}{(6.29082)^3 \times 1.9893} = 6.027 \times 10^{23}.$$

Example 6:

X-ray analysis shows that the unit cell length in NaCl is 5.628 Å. Calculate the density you expected on this basis. [Avogadro's Number = 6.203 × 10²³].

Solution:

Let a be the edge length of the unit cell of NaCl

$a = 5.628\ \text{Å} = 5.628 \times 10^{-8}$ cm.

Volume of unit cell = $[5.628 \times 10^{-8}]^3$.

Since NaCl has face centred lattice, there are 4 atoms per unit cell.

So, the volume occupied by one mole of NaCl

$$= \frac{\left[5.628 \times 10^{-8}\right]^3 \times 6.023 \times 10^{23}}{4}$$

$$\text{Density of NaCl} = \frac{\text{Mass of one mole}}{\text{Volume of one mole}}$$

$$= \frac{58.5 \times 4}{\left[5.628 \times 10^{-8}\right]^3 \times 6.023 \times 10^{23}}$$

$$= \frac{58.5 \times 4}{\left[5.628 \times 10^{-8}\right]^3 \times 10^{-1} \times 6.023} = \frac{58.5 \times 4 \times 10}{\left[5.628 \times 10^{-8}\right]^3 \times 6.023}$$

Hence the density of NaCl = 2.181 gm/cc.

Example 7:

Molybdenum forms body-centred cubic crystals whose density is 12.96 g cm^{-3}. Calculate :

(a) the edge length of unit cube.

(b) the distance between the (110) and between (111) planes.

Solution:

(a) Let 'a' Å be the edge length of the unit cell.

Volume of unit cell = $a^3 \times 10^{-24}$ cm^{-3}.

Since in body-centred cubic lattice, the number of atoms per unit cell = 2.

Volume occupied by one mole of molybdenum atoms

$$= \frac{a^3 \times 10^{-24} \times 6.023\ 10^{23}}{2}$$

Mass of one mole of molybdenum atom = 95.94.

$$\text{Density of molybdenum} = \frac{\text{Mass of one mole}}{\text{Volume of one mole}} = 12.96 \text{ g/cm}^{-3}$$

$$12.96 = \frac{95.94 \times 2}{a^3 \times 10^{-24} \times 6.023 \times 10^{23}}$$

$$\text{or} \quad a^3 = \frac{95.94 \times 2}{12.96 \times 10^{-1} \times 6.023}$$

$$a^3 = \frac{95.94 \times 20 \times 1}{6.023 \times 12.96} = 4.57 \text{ or } a = 2.907 \text{ Å}$$

(b) For body centred cubic lattice, $d_{110} = \frac{a}{\sqrt{2}}$

$$d_{111} = \frac{a}{2\sqrt{3}}$$

$$d_{110} = \frac{2.907}{\sqrt{2}} = \frac{2.907}{1.414} = 2.056 \text{ Å}$$

$$d_{111} = \frac{2.907}{2\sqrt{3}} = 0.8392 \text{ Å}.$$

Example 8:

Show that the maximum proportion of the available volume which may be filled by hard spheres in various structures is

Simple cubic : $\frac{\pi}{6}$ [= 0.52]

Body-centred cubic : $\frac{\pi\sqrt{3}}{8}$ [= 0.68]

Face-centred cubic : $\frac{\pi\sqrt{2}}{6}$ [= 0.7]

Hexagonal close packed : $\frac{\pi\sqrt{2}}{6}$ [= 0.72]

What is the percentage of void space in each structure ?

Solution:

Do yourself.

Example 9:

How many atoms are assigned to the unit cell of :

(a) primitive,

(b) face-centred, and

(c) body-centred cubic lattice?

Solution:

(a) In a primitive unit cell of a cubic lattice, one particle is present at each corner of the cube. Thus, there are in all eight particles. But each corner atom is contributing 1/8 to the unit cell. Therefore, the number of particles per unit cell of this type,

$$= 8 \times \frac{1}{8} = 1$$

(b) In a face-centred unit cell, there are eight corner atoms (one on each corner and six face particles (one on each face). Therefore, the total number of particles per unit cell of this type

$$= \left(8 \times \frac{1}{8}\right) + \left(6 \times \frac{1}{2}\right) = 1 + 3 = 4$$

(c) In a body-centred unit cell, there are eight corner atoms and one at the body centre. Therefore, the total number of particles in a such unit cell,

$$= \left(8 \times \frac{1}{8}\right) + 1 = 1 + 1 = 2.$$

2

MOLECULAR SPECTRUM AND ATOMIC SPECTRUM

SPECTROSCOPY

It is the branch of science that deals with the transitions that a molecule undergoes between its energy levels upon absorption of suitable radiation. Our knowledge about molecular structure is mainly derived indirectly from Spectroscopy.

Quantum mechanics reveals that the energy levels of all system: ,t quantized and are designated by the appropriate quantum numb .,s. These energy levels could be obtained by the solutions of the time-independent Schrodinger equation, the details of which are not regar :ed as essential to the understanding of Spectroscopy. We will now consider how a spectrum arises. Consider two energy levels E, and Em, as own in Fig. 2.1.

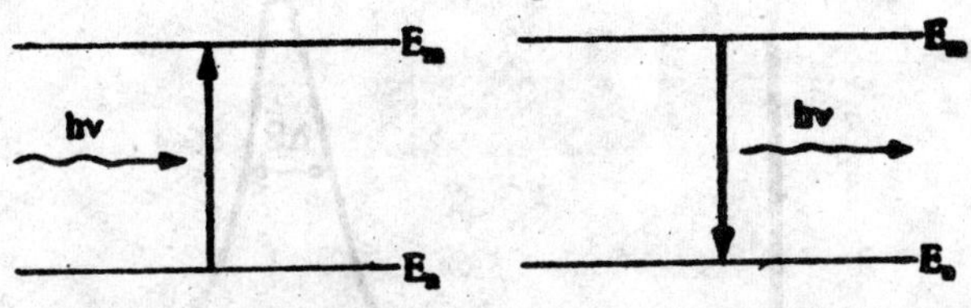

(a) Absorption spectrum **(b) Emission spectrum**

Fig. 2.1 : Spectroscopic transitions between energy levels.

If a photon of frequency v is incident on a molecule in the ground state and its energy Av is exactly equal to the energy difference ΔE ($= -E_m + E_n$) between the two molecular energy levels, then the molecule would undergo a *transition* from the lower energy level to the higher energy level with the absorption of a photon of energy hv. The spectrum thus obtained is termed as the *absorption spectrum*. If the molecule falls from the excited state to the ground state with the emission of a photon of energy Av, the spectrum obtained is termed as the *emission spectrum.*

The molecular spectra are governed by the so-called *selection rules*, which determine a spectrum. If there were no selection rules, the resulting spectrum would be very chaotic indeed 1. The selection rules are obtained from the quantum theory of interaction of radiation with matter.

The selection rules, are not always obeyed strictly. This is because certain approximations which have been used in the derivation of the selection rules are not valid strictly. The spectral transitions which obey a given selection rule are termed as *allowed transitions* while those which violate a selection rule are termed as *forbidden transitions.* In general, the allowed transitions are more intense (stronger) than the forbidden transitions which are weak.

Width and Intensity of Spectral Lines

When the spectrum of a molecule is analysed the first thing to know is how sharp and how intense (strong) is the spectral line. In other words, one is interested in the width and intensity of a spectral transition. These two quantities are common to all branches of Spectroscopy. Fig. 2.2(a) shows a sharp spectral line having no width while Fig. 2.2(b) shows a spectral line having a width ΔE at half-height. The chemist would require the spectral lines to be all very sharp and intense. In practice, this is no so.

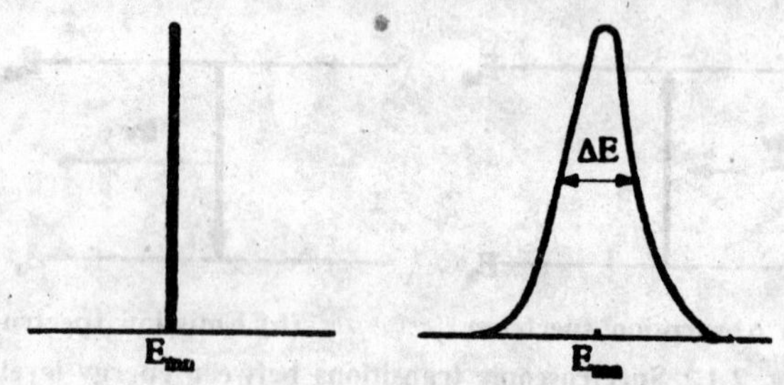

(a) Sharp spectral line; (b) Spectral line having a width

Fig. 2.2

Two factors contributing to broadening of a spectral line are:

(i) The *collision-broadening* and

(ii) The *Doppler-broadening.*

The *collision-broadening* is largely responsible for the width of spectral lines in the ultraviolet (UV) and visible regions. These transitions mostly occur between electrons in the outer shells. If molecules in the gaseous or liquid phase undergo collision with each other, they deform the charge-clouds of the outer electrons thereby slightly perturbing the energy levels of these electrons. Therefore, the spectral transitions between these perturbed energy levels get broadened.

The Doppler-broadening occurs if the molecule under investigation is having a velocity relative to the observer or observing instrument. This is generally the case with gaseous samples in which the molecules are undergoing random motions according to the postulates of the kinetic theory of gases. If the molecule is moving *towards* the measuring instrument with velocity u, then the frequency v' of radiation 'seen' by the molecule is given as follows :

$$v' = v\left(1 + \frac{u}{c}\right) \quad \text{...(1)}$$

where v refers to the actual radiation frequency. If, on the other hand, the molecule is moving away from the measuring instrument, the frequency of radiation 'seen' by the molecule is given as follows :

$$v' = v\left(1 - \frac{u}{c}\right) \quad \text{...(2)}$$

On rearranging Eq. (1). we get

$$\frac{v - v'}{v} = \frac{\Delta v}{v} = -\frac{u}{c} \quad \text{...(3)}$$

Similarly, rearranging Eq. (2), we get

$$\frac{v - v'}{v} = \frac{\Delta v}{v} = \frac{u}{c} \quad \text{...(4)}$$

The quantity Δv refers to the *Doppler broadening*. It can be seen that, depending upon the direction of the motion of the molecule relative to the instrument, the observed frequency becomes either higher or lower than the actual radiation frequency. It may be mentioned here that there is also the *natural line width*, a consequence of Heisenberg's uncertainty principle, which is always present in a spectral line.

Let us now turn to the *intensity of spectral transitions*. The intensity is determined by (i) the *Boltzmann population* of the energy levels and (ii) the *transition probability* between the energy levels.

According to Boltzmann, if, at a temperature T. N_0 refers to the number of molecules in the ground state, then the number of molecules, N. in the excited state is given as follows :

$$N = N_0 \, e^{-\Delta E/kT} \qquad ...(5)$$

where ΔE refers to the energy difference between the ground and excited states and k refers to the Boltzmann constant. The relative population at equilibrium is, thus, given as follows :

$$N/N_0 = e^{-\Delta E/kT} \qquad ...(6)$$

Evidently, if ΔE is large, N/N_0 would be small, *i.e.*, number of molecules in the excited state would be less than that in the ground state. In fact, at room temperature, most of the molecules are in the ground state. Therefore, the transitions originating from the *ground state* to a higher, say, third excited state wouid be more intense than those originating from the *first excited state* to the third excited state

The term transition probability implies the probability of transition between two given energy levels. It has been said earlier that a molecule, upon absorption of a photcn, does not just go anywhere it pleases; it ends up in an energy level determined by the *selection rule* for that particular transition. The selection rules, in turn, ascertain the allowed transitions and the forbidden transitions. Evidently, the allowed transitions are having greater intensity than the forbidden ones.

MOLECULAR SPECTRA

The atomic spectra arise from the transitions of an electron between the atomic energy levels whereas the molecular spectra arise from three types of energy changes, viz., *molecular rotation, molecular vibration and electronic transition.* Broadly speaking, the total energy of a molecule, according to Born-Oppenheimer approximation, is given as follows :

$$E = E_{tr} + E_{rot} + E_{vib} + E_{el} \qquad ...(7)$$

i.e., the energy refers to the sum of *transitional energy*, E_{tr}; *rotational energy*, E_{rot}; *vibrational energy*, E_{vib} and *electronic energy*, E_{et}. The translational energy does not get *quantized* while all the other energies get *quantized*. Furthermore,

$$E_{el} >> E_{vib} >> E_{rot} >> E_{tr}$$

As the translational energy is negligibly small, the *Born-Oppenheimer approximation* may be put as follows:

$$E = E_{rot} + E_{vib} + E_{el} \quad ...(8)$$

We will explain these energies with reference to a diatomic molecule. The molecule is having translational energy when, because of motion its centre of gravity changes. Rotational energy arises when the molecule rotates about an axis perpendicular to the internuclear axis and passing through the centre of gravity of the molecule. Vibrational energy gets associated with the to-and-fro motion of the nuclei of the molecule such that the centre of gravity does not change.

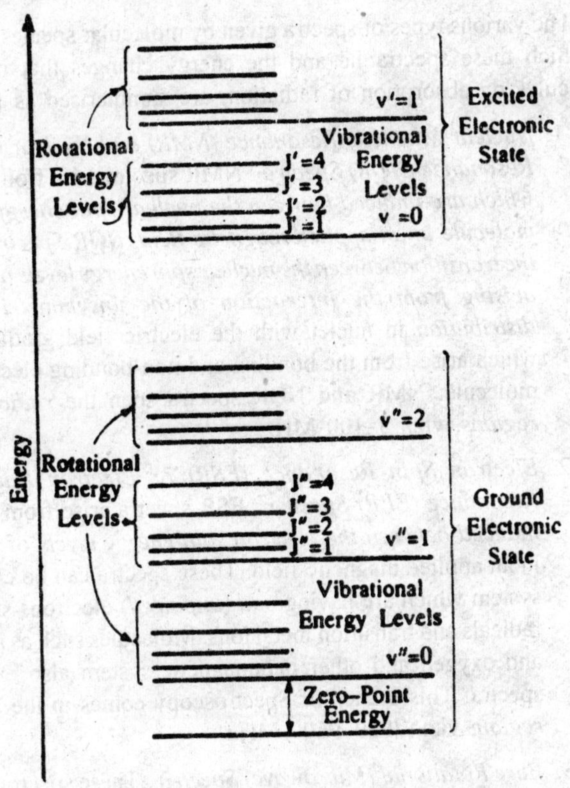

Fig. 2.3 : Energy level diagram depicting the various molecular energies.

And finally, electronic energy gets associated with the transition of an electron from the ground state energy level to an excited state energy level because of the absorption of a photon of suitable frequency. Associated with each electronic state there is a series of vibrational energy levels (designated by the vibrational quantum number, v) and associated with each vibrational energy level there is a series of rotational energy levels (designated by the rotational quantum number J), as depicted in Fig. 2.3. It is customary to use a single prime superscript for designating a higher energy level and a double prime superscript to designate a lower energy level.

The rotational spectra of a molecule can be observed in the microwave region; the vibrational spectra in the infrared (IR) region and the electronic spectra in the ultraviolet (UV) and/or visible regions of the electromagnetic spectrum.

The various types of spectra given by molecular species, the regions in which these spectra lie and the energy changes that occur in the molecules on absorption of radiation, are summarised as follows :

1. *Nuclear Magnetic Resonance (NMR) and Nuclear Quadrupole Resonance (NQR) Spectra :* NMR spectra arise from *transitions which are induced between the nuclear spin energy levels of a molecule in an applied magnetic field. NQR Spectra arise/row the transition between the nuclear spin energy levels of a molecule arising from the interaction of the unsymmetrical charge distribution* in nuclei with the electric field gradients (EFG) which arise from the bonding and non-bonding electrons in the molecule. NMR and NQR spectra span the *radio frequency regions*, viz., 5–100 MHz.

2. *Electron Spin Resonance (ESR) or Electron Paramagnetic Resonance (EPR) Spectra :* ESR spectra arise from *transitions induced between the electron spin energy levels* of a molecule in an applied magnetic field. These spectra can be exhibited by system which are having *odd* (*unpaired*) electrons such as free radicals and transition metal ions. Molecules such as nitric oxide and oxygen and other paramagnetic system also exhibit ESR spectra. This branch of Spectroscopy comes in the *microwave region*, viz., 2000–9600 MHz.

3. *Pure Rotational (Microwave) Spectra :* These spectra arise from transitions between the rotational energy levels of a *gaseous* molecule on absorbing radiations falling in the microwave region. These spectra are shown by molecules which are having *a permanent dipole moment*, *e.g.*, HCl, CO, H_2O vapour, NO, etc. Homonuclear diatomic molecules like H_2, Cl_2, O_2, etc., and linear polyatomic molecules such as CO_2, which are not having

a dipole moment, do not show microwave spectra. Microwave spectra come in the spectral range of 1–100 cm^{-1}.

4. *Vibrational and Vibrational-Rotational (Infrared) Spectra* : These spectra arise because of transitions induced between the vibrational energy levels of a molecule on absorbing radiations belonging to the infrared region. IR spectra are shown by molecules when vibrational motion is accompanied by a *change in the dipole moment of the molecule.* These spectra come in the spectral range of 500–4000 cm^{-1}.

5. *Raman spectra* : These are related to vibrational and/or rotational transitions in molecules but in a different manner. In this case, what we measure is the scattering and not the absorption of radiation. An intense beam of monochromatic radiation in the *visible region* is allowed to fall on a sample and the intensity of scattered light can be seen *at right angles to the incident beam*. Most of the scattered light is having the same frequency as the incident beam (this is called *Rayleigh scattering*). However, a small amount of light is having different frequencies than the incident beam. This is termed as *Raman scattering*. The energy differences between these weak lines and the main Rayleigh line correspond to vibrational and/or rotational transitions in the molecule under investigation. Raman spectra can be observed in the visible region, viz., 12,500–25,000 cm^{-1}.

6. *Electronic Spectra* : Electronic spectra arise due to electronic transitions in a molecule by absorbing radiations falling in the visible and ultraviolet regions. While electronic spectra in the visible region span 12.500–25.000 cm^{-1}, those in the ultraviolet region span 25,000–70,000 cm^{-1}. As electronic transitions in a molecule have been invariably accompanied by vibrational and rotational transitions, the electronic spectra of molecules would be highly complex.

7. *Photoelectron Spectra (PES)* : If a light photon falling on a molecule is having very high energy, it can bring about ionization of the molecule *i.e.*, the removal of the electron from the molecule. If the energy of the incident photon is greater than the ionization energy, the ejected electron will have excess kinetic energy. In PES, a beam of photons of known energy is made to fall on the sample and the kinetic energy of the ejected electrons is measured.

The difference between the photon energy and the excess kinetic energy would give the *binding energy* of the electron. PES provides one of the most accurate methods for determining the ionization energies of molecules. These values can provide a fairly good insight into the molecular electronic structures.

8. *Mossbauer Spectra or Gamma Ray Spectra :* Mossbauer spectra refers to a type of nuclear resonance spectra like nuclear magnetic resonance spectra. However, while NMR spectra result from absorption of low energy photons of frequency around 60 MHz, Mossbauer spectra arise from absorption of high energy γ-photons of frequency around 10^{13} MHz by the nuclei.

Gamma ray spectra have been used specifically for studying compounds of iron and tin. In this case γ-radiations from ^{57}Co source are allowed to fall on a sample in which the iron nuclei are in an environment identical with that of the source atoms. This causes in resonant absorption of γ-rays. The splittings in Mossbauer lines have been found to be of the same order as in NMR Spectroscopy. The transitions involved in the Mossbauer effect in case of ^{57}Fe nuclide have been depicted in Fig. 2.4.

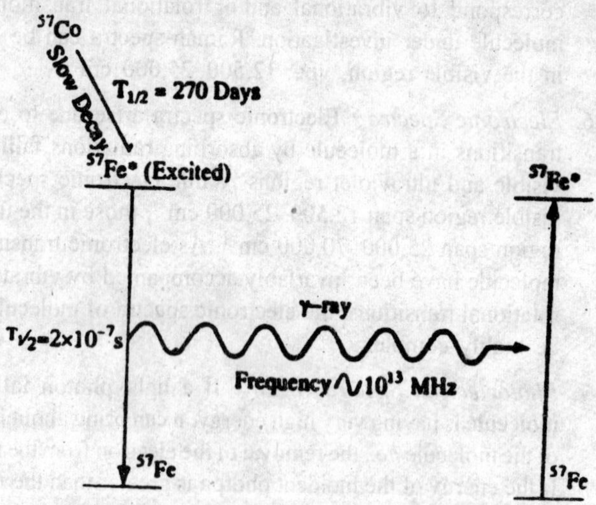

Fig 2.4 : The transitions involved in the Mossbauer spectrum of ^{57}Fe nuclide.

MICROWAVE SPECTROSCOPY (Rotational Spectroscopy)

The microwave Spectroscopy explores that part of the electromagnetic spectrum which is extending from 100 μm (3×10^{12} Hz) to 1 centimetre (3×10^{10} Hz). This region of electromagnetic spectrum is designated AS the *microwave region.* This lies between the far infrared and conventional radio frequency regions. Spectroscopic applications of microwaves consist almost exclusively of *absorption* type, rather than of the *emission type.* In most of the cases, absorption of microwave energy represents changes of the absorbing molecule from one rotational level to another. Therefore, *the microwave Spectroscopy deals with the pure rotational motion of molecules and is also known as rotational Spectroscopy.* The condition for observing resonance in that region is that a molecule must possess *permanent dipole moment.* When a molecule having dipole moment rotates, it generates an .electric field which can interact with the electric component of the microwave radiation. During the interaction energy can be absorbed or emitted and thus the rotation of the molecule gives rise to a spectrum. If molecules are not having dipole moments, interaction is not possible and these molecules are said to be "*microwave inactive*". Examples of such molecules are H_2, Cl_2, etc. On the other hand, the molecules like HCl CH_3Cl, etc. , are having dipole moments and their interaction will give rise to a spectrum. Such molecules are said to be "*microwave active*".

Generally, the microwave spectra obtained in most of the molecules are absorption spectra. Theoretical considerations reveal that the probability of microwave spectra is only about one per cent of the optical absorption.

Theory of Microwave Spectroscopy : We are aware of the simple fact that rotational energy, along with all other forms of molecular energy, is quantised, it means that rotational energy has certain permitted energy values, generally known as rotational energy levels which may be calculated for any molecule by solving the Schrodinger equation for the system represented by that molecule. For simple molecules the mathematics involved is simple whereas for complex molecules the mathematics involved is tedious and gross approximations have to be used.

We will now discuss each class of molecules one by one.

1. *Diatomic Molecule as a Rigid Rotator :* A rotating diatomic molecule whose nuclei are supposed to be separated by a definite

mean distance may be treated as a *rigid rotator* with free axis of rotation. In order to provide a suitable model for the interpretation of pure rotational spectra, consider a diatomic molecule in which masses m_1 and m_2 of atoms A and B are joined by a rigid bar whose length is r an is given by

$$r = r_1 + r_2 \qquad ...(1)$$

where r_1 and r_2 are the distances of atoms A and B from the centre of gravity G of the molecule AB about which the molecule rotates end-over-end as shown in Fig. 2.5. In this case, the moment of inertia about G is defined by

$$I = m_1r_1^2 + m_2r_2^2 \qquad ...(2)$$

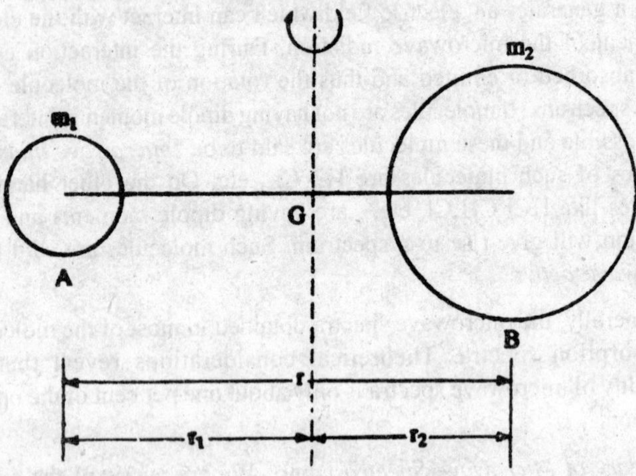

Fig. 2.5 : A Rotating diatomic molecule.

As the system is balanced about its centre of gravity G, one may write

$$m_1r_1 = m_2r_2 \qquad ...(3)$$

On substituting equation (3) in (2), we obtain the following expression :

$$I = m_2r_2r_1 + m_1r_1r_2 = r_1r_2\,(m_1 + m_2) \qquad ...(4)$$

But, from equations (1) and (3), we have

$$m_1r_1 = m_2r_2 = m_2\,(r - r_1) \qquad ...(5)$$

On solving equations (4) and (5), we get

$$r_1 = \frac{m_2}{m_1 + m_2} r \text{ and } r_2 = \frac{m_2}{m_1 + m_2} r \qquad ...(6)$$

On putting these values of r_1 and r_2 in equation (2), we get

$$I = \frac{m_1 m_2^2}{(m_1 + m_2)^2} r_2 + \frac{m_1^2 m_2}{(m_1 + m_2)^2}$$

$$r^2 = \frac{m_1 m_2^2 + m_1^2 m_2}{(m_1 + m_2)^2} r^2 = \frac{m_1 m_2 (m_1 + m_2)}{(m_1 + m_2)^2} r^2$$

$$I = \frac{m_1 m_2}{m_1 + m_2} r_2 = \mu r^2 \qquad ...(7)$$

where μ is the reduced mass of the diatomic molecule and its value is

$$\mu = \frac{m_1 m_2}{m_1 + m_2} \qquad ...(8)$$

Equation (7) defines moment of inertia in terms of atomic masses and bond length.

A rotating molecule having a permanent dipole or magnetic moment generates an electric field which can interact with the electric component of the microwave region. If it is assumed that a diatomic molecule behaves lake a rigid rotator, the rotational energy levels may in principle be calculated by solving the Schrodinger's equation for the system represented by that molecule.

$$E_J = \frac{h^2}{8\pi^2 I} J(J + 1) \text{ Joules where } J = 0, 1, 2, \qquad ...(9)$$

where h = Planck's constant, I = moment of Inertia,

J = Rotational quantum number ; it takes integral values from zero upwards, and

$$\frac{J(J + 1)h^2}{4\pi^2} = \text{The square of the rotational angular momentum.}$$

In the rotational region, spectra are generally expressed in terms of wave numbers, so it becomes useful to consider energies in these units. Thus, one may write

$$\epsilon_J = \frac{E_J}{hc} = \frac{h^2}{8\pi^2 Ihc} J(J + 1) \text{ cm}^{-1} = \frac{h}{8\pi^2 Ic} (J + 1) \text{ cm}^{-1} \qquad ...(10)$$

where c is the velocity of light expressed in cm per second. It is common to write B for $h/8\pi^2 Ic$ so that equation (10) reads as

$$\epsilon_J = J(J+1)\ B\ cm^{-1} \qquad ...(11)$$

where B is called the rotational constant and may be expressed in kMc per second or cm^{-1}, *i.e.*,

$$B = \frac{h}{8\pi^2 Ic}\ cm^{-1} \qquad ...(12)$$

From equation (11), we can show the allowed energy levels diagrammatically as in Fig. 2. When J = 0, equation (11) becomes as

$$\epsilon_J = BJ(J+1)\ \ cm^{-1} = B\ 0\ (0+1) = 0 \qquad ...(13)$$

From equation (13) it is evident that the molecule is not rotating at all. When J = 1, equation (11) becomes as

$$\epsilon_J = B.\ 1(1+1) = 2B\ cm^{-1} \qquad ...(14)$$

From equation (14), it follows that a ;rotating molecule ha its lowest angular momentum. Similarly, one can calculate the value of ϵ_J for J = 2, 3, 4 ... The allowed rotational energy levels of a rigid diatomic molecule are shown in Fig. 2.6.

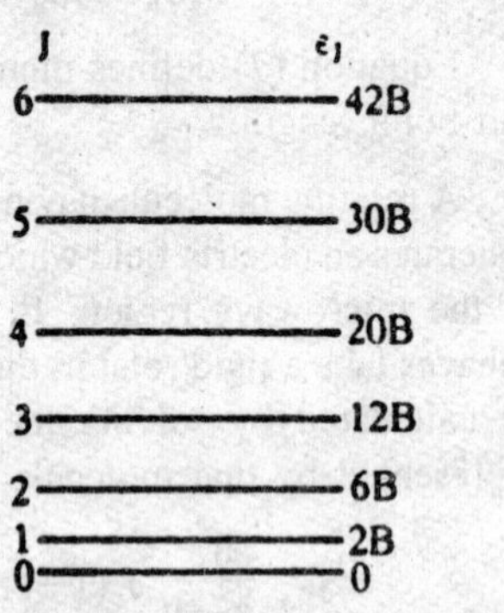

Fig. 2.6 : The allowed rotational energy levels of a rigid diatomic molecule.

Frequency or Rotational Spectral Lines

Consider there is a transition from rotational level of rotational quantum number J' to that of higher quantum number J. The energy difference between these two levels will be given by

$$\Delta E_J = E_J - E_{J'} = \frac{h^2}{8\pi^2 I} J(J+1) - \frac{h^2}{8\pi^2 I} J'(J'+1) \qquad ...15)$$

This energy, ΔE_J is evolved when the molecule returns to the original rotational level of quantum number J' from the excited rotational level of quantum number J According to quantum theory, the energy evolved is then given out in the form of spectral lines. The frequency of these spectral lines, expressed in wave number, is given by

$$\bar{v}_J = \frac{\Delta E_J}{hc} = \frac{h^2}{8\pi^2 Ihc}\left[J(J+1) - J'(J'+1)\right]$$

$$= \frac{h^2}{8\pi^2 Ic}\left[J(J+1) - J'(J'+1)\right]$$

$$= B[J(J+1) - J'(J'+1)] \text{ [using equation (12)]} \quad ...(16)$$

When J = 1, J' = 0, equation (16) becomes as

$$v_{J0\to1} = B[1(1+1) - 0(0+1)] = 2B \text{ cm}^{-1} \quad ...(17)$$

From equation (17) it follows ;that an absorption line will appear at 2B cm^{-1}. If the molecule is raised from J' = 1 to J = 2 level by the absorption of more energy in the microwave region, equation (16) becomes as

$$\bar{v}_{J1\to2} = B[2(2+1) - 1(1+1)] = B(6-2)\ 4B \text{ cm}^{-1} \quad ...(18)$$

It means that an absorption line will appear at 4bcm^{-1}. In general, when the molecule is raised from J to J + 1, equation (16) becomes as

$$v_{J\to J+1} = B[(J+1)(J+2) - J(J+1)] = B(J^2 + 3J + 2 - J^2 - J)$$

$$= 2B(J+1) \text{ cm}^{-1} \quad ...(19)$$

From equation (19), it is clear that a stepwise raising of the rotational energy gives rise to an absorption spectrum which consists of lines at 2B, 4B, 6B, ... cm^{-1} whereas a similar lowering would give rise to a similar identical emission spectrum. This is shown in Fig. 2.7.

Selection Rule for Rotational Spectra

In order for a molecule to give rise to rotational spectrum, it becomes essential that the molecule must have a dipole moment but all transitions are not permitted. There is a selection rule which is given as

$$\Delta J = \pm 1 \quad ...(20)$$

From the above rule, it is evident that only those transitions are permitted in which there is an increase or decrease by unity in the rotational quantum number. It means that J = 0 → J 2 → J = 4 ... transitions

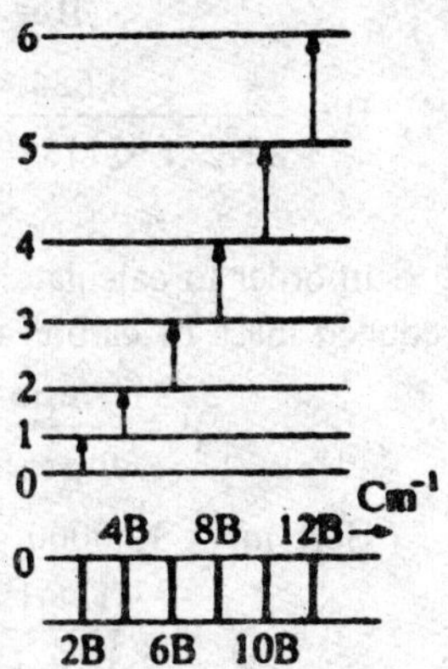

Fig. 2.7 : Allowed rotational transitions of a rigid diatomic molecule

are not possible. In other words, these transitions are spectroscopically forbidden.

Validity of the Theory

The theory of microwave spectroscopy can be confirmed by considering the following examples :

(i) Carbon Monoxide Molecule

Let us now apply equation (19) to the observed microwave spectrum of the molecule to calculate its moment of inertia and hence the bond length. The first line corresponding to J = 0 appears in the rotational spectrum of carbon monoxide at 3.84235 cm^{-1}, *i.e.*,

$$\overline{v_J} = 3.84235 \text{ cm}^{-1} \quad \text{...(21)}$$

From equation (19) we have

$$\overline{v_J} = 2B(J + 1)$$

But, $J = 0$

$$\therefore \overline{v_J} = 2B(0 + 1)\ 2B \quad \text{...(22)}$$

From equations (21) and (22), we have

$$2B = 3.84235 \text{ cm}^{-1} \text{ or } B = 1.92118 \text{ cm}^{-1} \quad \text{...(23)}$$

Also, from equation (12), we have

$$B = h/8\pi^2 Ic \text{ or } I = h/8\pi^2 Bc$$

$$B = \frac{6.624 \times 10^{-27}}{8\pi^2 \times 1.92118 \times 2.99976 \times 10^{10}} = 1.45673 \times 10^{-39} \text{ g cm}^2 \quad \text{...(24)}$$

In order to calculate bond length, one has to find out the value of reduced mass of carbon monoxide molecule which is given by

$$\mu = \frac{m_1 m_2}{m_1 + m_2} \times \frac{1}{N}$$

Here m_1 = 12.0000 g (carbon) and m_2 = 15.9949 (oxygen)

$$\mu = \frac{12 \times 15.9949}{12 + 15.9949} \times \frac{1}{6.024 \times 10^{23}} \text{g} \quad \text{...(25)}$$

From equation (7) we have, $I = \mu r^2$

On substituting equations (24) and (25) in (7), we get

$$1.45673 \times 10^{-39}\ \text{g cm}^2 = \frac{12 \times 15.9949}{12 + 15.9949} \times \frac{1}{6.024 \times 10^{23}} r^2$$

or $\quad r^2 = 1.28 \times 10^{-16}\ \text{cm}^2$

or $\quad r = 1.131\ \text{A}^o$

Gallium and coworkers have also found the rotational absorption of $^{12}C\ ^{16}O$ at 3.84235 cm^{-1} (C = 12 and O = 16) while that of $^{13}C\ ^{16}O$ was found to be at 3.67337 cm^{-1}. The values of B and B' from these figures come out to be

$$B = 1.92118\ cm^{-1} \text{ and } B' = 1.83669\ cm^{-1}$$

But, $\quad B = \dfrac{h}{8\pi^2 Ic}$ and $B' = \dfrac{h}{8\pi^2 I'c}$

$\therefore \quad B/B' = \dfrac{h}{8\pi^2 Ic} \times \dfrac{8\pi^2 I'c}{h} I'/I = \mu' r^2/\mu r^2 = \mu'/\mu$

or $\quad \mu'/\mu = 1.92118/1.83669 = 1.046$

If we take the mass of oxygen to be 15.9994 g an that of carbon to be 12.0000 g, we get

$$\mu'/\mu = 1.046 = \frac{15.9994 \times m'}{15.9994 + m'} \times \frac{12 + 15.9994}{12 \times 15.9994}$$

or $m' = 13.0007$

Thus, the atomic weight of carbon-13 comes out to be 13.0007. This value agrees with those values which are obtained by other methods.

From the above discussion, it follows that microwave spectroscopy can be successfully used to estimate the abundance of isotopes by comparison of absorption intensities.

(ii) HCl Molecule

From the microwave spectrum of HCl, it is observed that frequency difference between successive absorption lines is found to be 20.7 cm^{-1} and is identified with 2B, *i.e.*,

$$2 = 20.7 \text{ or } B = 10.35\ cm^{-1}$$

But, $\quad I_{HCl} = h/8\pi^2 Bc = 6.62 \times 10^{-27}/8 \times (3.14)^2 \times (10.35)^3 \times 10^{10}$

$\quad = 2.70 \times 10^{-40}\ \text{g cm}^2$

Reduced mass of HCI molecule is given by

$$\mu = \frac{m_1 m_2}{m_1 + m_2} \times \frac{1}{N}$$

$= 1.008 \times 35.46/(1.008 + 35.46) \times 1/6.024 \times 10^{-24}g$

But the bond length of HCl, *i.e.*, r_{HCl}, is given by

$r^2_{HCl}/\mu = I_{HCl}/\mu = 2.70 \times 10^{-40}/1.627 \times 10^{-24}$ or r_{HCl}

$= 1.29 \times 10^{-8}$ cm $= 1.20$ A°

The separation between energy levels J = 0 and J = 1 will be given by

$$\Delta E_J = 2\,\frac{h^2}{8\pi^2 I} = 2Bhc$$

$= 20.7 \times 6.624 \times 10^{-27} \times 2.99976 \times 10^{10}$ ($\because$ 2B = 20.7)

$= 0.405 \times 10^{-24}$ erg.

(iii) HI Molecule

The microwave spectrum of HI molecule consists of a series of equidistant lines with a spacing of 12.8 cm^{-1}, *i.e.*,

2B = 12.8 cm^{-1} or B = 6.4 cm^{-1}

$$I_{HCl} = \frac{h}{8\pi^2 BC} = \frac{6.624 \times 10^{-27}}{8 \times (3.14)^2 \times 6.4 \times 2.99976 \times 10^{10}}$$

$= 4.372 \times 10^{-40}$ g cm^2.

But, $\mu = 1 \times 127/(1 + 127) \times 1/6.024 \times 10^{23} = 1.65 \times 10^{-25}$g

But, bond length, $r^2_{HI} = I_{HI}/\mu = 4.372 \times 10^{-40}/1.65 \times 10^{-24}$ or r_{HI} = 1.63Å

From the above examples to is evident that the examination of microwave spectrum is an accurate method for calculating internuclear distance, at least for simple diatomic molecules.

2. The Diatomic Molecule as a Non-Rigid Rotator

So far we have considered the bond in a diatomic molecule to be a rigid bond. But this is only an approximation. However, all bonds are elastic to some extent. The elasticity results in some changes which are as follows :

(i) An elastic bond may have vibrational energy.

(ii) Another consequence of elasticity is that the quantities r and B vary during a vibration.

When the spectrum of a non-rigid rotator is considered, we must take into account the above two facts. For a non-rigid rotator, Schrodinger's wave equation yields the following rational terms :

$$E_J = \frac{h^2}{8\pi^2 I} J(J+1) - \frac{h^4}{32\pi^4 I^2 r^2 k} J^2 (J+1)^2 \text{ Joules.}$$

$$\text{or } \in_J = E_J/hc = BJ\,(J + 1) - DJ^2(J + 1)^2 \text{ cm}^{-1} \qquad ...(25)$$

where B is the rotational constant and its value is, as defined previously, as follows :

$$B = h/8\pi^2 Ic \qquad ...(26)$$

and D is the centrifugal distortion constant and its value is given by

$$D = h^3/32\pi^4 I^2 r^2 kc \text{ cm}^{-1} \qquad ...(27)$$

In equation (27) k is a force constant can its value is given by

$$k = 4\pi^{2^{-2}} wc^2 / \mu \qquad ...(28)$$

where $\overline{w}$ is the vibration frequency expressed in cm^{-1}. From the defining equations of B and D it may be shown directly that

$$D = 16B^3\pi^2 \text{ m } c^2/k = 4B^3/\overline{w^2} \qquad ...(29)$$

It is generally found that vibrational frequencies are of the order of 10^3 cm^{-1} and the value of B is found to be of the order of 10 cm^{-1}. It means that the value of D, according to equation (29), is of the order 10^{-3} cm^{-1} which is very small compared to B.

Selection Rule

The selection rule for the spectrum of a rotator ($\Delta J = \pm 1$) discussed earlier is valid whether it is rigid or not. The analytical expression for the transitions is as follows :

$$\overline{v_J} = B[(J + 1)(J + 2) - J(J + 1)] - D[(J + 1)^2 (J+2)^2 - J^2(J+1)^2]$$

$$= 2B\,(J + 1) - 4D(J + 1)^3 \text{ cm}^{-1} \qquad ...(30)$$

where $\overline{v_J}$ denotes the upward transition from J to J + 1 or the downward from J + 1 to J. From equation (30) and 919), it follows that the spectrum of a non-rigid rotator is almost similar to that of a rigid molecule except that each spectral line will undergo displacement towards the low frequency value and the displacement increases with $(J + 1)^3$.

From the value of D, two useful measurements can be made. These are :

(a) Firstly the value of vibrational frequency $\overline{w}$ can be found out from equation (29) provided the value of B is known. However, this method is not accurate.

(b) Secondly a knowledge of D allows us to determine the J value of lines in an observed spectrum.

INSTRUMENTATION FOR MICROWAVE SPECTROSCOPY

A microwave spectrometer consists of the following essential components :

(1) The Source and Monochromator

Reflex klystron valve is the main source of radiations in microwave region. As the klystron valve emits radiation of a very narrow frequency range (called monochromatic), it acts as own monochromator. Furthermore, the frequency of the emitted radiation depends on the voltage that is applied to the klystron valve. As the voltage is varied over a given range, the emitted radiation can thus be made to sweep through a region of the microwave range.

Klystrons are readily available from 3000 to 5000 Mc/second and weaker signals upto 250000 Mc/second may be obtained with harmonic generators.

One slight disadvantage of the klystron valve is that it radiates out very small energy which is of the order of 30 milliwatts. However, since the energy radiated is concentrated into very narrow frequency band, a sharply tuned detector may be activated to produce a strong signal.

(2) The Beam Direction

The radiation emitted by klystron cannot be handled with mirrors and lenses, but can be most advantageously transmitted through hollow metallic conductors of such geometry that the electric and magnetic fields can be utilised to the greatest extent. These are known as wave guides. These are hollow tubes of copper or silver of rectangular cross section inside which the radiation is confined. In order to maintain the direction of beam as well as its focussing, the wave guides may be bent or tapered the waveguides are generally evacuated because if air is present in them, considerable absorption of the radiation will occur. The waveguides used in microwave spectrometer is now commonly used in chemical research facilities.

(3) Sample and Sample Space

The sample is placed in a piece of evacuated waveguide which is closed at both ends by thin mica windows. Round holes are made in the tube for evacuation purposes and for introduction of the gas under test. The pressure of the gas is adjusted to make the absorption line sharp.

The sample must be in the gaseous state for studying in the microwave region. The pressures of the order of 0.01 mm mercury are generally required to give absorption spectrum. Many solid or liquid substances can be studied by the microwave techniques provided their vapour pressures are above the value of 0.01 mm of mercury.

(4) Detector

A quartz crystal is generally used as a detector. It is mounted on a cartridge made up of a tungsten whisker held in point contact with the crystal. In place of crystal detector, an ordinary superheterodyne radio receiver can be used provided it may be tuned to the appropriate high frequency. But a simple quartz crystal is more sensitive and easier to use.

(5) Spectrum Analyser

It consists of an amplifier of detected energy and an indicator which may be either a cathode ray oscillograph or a pen and ink-recorder. The vibrations emitted by the quartz crystal produce an electrical signal which is amplified an then displayed as a pattern on an oscilloscope screen or a recording on a chart by the pen-and-ink recorder.

A diagrammatic sketch of one of the simplest type of microwave spectrometers is shown in Fig. 2.8.

Working

Monochromatic radiations of various wave lengths in the microwave region emitted by klystron valve are allowed to pass through the sample space containing the gaseous sample of the substance under investigation. Then, the radiations are made to conduct along a rectangular tube called a waveguide. After this the radiations are received by the quartz crystal detector which is situated at the far end of the waveguide. After receiving the radiations from the wave guide it vibrates and produces an electrical signal which is amplified by amplifier and then displayed either as a recording on a chart or as a pattern on an oscilloscopic screen. The pattern obtained on the chart or on the screen or the oscillograph enables one to determine the frequencies of the detected microwave radiation.

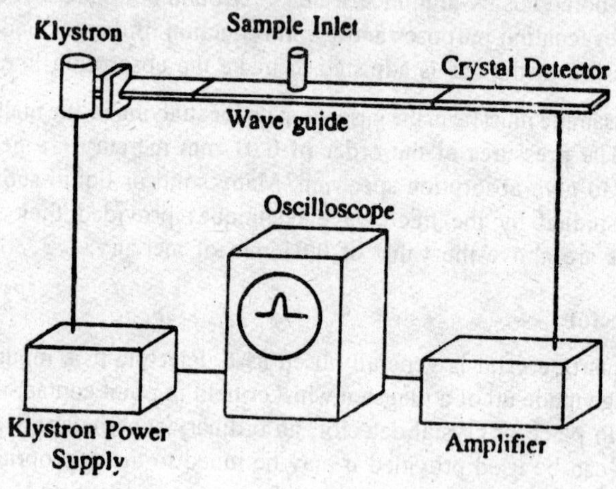

Fig. 2.8 : Simplified diagram of microwave spectrometer.

This microwave spectrometer described above is usually used for the measurements of the highest accuracy because the absorption lines are narrow an fairly regular in shape and relative intensities. The use of oscilloscope poses a serious problem that the amplifier bandwidth cannot be narrowed to remove noise and, thus, the sensitivity is not exceptionally high. At the same time the new lines for unknown substance cannot be obtained very easily unless their frequencies are known within narrow limits.

By changing the frequency of the oscillator and observing the intensity of transmitted beam, moment of inertia and internuclear distances upto ± 0002 A° can be calculate - Data obtained for bond lengths and bond angles calculated for linear molecules and symmetrical top molecules by microwave spectroscopy are given in Tables 2.1 and 2.2.

Table 2.1 : Some Molecular Data Determined by Microwave Spectroscopy

S. No.	*Molecules*	*Bond*	*r(A°)*	*Bond*	*r(A°)*
1.	HCN	C – H	1.06317	C – N	1.15535
2.	CICN	C – CI	1.629	C – N	1.163
3.	BrCN	C – Br	1.790	C – N	1.159
4.	NNO	N – N	1.126	N – O	1.191
5.	OCS	C – O	1.164	C – S	1.559
6.	CH_3CI	C – H	1.0959	C – CI	1.178
7.	CH_3F	C – H	1.109	C – F	1.385
8.	SiH_3Br	Si – H	1.570	Si – H	2.209

Table 2.2 : Bond Angles for Some molecules

S. No.	*Molecules*	*Bond Angle*	*Value of Bond Angle in Degrees*
1.	$CHCI^3$	CI – C – CI	110°24'
2.	CH^3CI	HI – C – H	108°0'
3.	CH^3F	H – C – H	110°0'
4.	SiH^3Br	H – Si – H	111°20'

It has been observed that microwave spectroscopy can also be used for finding the nuclear spin.

Applications of Microwave Spectroscopy

1. Structural Determination . From microwave studies, one can obtain information regarding molecular symmetry and molecular parameters. This can be illustrated by considering the following examples.

Structure of OCS Molecules

The problem of structural determination is of such importance in case of carbon oxysulphide, that is worth a more detailed discussion. This molecule is linear and has two different interatomic distances and is therefore confronted with the difficulty of solving two unknowns is one equations. This difficulty can be overcome by the isotopic technique which is as follows :

First of all consider the isotopic species $^{16}O\ ^{12}C\ ^{32}S$ (Fig. 2.9). If r_1, r_2, r_3 are the distances of C, O and S from the centre of gravity G of this isotopic molecule, it follows that

$$m_c r_1 + m_o r_2 = m_3 r_3 \qquad ...(1)$$

The moment of inertia of this molecule I is given by

$$I = m_c r_1^2 + m_o r_2^2 + m_3 r_3^2 \qquad ...(2)$$

It also follows that

$$r_2 = r_{co} + r_1 \qquad ...(3)$$

$$r_3 = r_{cs} - r_1 \qquad ...(4)$$

In equations (3) and (4) r_{co} and r_{cs} are the interatomic distances. On substituting equations (3) and (4) in (1), we obtain the following expression:

$$(m_c + m_o + m_s)\, r_1 = m_s r_{cs} - m_o r_{co} \text{ or } Mr_1 = m_s r_{cs} - m_o r_{co}$$

or $$r_1 = \frac{m_s r_{cs} - m_o r_{co}}{M} \qquad ...(5)$$

where $M = m_c + m_o + m_s$

On substituting equations (3) and (4) into (2), we obtain

$$I = m_o (r_{co} + r_1)^2 + m_c r_1^2 + m_s (r_{cs} - r_1)^2 \qquad ...(7)$$

On substituting the value of r_1 from equation (5) in (7), we get

$$I = \frac{m_s r^2_{cs} - m_o r^2_{co} + 2\, m_o m_s r_{co} r_{cs} - m_s r^2_{cs}}{M} \qquad ...(8)$$

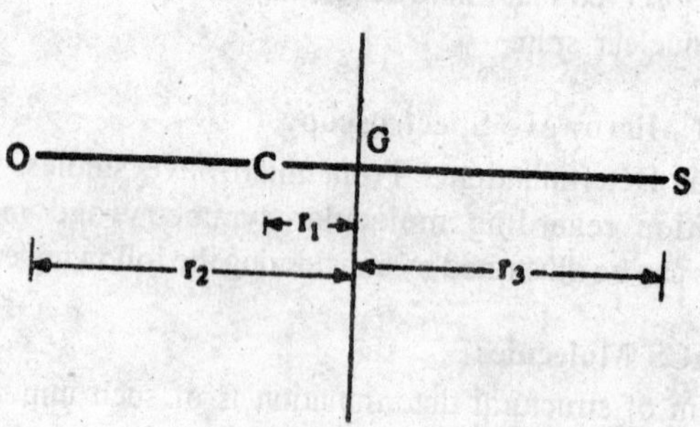

Fig. 2.9 : The $^{16}O\ ^{13}C\ ^{32}S$ Molecule.

In equation (8) r_{co} and r_{cs} are not known. In order to determine their values, a similar expression for the moment of inertia I' of another isotopic spices $^{16}O^{12}C^{34}S$ is obtained. This is as follows:

$$I' = \frac{m'_s r^2_{cs} - m_o r^2_{co} + 2\, m_o m'_s r_{co} r_{cs} - m'_s e^2_{cs}}{M} \quad ...(9)$$

In deriving equation (9) it is assumed that the isotopic substitution does not alter the interatomic distances. Tea and res. On solving equations (8) and (9), the values of r_{co}, and r_{cs} can be determined provided the values of I and I' are known which can be obtained from microwave spectra $^{16}O^{12}C^{32}S$. In Table 2.3, the interatomic distances are given which have been obtained from the various isotopic species.

Table 2.3 : Interatomic Distances in OCS

Pair of Isotopic Molecules	*O–C(A°)*	*C–S(A°)*
$^{16}O^{12}C^{32}S$, $^{16}O^{12}C^{34}S$	1.1647	1.5576
$^{16}C^{12}O^{32}S$, $^{16}O^{12}C^{32}S$	1.1629	1.5591
$^{16}O^{12}C^{34}S$, $^{16}O^{12}C^{34}S$	1.1625	1.5594
$^{16}O^{12}C^{34}S$, $^{16}O^{12}C^{34}S$	1.1552	1.5653

From Table 2.3, it can be seen that there are differences in interatomic distances when isotopes are changed. These differences may be attributed to zero-point vibrations which persist at all temperatures including absolute zero. The amplitude of zero-point vibrations depend on the mass of atoms and hence the bond lengths will change upon isotopic substitution.

2. The Inversion Spectrum of Ammonia

It was the first molecule to be studied by microwave Spectroscopy, by Bleaney and by Townes. In the spectrum of ammonia molecule, each of the lines is split into a doublet due to the inversion of the molecule.

The microwave spectrum of ammonia is concerned with the motion of the nitrogen atom between the centre of three hydrogen atoms as given in Fig. 2.10. When a graph is plotted between potential energy and position of nitrogen atom, the spectrum obtained is of the double minimum type shown with a central barrier of 2070 cm^{-1} (6 Kcal). The curvature of each minimum lies near 1000 cm^{-1} which is described as a NH bending fundamental. The splitting of the lower rotational levels (Fig. 2.11) occurs due to quantum mechanical tunnel effect which is due to finite barrier high and low mass of the hydrogen atoms.

In principle, the inversion of ammonia is a vibrational motion. As the inversion is hindered, the interaction will result in energy difference in microwave region. This interaction will cause the splitting of the vibrational levels. As the transitions are now between the two components of the vibrational levels, this will result in the rotational lines to split.

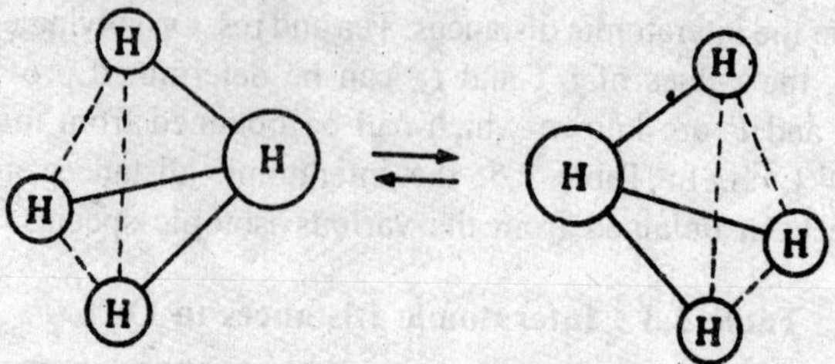

Fig. 2.10 : The inversion of ammonia molecule.

One may immediately raise the question that the molecule such as PH_3 and As H_3 should also show the splitting of the rotational levels. But they do not show such doublings. The main reason for this is that the inversion frequencies in these systems are much slower due to the increase in the barrier height.

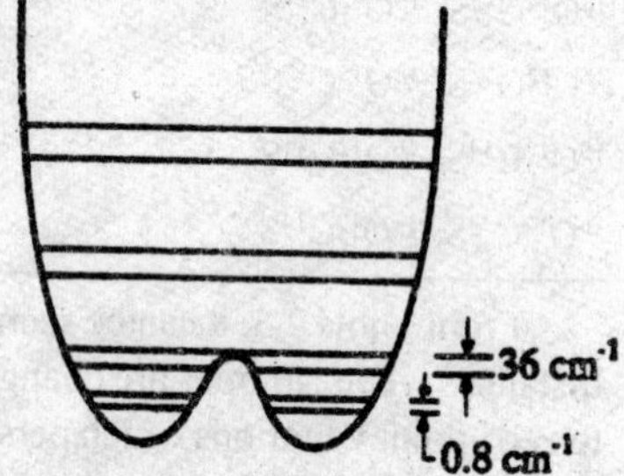

Fig. 2.11 : Energy diagram for the inversion coordinate of Ammonia.

INFRARED SPECTROSCOPY

Introduction : The technique is based upon the simple fact that a chemical substance shows marked selective absorption in the infrared region. After absorption of IR radiations, the molecules of a chemical substance vibrate at many rates of vibration, giving rise to closepacked absorption bands, called an IR *absorption spectrum* which may extend over a wide wavelength range. Various bands will be present in IR spectrum which will correspond to the characteristic functional groups and bonds present in a chemical substance. Thus, an IR *spectrum of a chemical substance is a fingerprint for its identification.*

Band positions in an infrared spectrum may be expressed conveniently by the wave number $\bar{v}$, whose unit is cm^{-1}. The relation between wave number $\bar{v}$, wavelength λ and frequency $\bar{v}$ is as follows

$$v = c/\lambda \quad \text{or} \quad \bar{v}/(cm^{-1}) = v/c = 1/\lambda$$

where c is the velocity of light. From the above relations, it follows that wave number is the reciprocal of wavelength.

Band intensities in IR spectrum may be expressed either as *transmittance* (T) or absorbance (A). Transmittance is defined as *the ratio of radiant power transmitted by a sample to the radiant power incident on the sample.* On the other hand, absorbance is defined as the logarithm, to the base 10, of the reciprocal of the transmittance, *i.e.*,

$$A = \log_{10}(1/T)$$

The Range of Infrared Radiation : The inferred radiation refers broadly to that region of electromagnetic spectrum which lies between the visible and microwave regions. However, this region may be divided into four sections :

(a) *The photographic region* : This ranges from visible to 1.2 (A.

(b) *The Very Near Infrared region* : This is also known as overtone region and ranges from 1.2 to 2.5μ.

(c) *The Near Infrared Region* : This is also known as vibration rotation region and ranges from 2.5 to 15μ.

(d) *The Far Infrared Region* : The is known as the rotation region. This ranges from 25 to 300-440μ.

When an analytical chemist speaks of infrared spectroscopy, he usually means the range from 2.5 to 25μ (microns) or 4000 to 400 wave numbers (waves per cm or cm^{-1}). This range gives him the important information about the vibrations of molecules, and hence about the structure of molecules.

The Requirements for Infrared Radiation Absorption

For a molecule to absorb IR radiation, it has to fulfil certain requirements which are as follows :

(a) Correct Wavelength of Radiation

A molecule absorbs radiation only when the natural frequency of vibration of some part of a molecule (*i.e.* atoms or group of atoms comprising it) is the same as the frequency of the incident radiation. For example, the natural frequency of vibration of HCl molecule is about 8.7×10^{13} sec^{-1} (2890 cm^{-1}). When IR radiation is permitted to pass through a sample of HO and the transmitted radiation is analysed by the

IR spectrometer, it is observed that part of the radiation which has a frequency of 8.7 × 10" sec^{-1} has been absorbed by HCl molecule whereas the remaining frequencies of the radiation are transmitted. Thus, the frequency 8.7×10^{18} sec^{-1} is characteristic of HCl molecule.

After absorbing the correct wavelength of radiation, the molecule vibrates at an increased amplitude. This occurs at the expense of the energy of the IR radiation which has been absorbed.

(b) Electric Dipole

This is another condition for a molecule to absorb IR radiation. A molecule can only absorb IR radiation when its absorption causes a change in its electric dipole (dipole moment). A molecule is said to have electric dipole when there is a slight positive and a slight negative electric charge on its component atoms.

Wnen the molecule having electric dipole is kept in the electric field (as in the case when the molecule is kept in a beam of IR radiation), this field will exert forces on the electric charges in the molecule. Opposite charges will experience force in opposite direction. This tends to decrease or increase separation in. A the electric field of the IR radiation is changing its polarity periodically, it means that the spacing between the charged atoms (electric dipoles) of the molecules also changes periodically. When these charged atoms vibrate, they absorb IR radiation from the radiation source. If the rate of vibration of charged atoms in a molecule is fast, the absorption of radiation is intense and, thus, the IR spectrum will have intense absorption bands. On the other hands, when the rate of vibration of the charged atoms in a molecule is slow, there will be weak bands in the IR spectrum.

As the symmetrical diatomic molecules, like O_2 and N_2, do not possess electric dipole, they cannot be excited by infrared radiation and, thus, they do not give rise to IR absorption spectra. This is very fortunate, for otherwise one would have to evacuate the air from infrared spectrometers. However, carbon dioxide and water vapour in air do absorb in the molecule, but these do not affect IR spectra taken on a double-beam instrument, as they are fairly weak and cancel out between the sample beam and the reference beam.

The Theory of IR Absorption Spectroscopy

In order to understand the theory of IR absorption spectroscopy, one has to understand the phenomenon of vibrational-rotational spectra. Let us consider a diatomic molecule associated with a dipole moment.

The vibratory motion of the nuclei of diatomic molecule may be similar to the vibration of a linear harmonic oscillator. In such an oscillator, the force tending to restore an atom to its original state is proportional to the displacement of the vibratory atom from the original position Hooke's Law). Suppose the bond between the two nuclei of a diatomic molecule is distorted from its equilibrium length r_e to a new length r. Then the restoring forces on each atom of the diatomic molecule will be given by

$$m_J \frac{d^2 r_1}{dt^2} = K(r - r_e) \qquad ...(1)$$

$$m_2 \frac{d^2 r_2}{dt^2} = K(r - r_e) \qquad ...(2)$$

where K is the proportionality constant and known as force constant : it is regarded as a measure of the stiffness of the bond r_1 and r_2 are the positions of atoms 1 and 2 relative to the centre of gravity of the molecule. We known

$$r_1 = \frac{m_2}{m_1 + m_2} r \qquad ...(3)$$

But
$$r_2 = \frac{m_1}{m_1 + m_2} r \qquad ...(4)$$

where m_1 and m_2 are the masses of two atoms of a vibrating dia-atomic molecule. On substituting equation (3) in (1), we get.

$$\left(\frac{m_1 m_2}{m_1 + m_2}\right), \frac{d^2 r}{dt^2} = -K(r - r_e) \qquad ...(5)$$

As r_e is a constant, its differentiation with respect to time t will be zero, *i.e.*,

$$\frac{d^2 r}{dt^2} = \frac{d^2(r - r_e)}{dt^2} \qquad ...(5A)$$

Substitution of the above equation in equation (5) yields

$$\left(\frac{m_1 m_2}{m_1 + m_2}\right), \frac{d^2(r - r_e)}{dt^2} = -K(r - r_e) \qquad ...(6)$$

put
$$r - re = x \text{ and } \frac{m_1 m_2}{m_1 + m_2} = \mu \qquad ...(7)$$

On substituting equation (7) in (6), we get

$$\mu \frac{d^2x}{dt^2} = -kx$$

or $$\frac{d^2x}{dt^2} + \frac{k}{\mu} x = 0 \qquad ...(8)$$

or $$\frac{d^2x}{dt^2} + \omega^2 x = 0 \qquad ...(9)$$

where $\omega^2 = k/\mu$ or $\omega = \sqrt{k/\mu}$ ...(10)

Equation (9) is the expression of a simple harmonic motion with frequency of vibration as follows

$$v = \omega/2\pi = 1/2\pi\sqrt{k/\mu} \qquad ...(11)$$

where K is a force constant expressed in dynes per cm and μ is the reduced mass of the system (Eq. 7).

Energy Levels

Vibrational energies, E_v, like all other energies for any particular vibrational system may be evaluated from Schrodinger's equation. The energy values for the energy of a linear harmonic oscillator are of the following type :

$$Ev = (v + 1/2)\, hc\overline{w} \text{ where } \overline{w} = v/c = w/2\pi c \qquad ...(12)$$

In equation (12), v is the vibrational quantum number and can have values of 0, 1, 2, ... etc. and $\overline{w}$ is the vibrational frequency of the oscillator in terms of wave numbers.

One can deduce a very important result from equation (12) when the oscillating molecule is in its lowest vibrational state, *i.e.*, when v = 0, the energy E_0 may be given by

$$E_0 = 1/2\, hc\overline{w} \qquad ...(13)$$

where E_0 is know as zero-point energy of the molecules. Thus, even at absolute zero, when the translational and rotational motion cease, vibrational motion still exists.

Suppose we consider a diatomic molecule which under goes transition from an upper vibrational level v' to lower vibrational level v" The change in vibrational energy will then be given as

$$\Delta E_v = E_v' - E_v'' = (v' + 1/2)\, hc\overline{w} - (v'' - 1/2)\, hc\overline{w}$$

or $$\Delta E_v = (v' - v'')\, hc\overline{w} \qquad ...(14)$$

The frequency of transition in wave numbers may be written as

$$v_V = \frac{\Delta E_V}{hc} = \frac{(v' - v'')hc\overline{w}}{hc} = (v' - v'')\overline{w} \quad ...(15)$$

so that the frequency of spectral line arising due to transition between v = 1 and v = 0 will be given by

$$\overline{v}_{1,0} = \overline{w} = v / c = \frac{1}{2}\pi c \sqrt{\frac{k}{\mu}} \text{ cm}^{-1} \quad ...(16)$$

Let us consider a diatomic molecule which possesses a dipole moment. Then the change in the vibrational energy may be obtained as follows :

$$v = \frac{E_v' - E_v''}{h} \text{(Bohr – frequency condition)}$$

$$\overline{w}_B = v/c = Ev' - Ev''/hc \quad ...(17)$$

where $\overline{w}_B$ is the wave number of the radiation emitted. On substituting equation (14) in (17), we get

$$\overline{w}_B = \frac{\left(v' + \frac{1}{2}\right) hv\overline{w} - \left(v'' + \frac{1}{2}\right) hc\overline{w}}{hc}$$

$$= (v' + 1/2)\ \overline{w} - (v'' + 1/2)\ \overline{w} \quad ...(18)$$

But $\quad v' - v'' = 1$ or $v' = v'' + 1$

$$\therefore \quad \overline{w}_B = (v'' + 1 + 1/2)\ \overline{w} - (v'' + 1/5)\overline{w} \quad ...(19)$$

or $\quad \overline{w}_B = \overline{w}$

or $\quad \dfrac{\overline{w}_B}{c} = \dfrac{\overline{w}}{c}$, or $\overline{v}_B = \overline{v} \quad ...(20)$

From the above equation it follows that the vibration spectrum should consist of single lines provided there are no accompanying rotational changes. From equation (20) it can be concluded that the frequency of radiated light is equal to the frequency of oscillator, NO matter, whatsoever is the v value of the initial state may be. It means that all the vibrational lines obtained for many harmonic oscillator are of the same frequency (Fig. 2.12).

The selection rule for vibrational quantum number of the harmonic oscillator is restricted to $\Delta v = \pm 1$ and hence each mode of vibration yields one band only. The plus sign applies to the absorption spectra and minus sign to emission spectra. As vibration spectra are usually determined by absorption phenomena, the selection rule becomes as $\Delta v = \pm 1$.

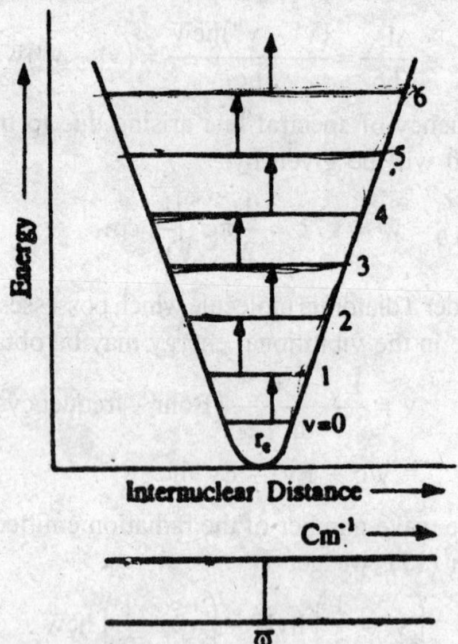

Fig. 2.12 : The allowed-vibration energy levels and transitions between them for a diatomic molecule undergoing simple harmonic motion.

Discussion

From the above treatment we observe the following important points:

(i) When one considers a diatomic molecule as harmonic oscillator, this describes lowest vibrational states, but fails for higher vibrational states Therefore, this requires correction.

(ii) Another weakness of the harmonic oscillator is that one observes experimentally that vibrational energy levels are less and less separated as v increases whereas the harmonic oscillator results in even spacing.

(iii) Overtones and combination terms are also observed in vibrational spectra. But the harmonic oscillator does not explain this.

Due to the above mentioned reasons, one should consider the diatomic molecule as an harmonic oscillator.

Vibrating Diatomic Molecule as Anharmonic Oscillator

Real molecules do not obey exactly the laws of simple harmonic motion. The bonds in real molecules are known as real bonds. Although, for small expansions, and extensions, real bonds may be considered to be perfectly elastic, obeying Hooke's law, yet at larger distortions they deviate from the behaviour of Hooke's law.

Fig. 2.13. shows the energy curve for a typical diatomic real molecule which behaves as anharmonic oscillator and undergoes extensions and compressions.

In the same figure, a dotted parabola is there which is due to an ideal diatomic molecule obeying ideal simple harmonic motion.

In order to explain the energy curve due to a real molecule, P.M. Morse invented a purely empirical expression known as Morse function which is as follows :

$$E = D_{eq}[\,1 - \exp\{a(r_{eq} - r)\}]^2 \quad ...(21)$$

where D_{eq} = the dissociation energy for a particular molecule

$\propto$ = a constant for a particular diatomic molecule and

r_{eq} = the value of internuclear distance which corresponds to a minimum of Morse curve.

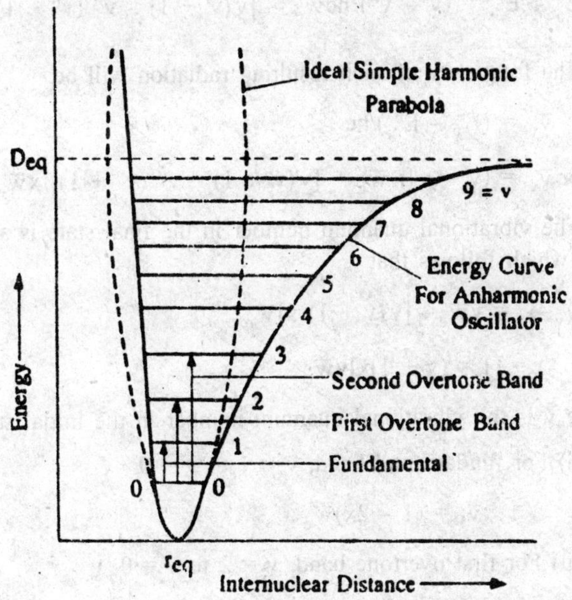

Fig. 2.13

When equation (21) is used in the Schrodinger's equation, the values of allowed vibrational energy levels are as follows :

$$E_v = (v+1/2)\, hc\overline{w}_e - (v+1/2)^2 hcx\, \overline{w}_e + (v + 1/2)^2\, hcy\, \overline{w}_e + ... \quad ...(22)$$

where $\overline{w}_e$ = equilibrium frequency of the molecule expressed in wave numbers, and x, y, ..., = the anharmonic constants.

The selection rules for all the transitions in anharmonic oscillator may be given as :

$$\Delta v = \pm 1, \pm 2, \pm 3, ...$$

for these transitions in which

(i) v = 1 to v = 0 gives fundamental band

(ii) v = 2 to v = 0 gives first overtone (second anharmonic)

(iii) v = 3 to v = 0 gives second overtone (third ;anharmonic), etc.

The energy change when a transition results from an upper level v' to the lower level v" will be given by

$$E'_v - E''_v = (v' - v'')\, hc\overline{w}_e - [v'(v' + 1) - v''\, (v'' + 1)]\, xhc\overline{w}_e \quad ...(23)$$

The frequency of such resulting radiation will be

$$\overline{v}_v = (E'_v - E''_v)/hc \quad ...(24)$$

or $\overline{v}_v = (v' - v'')\, \overline{w}_e - [v'(v' + 1) - v''\, (v'' + 1)]\, x\overline{w}_e$.

The vibrational quantum number in the final state is always zero from which follows that

$$\overline{v}_{v \to o} = v\overline{w}_e - [v'(v + 1)\, x]\overline{w}_e \quad ...(25)$$

$$= [1 - (v + 1)x]v\overline{w}_e \quad ...(26)$$

where v is the vibrational quantum number in the initial state. Thus,

(i) For fundamental band, v = 1 to v = 0

$$\overline{v}_1 = (1 - 2x)\overline{w}_e \quad ...(27)$$

(ii) For first overtone band, v = 2 to v = 0

$$\overline{v}_2 = (1 - 3x)2\overline{w}_e \quad ...(28)$$

(iii) For second overtone band, v = 3 to v = 0

$$\overline{v}_3 = (1 - 4x)3\overline{w}_e \qquad ...(29)$$

where $\overline{v}_1$, $\overline{v}_2$, $\overline{v}_3$ are the frequencies of the origins or centres of fundamental, first and second overtone respectively. As the frequencies of the first and second overtone bands are 2 to 3 times the frequency of the fundamental, they appear in the regions of shorter wave lengths as compared to the fundamental band. These have been shown in Fig. 2.13.

Diatomic Vibrating Rotator

In the earlier discussion it was assumed that a diatomic molecule behaves as harmonic or anharmonic oscillator. But it seems natural to assume that the rotation and vibration must take palace simultaneously and in fact the observed fine structure of rotation bands reveals that a simultaneous rotation and vibration do occur in such molecules. For this reason, a diatomic molecule will be considered which can execute rotations and vibrations simultaneously. Such a system is termed as a rotating vibrator or a rotating oscillator.

If there is no interaction of vibration and rotation, the total energy of the vibrating rotator would be given by the sum of vibrational and rotational energies.

$$E_{total} = E_{rot} + E_{vib} \qquad ...(30)$$

Again, we will have to assumed whether the diatomic molecule is behaving as a harmonic oscillator and a rigid rotator or anharmonic oscillator and nonrigid rotator. Let us discuss these cases one by one.

(1) Diatomic Molecule as a Harmonic Oscillator and a Rigid Rotator

The combined energy of such a system is given by

$$E_{vr} = E_{total} = (v + 1/2)\, hc\overline{w} + \frac{h^2}{8\pi^2 I} J(J+1) \qquad ...(31)$$

It can be proved very easily that the selection rules for combined motions are the same as those for each separately. Therefore,

$$\Delta v = \pm 1, \pm 2, \pm 3, \text{ and } \Delta J = \pm 1.$$

The frequency of spectral lines arising due to transition between the two states can be calculated by using equation (31). Suppose a simultaneous transition from the vibrational level v' to v" and from rotational level J' to J" occur. Now the change in energy, on using equation (31), may be put as

$$E'_{vr} - E''_{vr} = (v' - v'')\, hc\overline{w} + \frac{h^2}{8\pi^2 I}\,[j''+1)-J''\,(J''+1)] \quad ...(32)$$

and the corresponding frequency of radiation, which arises due to this transition, will be given by

$$\overline{v} = \frac{E'_{vr} - E''lvr}{hc} = \frac{(v' - v'')hc\overline{w} + \frac{h^2}{8\pi^2 I}\left[J'(J'+1) - J''(J''+1)\right]}{hc}$$

$$= (v' - v'')\,\overline{w} + B[J'\,(J'+1) - J''\,(J''+1)] \quad ...(33)$$

where $B = \frac{h^2}{8\pi^2 I}$

Now $\overline{v} = (v' - v'')\,\overline{w} + B[J'^2 + J' - J''^2 - J'']$

$$= (v' - v'')\,\overline{w} + B(J' - J'')\,(J' + J'' + 1) \quad ...(34)$$

As the vibrations are assumed to be harmonic, the possible vibrational transitions are those in which v' – v" = 1. Thus, equation (34) becomes as

$$\overline{v} = \overline{w} + B(J' - J'')\,(J' + J'' + 1) \quad ...(35)$$

Form Eq. (35) it is evident that the frequencies of the lines will evidently vary with the values of J' and J". Thus,

(i) if $\Delta J = +1$, *i.e.*, $J' = J'' + 1$ or $J' - J'' = 1$,

so that equation (35) becomes as

$$\overline{v}\,(R) = \overline{w} + 2B(J''+1)\ cm^{-1} \text{ where } J'' = 0, 1, 2, \text{etc.,} \quad ...(36)$$

(ii) if $\Delta J = -1$, $J'' = J' + 1$ or $J' - J'' = 1$, so that equation (35) becomes as

$$\overline{v}\,(P) = \overline{w} - 2B(J'+1)\ cm^{-1} \text{ where } J' = 0, 1, 2, \text{etc,} \quad ...(37)$$

On combining equations (36) and (37), we get

$$\overline{v} = \overline{w} + 2Bm\ cm^{-1} \quad ...(38)$$

where m = ± 1, ±2, ±3, ... etc. The value of m cannot be zero because this would imply value of J' or J" to be – 1 which is not possible. As can be seen that the spectrum given by equation (38) will consist of equally spaced lines with a spacing of 2B on each side of band origin.

If v' – v" = 1 and J' – J" = 0 equation (35) becomes as

$$\bar{v} = \bar{w} \quad ...(39)$$

From equation (39) it follows that the frequency (in wave numbers) of the corresponding spectral line is equal to the vibration frequency of the molecule expressed in cm^{-1} units. However, the spectral line for J = 0 is generally missing.

Limitations

The theoretical predictions made by equation (38) are not in agreement with experimental results Therefore, we will now drop this concept and consider a diatomic molecule as anharmonic oscillator an a non-rigid rotator.

(2) Diatomic Molecule as Anharmonic Oscillator and a Non-rigid Rotator

Suppose we assume the diatomic molecule to be anharmonic oscillator and non-rigid rotator. Again, it is assumed that there is no interaction of vibration and rotation. In such a situation it may be put as

$$E_{vr} = E_{total} = E_{rot} + E_{vib}$$

The combined energy of a diatomic molecule will be given by

$$E_{vr} = hc[\{B(J + 1) - DJ^2(J + 1)^2 + ...\}+$$

$$\left\{\left(v + \frac{1}{2}\right)\bar{w}_e - x\left(v + \frac{1}{2}\right)\bar{w}_e\right\}\Bigg] \quad ...(40)$$

It can be proved that the selection rules for the combined motions remain same as those for each separately, thus,

$$\Delta v = \pm 1, \pm 2, ... \text{ and } \Delta J = \pm 1$$

The frequency of spectral line arising due to transition between the two states may be evaluated by using equation (40). For a good infrared spectrometer possessing a resolving power of about 0.5 cm, the value of D can be neglected. Thus, equation (40) becomes as

$$E_{vr} = hc\ BJ(J + 1) + (V + 1/2)\ \bar{w}_e\ hc - x(v + 1/2)^2\ \bar{w}_e hc \quad ...(41)$$

Suppose we consider only v = 0 to v = 1 transition. Now designate rotational quantum number in v = 0 state as J' and v = 1 state as J". Thus one can write for the frequency of transition by using equation (41).

$$\bar{v} = \frac{E' - E''}{hc}$$

or $\bar{v} = 1/hc[\{hcBJ'(J' + 1) + (1 + 1/2)\,\bar{w}_e\, hc - x(1 + 1/2)^2\,\bar{w}_e hc\}$

$$- \{hcBJ''(J'' + 1) + (0 + 1/2)\,\bar{w}_e\, hc - x(0 + 1/2)^2\,\bar{w}_e hc\}]$$

$$\bar{v} = BJ'(J' + 1) + 3/2\,\bar{w}_e - 9/4\, x\,\bar{w}_e - BJ''(J'' + 1)$$

$$- 1/2\,\bar{w}_e + 1/4\,\bar{w}_e x$$

$$= BJ'(J' + 1) + (1 - 2x)\,\bar{w}_e - BJ''(J'' + 1) = \bar{v}_o$$

$$+ B[J'(J' + 1) - J''(J'' + 1)]$$

$$= \bar{v}_o + B(J' - J'')(J' + J'' + 1)\text{cm}^{-1} \qquad ...(42)$$

where $\bar{v}_o = \bar{w}_e\,(1 - 2x)$. The value of B remains the same in upper and lower vibrational states because we have taken rotation to be independent of the vibrational changes. Thus, one can write

(i) When $\Delta J = +1$, *i.e.*, $J' = J'' + 1$, or $J' - J'' = +1$ so that equation (42) becomes as

$$\bar{v}(R) = \bar{v}_o + 2B(J'' + 1)\text{ cm}^{-1} \qquad ...(43)$$

where $J'' = 0, 1, 2, \ldots \ldots$ etc,

(ii) When $\Delta J = -1$, *i.e.*,

$J'' = J' + 1$ or

$J' - J'' = -1$ so that equation (42) becomes as

$$\bar{v}(P) = \bar{v}_o - 2B(J' + 1)\text{ cm}^{-1} \qquad ...(44)$$

Where $J' = 0, 1, 2, \ldots \ldots$ etc.

On combining equations (43) and (44), we get

$$\bar{v} = \bar{v}_o + 2Bm\text{ cm}^{-1}$$

where $m = \pm 1, \pm 2, \ldots$

The value of m cannot be zero because this will mean that the values of J' or J'' be negative which is not possible. The frequency $\bar{v}_o$ is generally termed as band origin.

Equation (44) represents the vibration-rotation spectrum. Such a spectrum would contain equally spaced lines with a spacing of 2B on either side of the bond origin $\bar{v}_o$. As the value of m cannot be zero, no line will appear at $\bar{v}_o$ and thus Q branch will be absent (Fig. 2.14).

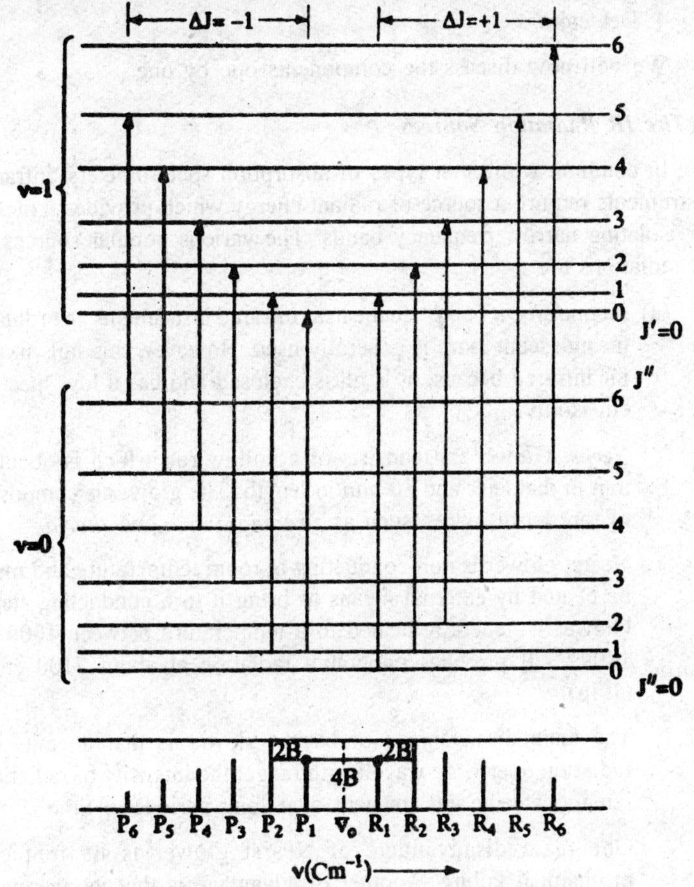

Fig. 2.14 : Transitions between the rotational vibrational energy levels of a diatomic molecule (for the states v = 0 and v = 1d). Note that first line in P branch is P_1 where in R branch it is R_0.

Instrumentation

The usual optical materials, glass or quartz absorb strongly in the infrared region, consequently the apparatus for measuring infrared spectra is appreciably different from that for the visible and ultraviolet regions. The main parts of n IR spectrometer are as follows :

1. The IR radiation sources.
2. The monochromators.

3. The sample cells and sampling of substances.

4. Detectors.

We will now discuss the components one by one.

1. The IR Radiation Sources

In common with other types of absorption spectrometers, infrared instruments require a source of radiant energy which provides a means for isolating narrow frequency bands. The various popular sources of IR radiations are :

(a) *Incandescent Lamp* : In the near infrared instruments an ordinary incandescent lamp is generally used. However, this fails in the far infrared because it is glass enclosed and has a low spectral emissivity.

(b) *Nernst Glower* : It consists of a hollow rod which is about 2 mm in diameter and 30 mm in length. The glower is composed of rare earth oxides such as *zirconia*, *yttria* and *thoria*.

Nearst glower is non-conducting at room temperature and must be heated by external means to bring it to a conducting state. Glower is generally heated to a temperature between 1000 to 1800°C. It provides maximum radiation at about 7100 cm^{-1} (1.4μ).

The main disadvantage of Nernst glower is that it emits IR radiation over wide wavelength range; the intensity of radiation remains steady and constant over long periods of time.

One main disadvantage of Nearst glower is its frequent mechanical failure. Another disadvantage is that its energy is also concentrated in the visible and near infrared regions of the spectrum.

(c) *Globar Source* : It is a rod of sintered silicon carbide which is about 50 mm in length and 4 mm in diameter. When it is heated to a temperature between 1300 and 1700°C, it strongly emits radiation in the IR region. It emits maximum radiation at 5200 cm^{-1}.

Unlike the Nernst glower, it is self-starting. As its temperature coefficient is positive, it can be conveniently controlled with a variable transformer.

Unlike the Nearst glower, it is more satisfactory, for it works at wavelengths longer than 650 cm^{-1} (0.15μ).

The main disadvantage is that it is a less intense source than the Nearst glower.

(d) *Mercury Arc :* In the far infrared instruments, high pressure mercury arc is generally employed. Beckmann devised the quartz mercury lamps for the same region in a unique manner. At the shorter wavelengths, the heated quartz envelope emits the radiation whereas at longer wavelengths the mercury plasma provides radiation through the quartz.

Monochromators

The radiation source emits radiations of various frequencies. As the sample in IR Spectroscopy absorbs only at certain frequencies, it therefore becomes necessary to select desired frequencies from the radiation source and reject the radiations of other frequencies. This selection has been achieved by means of monochromator which are mainly of two types:

(a) *Prism Monochromator :* A single pass monochromator has been illustrated in Fig. 2.15. The sample is kept at or near the focus of the beam, just before the entrance slit 'A' to the monochromator. The radiation from the source after passing through the sample and entrance slit, strikes the off-axis parabolic Littrow mirror B which renders the radiation parallel and sends it to the prism C. The dispersed radiation after reflecting from a plane Mirror D returns through the prism a second time and focusses into the exist slit of the monochromator, through which it finally passes into the detector section.

The double-pass monochromator has been illustrated in Fig. 5. In this there occurs a total of four passes of radiation through the prism as shown (1), (2), (3) and (4) in Fig. 2.15. The double pass monochromator produces more resolution than the monochromator in the radiation, before it finally passes on to the detector.

In both mono-pass and double-pass monochromators, sodium chloride (rock-salt) prism is employed for the entire region from 4000 to 650 cm^{-1} (2.5 to 15.4μ). Prisms of lithium fluoride or calcium fluoride give more resolution in the region where the significant stretching vibrations are located.

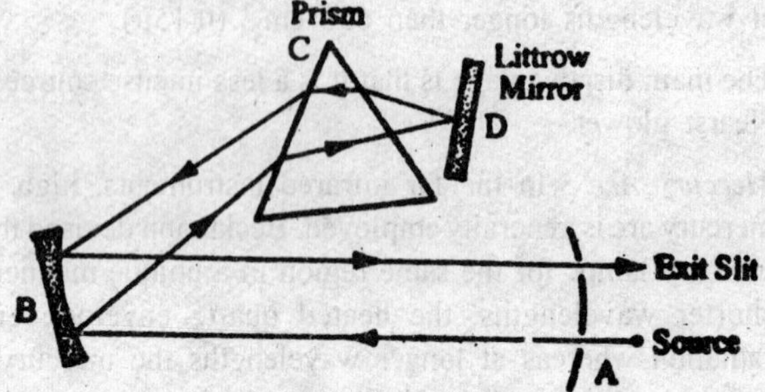

Fig. 2.15 : Single-pass monochromator.

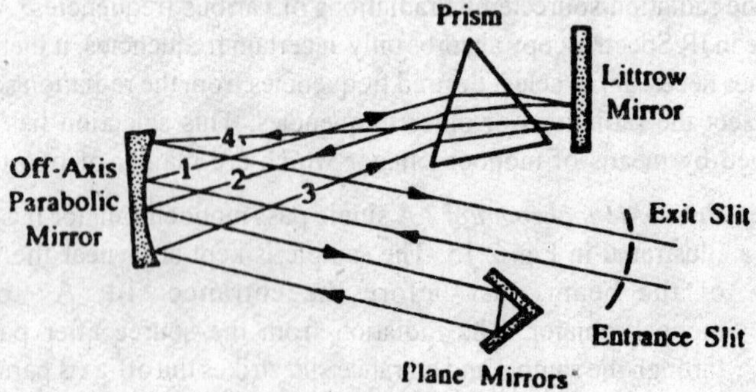

Fig. 2.16 : Double-pass monochromator.

(b) *Grating monochromator :* If a prism in a prism monochromator is replaced by a grating, higher dispersion can be achieved.

The grating is essentially a series of parallel straight lines cut into a plane surface (Fig. 2.17). Dispersion by a grating follows the law of diffraction (Fig. 2.18). It follows the following mathematical relation,

$$n\lambda = d(\text{Sin } i \pm \text{Sin } \theta)$$

where n is the order (a whole number), λ the wavelength of the radiation, d the distance between grooves, the angle of incidence of beam of IR radiation and θ the angle of dispersion of light of a particular wavelength.

For radiations of different wavelengths (λ), the angle of dispersion (θ) is different. At a grating, separation of light occurs because light of different wavelengths is dispersed at different angles.

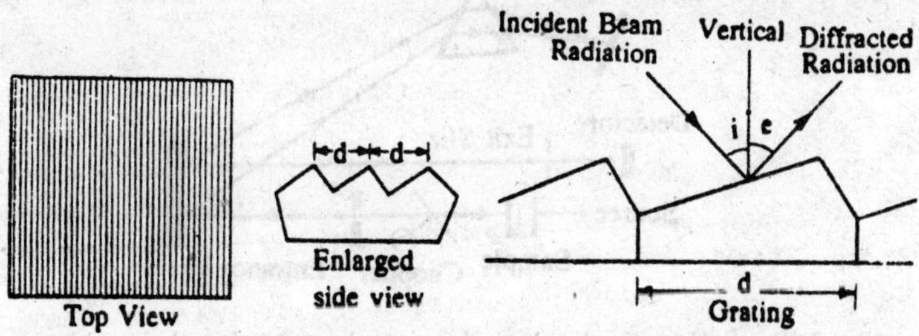

Fig. 2.17 : Magnified view of grating monochromator.

Fig. 2.18 : Path of IR Radiation diffracted by a grating monochromator.

Grating monochromator possesses the following advantages over prism monochromator :

(i) Grating can be made with materials like aluminium which are not attacked by moisture. On the other hand, metal salt prisms are subject to etching from atmosphere moisture.

(ii) Grating monochromators can be used over considerable wavelength ranges.

A grating is generally used in combination with a small prism which acts as an *order sorter*. Sometimes filters transparent over limited wavelength ranges can be used in combination with gratings.

Single Beam and Double Beam Spectrophotometers

A diagram of the optical system of a single beam infrared spectrophotometer is shown in Fig. 2.19.

In the single-beam system, the radiation is emitted by the source through the sample and then through a fixed prism and a rotating Littrow mirror. Both prism and Littrow mirror select the desired wave length and then allow it to pass on to the detector. The detector measures the intensity of radiation after it passes through the sample. Knowing the original intensity of radiation, one can measure how much radiation has been absorbed. By measuring the degree of absorption at wavelengths, the absorption spectrum of the sample can be obtained.

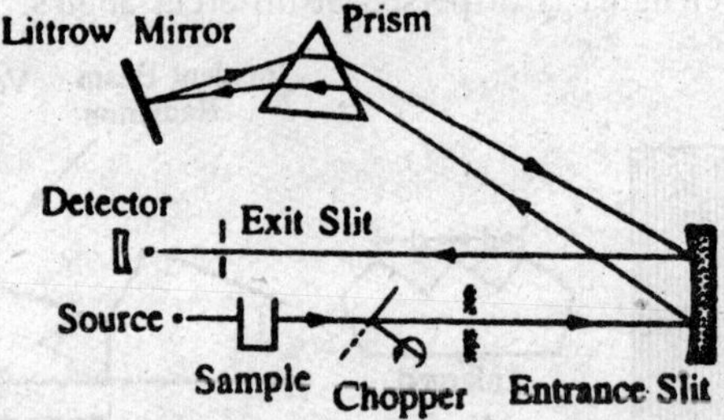

Fig. 2.19 : Schematic diagram showing the optical path in a single beam infrared spectrophotometer.

Disadvantages : The various disadvantages of a single-beam are as follows :

(i) This type of instrument has toe basic disadvantage that the intensity of the emission of the radiation source varies from point to point in IR absorption spectrum; therefore, the resulting spectrum is considerably deformed. The necessary correction by the continuous variation of slit is cumbersome.

(ii) When the sample is analyzed in solution, the bands of solvent appear in the spectrum. In this case, the spectrum of the sample is obtained by subtracting the spectrum of the solvent from the resultant spectrum, the former must be recorded under identical conditions (thickness of layer, etc.)

In order to overcome the above mentioned difficulties, a double-beam spectrophotometer is used. This is shown in Fig. 2.20.

The energy emitted by the radiation source is split by the instrument into two beams, which are energetically and optically identical. One of the beams passes through the sample and the other through the reference sample. The sample is placed in the sample beam and a reference material, such as the solvent used in the sample, is placed in the reference beam. The two half-beams are recombined and pass along the optical path to the detector.

When there is no sample in the reference beam, it arrives at the detector unabsorbed. When there is no sample in the sample cell, the half-beam travelling along the sample beam is not absorbed and is equal to the reference beam. When these two equal half beams recombine, a steady signal reaches the detector.

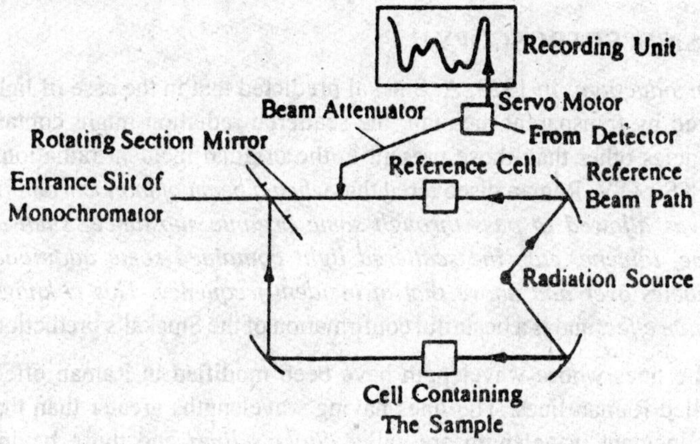

Fig. 2.20 : Schematic Diagram of a Double-beam Infrared Spectrophotometer.

When the sample cell contains the sample, the half-beam travelling through it undergoes a decrease in intensity. After the two half-beams are recombined, they produce an oscillating signal which is measured by the detector. The signal from the detector is passed on to the recording unit through a servo motor.

Limitations of Infrared Spectroscopy

Although infrared Spectroscopy has proved to be one of the most valuable methods for characterizing both qualitatively and quantitatively the multitude of inorganic compounds encountered in research as well as in industry yet it suffers from some shortcomings which are outlined as below:

1. By IR Spectroscopy, it is not possible to know molecular weight (except under special circumstances) of a substance.

2. Generally, the IR Spectroscopy does not provide information of the relative positions of different functional groups on a molecule.

3. From the single IR spectrum of an unknown substance, it is not possible to know whether it is a pure compound or a mixture of compounds. An interesting example is that a mixture of

paraffins and alcohols will give the same IR spectra as by higher molecular weight alcohols

RAMAN SPECTROSCOPY

Introduction : In 1923, A. Smekal predicted that in the case of light scattered by transparent medium, the scattered radiation might contain frequencies other than those present in the original incident radiations. In 1928 Sir C.V. Raman discovered that *when a beam of monochromatic light was allowed to pass through some organic substances such as benzene, toluene, etc., the scattered light contained some additional frequencies over and above that of incident frequency. This is known as Roman effect* and is a beautiful confirmation of the Smekal's prediction.

The lines whose wavelength have been modified in Raman effect are called Raman lines. The lines having wavelengths greater than that of the incident wavelength are called *Stake's lines* and those having shorter wavelengths are called *anti-Stoke's lines.*

If vi represents the frequency of the incident radiation and v½ that of the light scattered by a given molecular species, then the *Roman Shift*, Δv, is defined by $\Delta V = v_i - ve$

The Raman shift does not depend upon the frequency of the incident light but it is regarded as a characteristic of the substance causing Raman effect. For *Stoke's lines*, AV is positive and for anti-Stoke's Δv is negative (Fig. 2.22).

Characteristic properties of Raman lines : The lines observed in Raman effect exhibit a number of characteristics which are summarised as given below :

(i) The intensity of the *Stoke's lines* is always greater than the corresponding *anti-Stoke's lines.* With the rise of temperature, the intensity of anti-Stoke's lines increases.

(ii) Raman shift, Δv, generally lies within the range of 100 to 3000 cm^{-1} which lies in far and near infra-red regions of the spectrum.

(iii) The Raman lines are symmetrically displaced about the parent line. When the temperature rises, their individual separations from the parent lines decrease. The intensity of the anti-Stokc's lines falls off much rapidly with the increase in separation from the parent line.

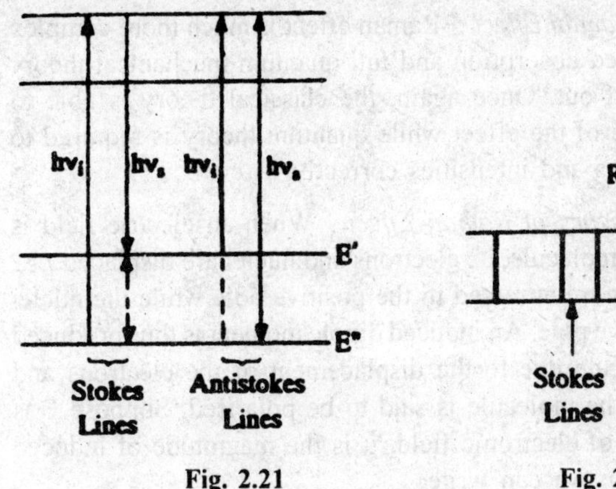

Fig. 2.21

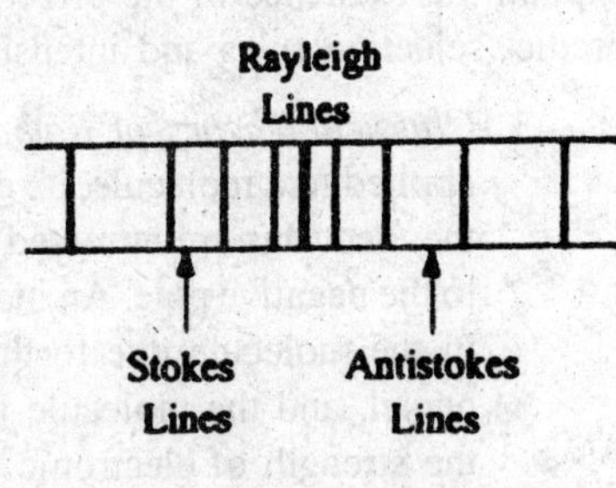

Fig. 2.22

(iv) The frequency difference between the modified and parent line represents the frequency of the absorption band of the material.

Differences between Raman Spectra and Infra-red Spectra

Though given in the above characteristics that Raman shift lies in far and infrared regions of the spectrum, yet Raman spectrum is quite different from infra-red spectra as pointed out below :

Table 2.4

Raman Spectrum	*Infra-red Spectrum*
1. It is due to the scattering of light by the vibrating molecules.	1. It is the result of the absorption of light by vibrating molecules.
2. Polarizability of the molecule will determinewhether the Raman spectrum will be observed or not.	2. The presence of a permanent dipole moment in a molecule is regarded as a criterion of infra-red spectrum.
3. As Raman lines are of weaker intensity, con-centrated solution must be utilised to increase the intensity of Raman lines.	3. Generally dilute solutions are preferred.
4. Water can be used as a solvent.	4. Water being opaque, cannot be utilised as a solvent.
5. Method is most accurate but is less sensitive.	5. Method is accurate as well as sensitive.
6. Optical systems are made of glass or quartz.	6. Optical systems are made up of special crystals such as CaF_2, NaBr.

Mechanism of Raman Effect : Raman effect is much more complex than ordinary infra-red absorption and full quantum mechanical theory is still being worked out. Once again, the classical theory is able to explain the existence of the effect while quantum theory is required to predict selection rules and intensities correctly.

1. *Classical Theory of Raman Effect :* When an electric field is applied to a molecule, its electrons and nuclei are displaced *i.e.*, the electrons are attracted to the positive pole while the nuclei to the negative pole. An induced dipole moment is thus produced in the molecule due to the displacement of the electrons and nuclei, and the molecule is said to be polarised. Suppose F is the strength of electronic field, μ is the magnitude of induced moment, then we can write

$$\mu = \alpha F \qquad ...(1)$$

where a denotes the polarizability of the molecules. The strength F of the electronic field of an electromagnetic wave of frequency, v, may be put as

$$F = F_0 \sin 2\pi Vt \qquad ...(2)$$

where F_0 is the equilibrium value of field strength. On combining equations (1) and (2), we get

$$\mu = \alpha\ F_0 \sin 2\pi evt \qquad ...(3)$$

From equation (3), it follows that the interaction of electro-magnetic wave of frequency v induces in the atom or molecule a dipole which oscillates with the same frequency. According to the classical theory, this oscillating dipole would emit radiation of the same frequency, v, i e., incident and scattered frequencies will be same; this is the case of Rayleigh's scattering. While deducing equation (3), vibration and rotation of molecules have not been considered. We will now consider the effect of vibration and rotation on equation (3).

(1) *Effect of vibration :* Suppose we are considering the diatomic molecule. As the two nuclei of the diatomic molecule vibrate along the line joining them, the polarizability of the molecule will change. If x is the small displacement from the equilibrium position, the vibration in polarizability, a is given by

$$\alpha - \alpha_0 - \beta \frac{x}{A} \qquad ...(4)$$

where αo is known as equilibrium polarizability, β denotes the rate of vibration of the polarizability with distance and A is the vibration amplitude. Suppose the molecule executes simple harmonic motion. Then, the displacement, x, can be put as

$$x = A \sin 2\pi v_V t \qquad ...(5)$$

where vy denotes the frequency of the vibration of the molecule. On substituting equation (5) in (3), we get

$$\mu = \alpha_0 F_0 \sin 2\pi vt + \beta F_0 \sin 2\pi vt \sin 2\pi v_V t$$

$$= \alpha_0 F_0 \sin 2\pi vt + 1/2\, \beta F_0 [\cos 2\pi (v - v_V) (-\cos 2\pi (v + v_V) t] \quad ...(6)$$

Thus, the induced dipole oscillates with frequencies $(v + v_V)$ and $(v - v_V)$ which are more and less than the frequency of incident radiation and predicts the existence of Raman scattering. Thus, the Raman shift will be as follows :

$$\text{Raman shift} = (v - v_V) - v = v_V \qquad ...(6A)$$

From equation (6A) it follows that the *Raman shift will be equal to the frequency of vibration of the diatomic molecule.*

(2) *Effect of rotation* : We will now discuss the effect of rotation of molecule on polarizability. When a diatomic molecule rotates, the orientation of a molecule varies with respect to the electric field of rotation. If the molecule is not optically isotropic, *i.e.*, it exhibits different polarizabilities in different directions then its polarization will vary with time. If we express variation of a by an equation identical to equation (5), we have

$$\alpha = \alpha_0 + \beta' \sin 2\pi (2v_r) t \qquad ...(7)$$

where v_r is the frequency of rotation. In the above equation, 2i'r has been put instead of v_r because a rotation through n angle will bring the diatomic molecule in a position in which its polarizability becomes same as initially. On substituting equation (7) into (3), we get

$$\mu = \alpha_0 F_0 \sin 2\pi vt + \beta' F_o \sin 2\pi vt \sin 4\pi v_r t$$

$$= \alpha_0 F_0 \sin 2\pi vt + 1/2\beta' F_o [\cos 2\pi (v - 2v_r) t$$

$$- \cos 2\pi (v + 2v_r) t] \qquad ...(8)$$

From the above equation it follows that the frequency of Raman lines will be $(V + 2v_r)$ and $(v - 2v_r)$. In this case, the Raman shift would be

$$\text{Raman shift} = (v + 2v_r) - v = 2v_r \quad ...(8A)$$

From equation (8A) it follows that the Raman shift would be equal to twice the frequency of rotation of molecule.

Quantum Theory

Raman effect cannot be explained on the basis fo classical electromagnetic theory. It becomes necessary to apply the quantum principles for its proper explanation. The Raman effect may be regarded as the outcome of the collisions between the light photons and molecule of the substance.

Suppose a molecule of mass m in the energy state E_p (the sum of electronic, vibrational and rotational energy) is moving with a velocity u and is colliding with a light photon hv. Suppose this molecule undergoes a change in its energy state as well as in its velocity. Let the new energy state be E_q and the velocity be u' after suffering a collision. If we apply the principle of conservation of the energy, we can write

$$E_p + 1/2mu^2 + hv = E_q + 1/2mu^2 + hv' \quad ...(1)$$

It can be easily proved that the change in velocity of the molecules is practically negligible. Thus, equation (1) can be written as

$$E_p + hv = E_q + hv' \text{ or } v' = v + \frac{E_p - E_q}{h} \quad ...(2)$$

or $$v' = v + \Delta v \quad ...(3)$$

From Eq. (2), three cases may arise :

(i) If $E_p = E_q$, then the frequency difference (Raman shift) Δv $\left(\text{i.e., } \frac{E_p - E_q}{h}\right)$ is zero. Hence, v' = v and this refers to the unmodified line where the molecule simply deflects the photon without receiving any energy from it. The collision thus being elastic is analogous to Rayleigh scattering.

(ii) If $E_p > E_q$ then v' < v which refers to anti-stoke's lines. It means that the molecule was previously in the excited state and it handed over some of its intrinsic energy to the incident photon the scattered photon thus has grater energy.

(iii) If $E_p < E_q$ then v' < v which corresponds to the Stoke's lines. The molecule has absorbed some energy from the incident photon and consequently the scattered photon will possess lower energy.

As the change in the intrinsic energy of the molecule is governed by quantum rules, we can write

$$E_p - E_q = \pm\ nh\nu_c \qquad ...(4)$$

where n = 1, 2, 3, etc. and ν_c the characteristic frequency of the molecule. IN the simplest case when n = 1, equation (4), reduces to the form

$$\nu' = \nu \pm \nu_c \qquad ...(5)$$

It follows from equation (5) that the frequency difference ($\nu - \nu'$) between the incident and scattered photon in the Raman effect corresponds to the characteristic frequency ν_c of the molecule. The Raman lines are equispaced from the unmodified parent line on the either sidè, at distance equal to the characteristic frequency lines on the scatterer.

Explanation of the intensity of Raman lines : The molecules in a medium are supposed to be distributed among the different quantum states of energies E_1, E_2, ..., etc. Assuming their statistical distribution to be according to Boltzmann's law, the number of molecules N_p having a particular energy state E_p is given by

$$N_p = CNg_p\ e^{-Ep/kT}$$

where C is a constant, N the total number of molecules, g_p the statistical weight of the state, k the Boltzmann's constant and t the absolute temperature.

Now at the room temperature, the energy E_p of a molecule will be small and therefore the $e^{-Ep/kT}$ will be high. Hence the number of molecules possessing this low energy E_p will be numerous, AS a consequence of this, the Stoke's transitions will occur more frequently than the anti-Stoke's. This will make the Stoke's line more intense than the anti-Stoke's lines.

As the temperature is raised, the kinetic energy of the molecules increases and more molecules are raised to the higher energy states. As a result the anti-Stoke's lines will gradually grow in their intensities and become more prominent.

(3) *Pure Rotational Raman Spectra* : The selection rule for rotational Raman spectra is $\Delta J = 0, \pm 2$ in contrast to the corresponding selection rule in infrared spectroscopy, $\Delta J = \pm 1$.

$$\Delta J = 0, \pm 2.$$

When ΔJ is zero, *i.e.*, $\Delta J = 0$, the scattered Raman radiation will be of the same frequency as that of incident light (Rayleigh Scattering). The transition $\Delta J = +2$ gives Stokes lines (linger wave lengths) where $\Delta J = -2$ gives two anti Stoke's lines (shorter wave lengths) Utilising the relation for the energy of a rigid rotator,

$$E_r = \frac{h^2}{8\pi^2 I} J(+1) \qquad r...(1)$$

when $\Delta J = +2$, the values of rotational Raman shifts (Stoke's line) will be given by

$$\Delta\bar{v} = \frac{h^2}{8\pi^2 I_C}\{(J + 2)(J + 3) - J(J + 1)\}$$

$$= 2B(2J + 3) \text{ where } B = \frac{h^2}{8\pi^2 I_C} \qquad ...(2)$$

When $\Delta J = -2$, the values of rotational Raman shifts (Anti-Stoke's lines) will be given by

$$\Delta\bar{v} = -2B(2J + 3) \qquad ...(3)$$

On combining equations (2) and (3), the Raman shift can be put in the form

$$\bar{v} = \pm 2B(2J + 3) \text{ where } J = 0, 1, 2, ...$$

The wave numbers of the corresponding spectral lines will be thus, given by

$$\bar{v} = \bar{v}_{ex} + \Delta\bar{v} \qquad ...(4)$$

where $\bar{v}_{ex}$ is the wave number of exciting radiation.

The transitions and the Raman spectrum arising from there are drawn schematically in Fig. (3). From the Fig. (3), it can be seen that the frequency separation of successive lines is $2B cm^{-1}$ where as it is $4B$ cm^{-1} in the far infrared spectra. While on substituting $J = 0$ in equation (4), we observe that the separation of the first line from the exciting line will be $6B$ cm^{-1}.

The homonuclear diatomic molecules (*e.g.*, O_2 and H_2) do not exhibit infra-red or microwave spectra but they do exhibit rotational Raman spectra. Therefore, their structures can be ascertained. If a molecule possesses a centre of symmetry (*e.g.* H_2, O_2 and CO_2), then the effect of nuclear spin will be observed in Raman as well as in infra-red spectra. Thus, in O_2 ,CO_2 etc. every alternate rotational level will be missing from the spectrum (Fig. 3), in the case of O_2, every level with even J values in missing (J = 0, 2, 4, ...).

(4) *Vibrational Rotational Raman Spectra :* It is possible theoretically for vibrational and rotational transitions to occur simultaneously in a Raman transition. In such a case, the selection rules governing such simultaneous changes in rotational and vibrational energy will be

$$\Delta J = 0, \pm 2 \text{ and } \Delta v = \pm 1$$

For a diatomic molecule vibration-rotation energy levels are given by

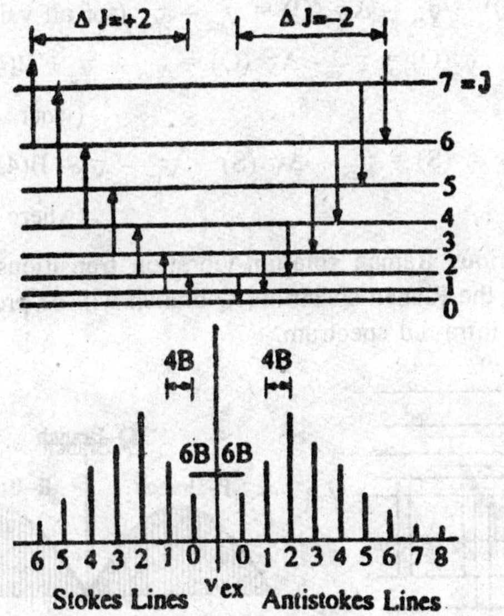

Fig. 2.23 : Rotational energy levels of a diatomic molecule and the rotational Raman spectra.

$$Erv = hv\,[\overline{w}_e(v+1/2) - \varpi_e\, x\,(v+1/2)^2 + Bhc\, J(J+1)]\ cm^{-1} \quad ...(1)$$

where v = 0, 1, 2, ... and = 0, 1, 2, ... In terms of frequency equation (1) modifies to

$$\overline{v} = \varpi_e(v + 1/2) - \varpi_e\, x\,(v + 1/2)^2\, BJ\,(J + 1) \quad ...(2)$$

We will now apply selection rules as stated earlier to equation (2) and get the following results :

(i) When $\Delta J = 0$, $\Delta\bar{v}\ (Q) = \bar{v}_o$ (for all values of J).

(ii) When $\Delta J = +2$, $\Delta\bar{v}\ (S) = \bar{v}_o + B\ (4J + 6)$

(where J = 0, 1, 2, ...)

(iii) When $\Delta J = -2$, $\Delta\bar{v}\ (O) = \bar{v}_o - B(4J + 6)$

(where J = 2, 3, 4, ...)

In the results (i) , (ii) and (iii), $\bar{v}_o$ is write for $\varpi_e\ (1 - 2x)$; O, Q, and S denote O branch lines, Q branch lines and S branch lines respectively. The corresponding Stoke's lines will occur at wave number represented by

(i) $\bar{v}\ (Q) = \bar{v}_{ex} - \Delta\bar{v}\ (Q) = \bar{v}_{ex} - \bar{v}_o$ (for all values of J)

(ii) $\bar{v}\ (O) = \bar{v}_{ex} - \Delta\bar{v}\ (O) = \bar{v}_{ex} - \bar{v}_o + B(4J + 6)$

(wnere J = 2, 3, 4, ...)

(iii) $\bar{v}\ (S) = \bar{v}_{ex} - \Delta\bar{v}\ (S) = \bar{v}_{ex} - \bar{v}_o - B(4J + 6)$

(where J = 0, 1, 2, ...)

The various Raman rotation-vibration transitions are shown in Fig. 2.24. In the Raman spectrum, Q branch will be present whereas it is absent in infra-red spectrum.

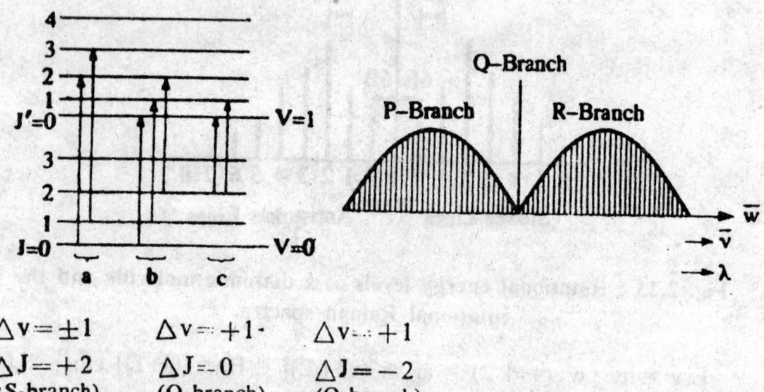

Fig. 2.24

Fig. 2.25

All the lines in Q branch are very close to one another an dare not resolved. On the other hand the S and O branches are very weak and their lines are not superimposed. Both S and O branches are somewhat similar to the R and P branches in infrared spectroscopy. The three branches O, S and Q are illustrated in Fig. 2.24. As all the lines of Q branch all almost in the same position for all values of J whereas those in the two wings are separated. The branch on the frequency side for which ΔJ is – 2 is termed as the O-branch whereas that on the high frequency side for which $\Delta J = + 2$ is termed as S-branch.

Experimental Study

The original simple arrangement of raman was not quite efficient and required very long exposures of about 100 hours and more to obtain good records of the Raman spectrum. Hence, improvements were made as regards the container of the substance, the source of radiation, filter, spectrograph, etc. The apparatus shown in Fig. 2.26 is the one developed by Wood and now ordinarily used in the study of the Raman effect in liquids.

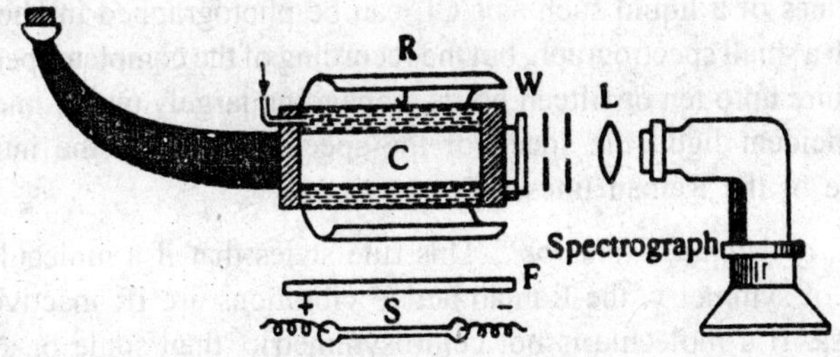

Fig. 2.26

Container

The container C of the liquid to be investigated, called the Raman tube, consists of a glass tube of about 1 – 2 cm in diameter and 10 to 15 cm. long, one end of which is drawn out into the shape of a horn and blackened outside to provide a suitable background, the other end being closed with optically plane glass plate constituting the window W through which the scattered light emerges. The container is surrounded by a water jacket J in which clod water is circulated to prevent over-heating of the liquid due to the proximity of the exciting are.

Source of Light

S is an ideal source which was previously a helium discharge tube filtered by nickel oxide glass, giving a strictly monochromatic line of wavelength 3888 Å but now used is the mercury are which is placed as close to the Raman tube as possible, which results in the large intensity of the incident light. A semi-cylindrical aluminium reflector R enhances the intensity of illumination still further.

In order to get lines of particular wave length from a mercury lamp, suitable filters, for example, to obtain the 4358Å line acidulated quinine sulphate solution contained in a novial glass vessel is used as a filter which cuts off all the other lines except 4358Å. The filter solutions may be arranged either to surround the Raman tube or in front of the arc.

Spectrograph : The chief features of a spectrograph, suited for the study of Raman spectra are (i) large light gathering power, (ii) special prism of high resolving power and (iii) a short-focus camera. A lens L in front of the plane window W directs the scattered radiation upon the slit of the spectrograph which is carefully aligned along the axis of the Raman tube and screened from the direct rays of the arc. The intense Raman lines of a liquid such as CCl_4 can be photographed in about an hour with a small spectrograph, but the recording of the complete spectrum may require upto ten or fifteen hours, depending largely on the intensity of the incident light, the speed of the spectrograph and the intrinsic brilliance of the Raman lines.

Rule of Mutual Exclusion : This rule states that if a molecule has a centre of symmetry, the Raman active vibrations are IR inactive and vice-versa. If a molecule is not centrosymmetric, then some or all the vibrations are simultaneously Raman and IR active.

By the above rule it means that for a molecule the Raman and IR spectra show no common line, the molecule must have a centre of symmetry while if one or more line happen to be present in both the spectra, the molecule has no centre of symmetry.

Rule of mutual exclusion provides a valuable information about the structure of any molecule.

Applications of Raman Spectroscopy : The applications to which Raman Spectroscopy can be put are very nearly the same as infrared Spectroscopy because both types of measurement deal with molecular vibrational transitions. However, because of fundamental differences in the two effects they often provide complementary information, with one type of measurement yielding data not supplied by the other.

1. *Identification of unknown compound* : Perhaps the most common application of molecular Spectroscopy is the identification of an unknown compound by comparison of its spectrum with spectra of known substances. Raman Spectroscopy is useful in this work, just as is infra-red Spectroscopy, because of the 'finger-print' property of vibrational spectra. Each molecule has its own unique spectrum, although differences can be quite appreciable for similar compounds. A vast number of Raman spectra have been recorded and Raman data are found to be scattered throughout the literature.

2. *Quantitative analysis* : Raman Spectroscopy can be used for quantitative analysis with about the same degree of success as infra-red and with the same limitations due to interferences by band overlap. Multi-component analysis is performed by measuring the lines on the pure constituents separately and developing a set of equations to be used in analyzing mixtures. Raman analysis of colourless solutions shares with emission Spectroscopy an advantage over absorption Spectroscopy, viz. signals are directly proportional to concentrations, Raman analysis of coloured solutions is more complicated but nontheless quite feasible.

3. *Determination of molecular structure* : When a plausible model for a not-too-complicated molecule is examined, it is possible to predict theoretically the number of Raman lines it should have. If the predicted numbers are of significantly different from those observed, the model is unsatisfactory and it is discarded; when agreement is reached for some model, some of the observed Raman frequencies can be used in a theoretical treatment to calculate other vibrational frequencies. If there is agreement between the calculated and observed frequencies the model is a good representation of the molecule and the relative positions of the atoms in the molecule are known. At the same time information is obtained about the magnitude of the forces holding the atoms together in the molecule and conclusions can be drawn about the chemical bonds in the molecule. Few examples are :

(a) *Diatomic molecules* : Let us consider molecules like H_2, N_2, O_2, HC1, HBr, HI. The first three molecules are composed of identical atoms and, therefore, are known as homonuclear molecules. The last three belong to heteronuclear molecules since they are composed of non-identical atoms. Moreover, the first three molecules are non-polar and their vibrational energy is not influenced by alternative electric field of light. Hence they do not exhibit any vibration-rotation absorption band in the infra-red which are necessarily found in polar molecules such as HC1, HBr and HI. Each of those has only one vibrational frequency and its value can be evaluated by studying the Raman spectra. The following table gives the natural frequency of vibration and restoring force per unit displacement.

Table 2.5

	H_2	N_2	O_2	HCl	HBr	HI
v_0 (cm^{-1})	4156	2331	1556	2880	2558	2233
$f \times 10^{-5}$	5.1	22.4	11.4	4.8	3.8	2.9

It is evident from the above table that lighter the molecule, greater is the vibration frequency. The values of the restoring force per unit displacement are approximately in the ratio of 3 ; 2 : 1 for Ng. Og and Hg, this suggests that the atoms in their respective molecules are held together by triple, double and single bonds.

(b) *Triatomic molecules* : These include the following examples :

(i) *Carbon dioxide* : Theoretically, one should expect two very strong bands in the infra-red absorption spectrum at 668 and 2349 cm^{-1} and one strong band in the Raman effect at 1389 cm^{-1} if one tries to study the infra-red and Raman spectrum of carbon dioxide. As none of these occurs both in the Raman and infra-red spectrum, it follows from the rule of mutual exclusion that the molecule has a centre of symmetry. Since it is a triatomic molecule, its structure should, therefore, be linear as well. In other words the structure should be O–C–O. This conclusion has been further confirmed by the rotational Raman spectrum of CO_2.

Carbon disulphide resembles a similar structure given by S–C–S, as obtained from the study of the infra-red absorption and Raman Spectrum.

(i) *Nitrous oxide* : This molecule has the same number of electrons as CO_2 and one might expect that it should have a linear symmetric structure. But the analysis of the infrared and Raman spectrum of N_2O has revealed that this molecule, though linear, is not symmetrical.

The three fundamental frequencies of N_2O are 2224, 1285, 589 cm^{-1}. All the three lines appear in the infra-red absorption and two of them viz. 2224 and 1285 cm^{-1} appear in the Raman spectrum. The third line 589 has, however, not been recorded due to weak intensity. As the two Raman lines appear in the infra-red which means the absence of the centre of symmetry in molecule. If there has been a centre of symmetry in the N_2O molecule, only one fundamental line should appear in the infra-red, according to the rule of mutual exclusion. It means that the molecule has unsymmetrical structure N–N–O. This is also confirmed by the rotational Raman spectrum of N_2O which consists of odd and even sets of lines without any alteration in intensity.

Other molecules having similar structures as N_2O are HCN, ClCN, BrCN, ICN, etc.

(iii) *Sulphur dioxide* : The results of SO_2 are

Table 2.6

$\bar{v}$ *cm*$^{-1}$	***Infra-red***	***Raman***
519	active (‖, PQR)	active (polarised)
1151	active (‖, PQR)	active (polarised)
1361	active (⊥, PQR)	active (depolarised)

The absence of PR branch in the IR spectrum indicates a non-linear structure for SO_2, Since all the vibrations are simultaneously active in the IR and Raman spectra, molecule does not have a centre of symmetry. These informations when coupled with the dipole results suggest the bent structure for the molecule :

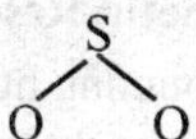

NUCLEAR MAGNETIC RESONANCE

Nuclear Magnetic Resonance is a branch of Spectroscopy in which radio frequency waves induce transitions between magnetic energy levels of nuclei of a molecule. The magnetic energy levels are created by keeping the nuclei in a magnetic field.

Without the magnetic field the spin states of nuclei are degenerate; *i.e.*, possess the dame energy, and energy level transition is not possible. When a magnetic field is applied, the separate levels and radio frequency radiation can cause transitions between these energy levels.

NMR Spectroscopy is most often concerned with nuclei with $1 = 1/2$. Examples of such nuclei are ^1H, ^{31}P and ^{19}F. Spectra cannot be obtained from nuclei with $I = 0$. In special cases, spectra can be obtained from nuclei where $1 \geq 1$.

Nuclear magnetic resonance is a powerful tool for investigating nuclear structure. The importance of NMR can be seen from the fact that the first observations of NMR signals were observed independently by Purcell at Harvard Bloch at Stanford In 1945, and the first application to study the structure of ethyl alcohol was made in 1951. In 1952 Purcell and Bloch won the Nobel Prize in Physics for their discovery.

Radio waves are regarded as the lowest-energy form of electromagnetic radiation that find valid applications in analytical chemistry. The frequency of radio waves lies between 10^7 and 10^8 cps. The energy of radio frequency (rf) radiation can be calculated by using the equation :

$$E = hv \qquad ...(1)$$

where h is the Planck's constant and v the frequency. Here h is 6.6×10^{-17} erg sec and v is between 10^7 and 10^8 cps.

$E = 6.6 \times 10^{-27}\ 10^7$ (or 10^8 ergs). $= 6.6 \times 10^{-20}$ (or 6.6×10^{-19} ergs).

From the above, it can be seen that the quantity of energy involved in rf radiation is very small which is too small to vibrate, rotate or excite an atom or molecule. But this energy is sufficient to affect the nuclear spin of the atoms of a molecule. Therefore, the nuclei of atoms in a molecule on absorbing rf radiation may change their direction of spin.

Quantum Description of Nuclear Magnetic Resonance

According to the quantum theory, a spinning nucleus can only have values for the spin angular momentum given by the equation.

Spin angular momentum = [I (I + l)]1/2 h/2π ...(2)

where I is the spin quantum number of the nucleus and h the Planck's constant.

But μ = γ × spin angular momentum

$$\because \quad \mu = \gamma \times [I\ (I + 1)]^{1/2}\ h/2\pi \qquad ...(3)$$

where μ is the magnetic moment of the nucleus and y is the gyromagnetic ratio.

If a nucleus having a magnetic moment is introduced into a magnetic field, H_0, the two energy levels become separate corresponding to m_I = −1/2 (antiparallel to the direction of magnetic field) and to m_I = +1/2 (Parallel to the direction of magnetic field). For a nucleus with 1 = 1/2. The energies E_1 and E_2 for the two states with m_I = +1/2 and m = −1/2i. respectively, are

$$E_1 = -\frac{1}{2}\left(\frac{\gamma^h}{2\pi}\right)H_0 \qquad ...(4)$$

and

$$E_2 = +\frac{1}{2}\left(\frac{\gamma^h}{2\pi}\right)H_0 \qquad ...(5)$$

The energies of these two states are represented in Fig. 2.27.

When the nucleus absorbs energy the nucleus will be promoted from the lower energy state E_1 to the higher energy state E, by absorption of energy, ΔE, equal to the energy difference, $E_2 - E_1$. It means that the absorption of energy ΔE changes the magnetic moment from the parallel state m_I = +1/2 to the antiparallel state (m_I = −1/2).

If the nucleus lies in the upper energy state. E_2 and radiation of energy ΔE is incident upon the system, the nucleus will come to lower energy level E_1 emitting energy corresponding to ΔE.

The frequency v at which energy is absorbed or emitted is given by Bohr's relationship

$$v = \frac{E_2 - E_1}{h} \qquad ...(6)$$

On substituting equations (4) and (5) in (6), we get

$$v = \frac{\frac{1}{2}\left(\frac{\gamma h}{2\pi}\right)H_0 + \frac{1}{2}\left(\frac{\gamma h}{2\pi}\right)H_0}{h}$$

or $$v = \frac{\gamma}{2\pi} H_0 \qquad ...(7)$$

From equation (7) it follows that the frequency absorbed or emitted by a nucleus in moving from one energy level to another is directly proportional to the applied magnetic field. In NMR it is the absorption of energy which is detected.

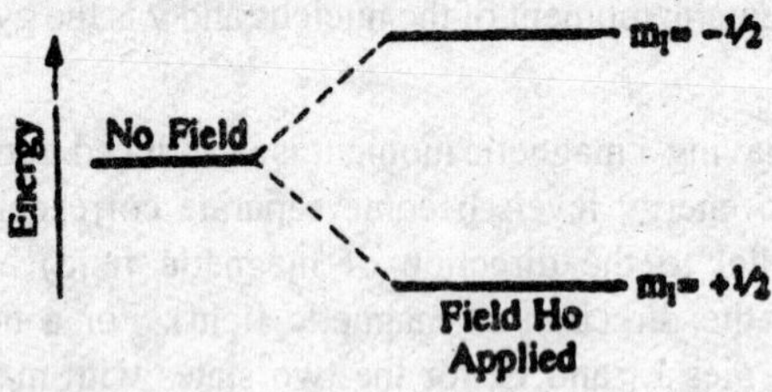

Fig. 2.27 : Energy level of a nucleus with I = 1/2 in the absence and in the presence of a magnetic field.

When a nucleus is placed in a system where it absorbs energy, it becomes excited. It then loses energy to return to the unexcited state. It absorbs energy and again enters an excited state. This nucleus which alternately becomes excited and unexcited is said to be in a state of resonance.

In order to determine the resonance frequency, the energy absorbed by nuclei is measured as the magnetic field H_0 is varied. As the field H_0 is increased so that precessional frequency of the nucleus increases and when this frequency becomes equal to the frequency of oscillation field, transitions occur between nuclear energy states.

The energy absorbed in this process produces a signal at the detector and this signal is amplified and recorded as a band in the spectrum.

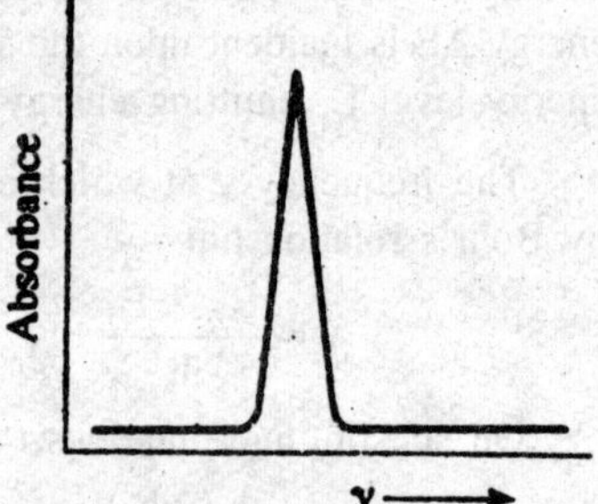

Fig. 2.27(a) : Absorption v versus the frequency of rf radiation.

As NMR spectrum is plotted between absorption signal at the detector and the strength of the magnetic field H_0. It is important to mention here that the NMR spectrum is generally calibrated in units of frequency rather than in units of magnetic field strength Fig. 2.27(a).

Instrumentation

A high resolution spectrometer contains a complex collection of electronic equipments. It is to be remembered that in the technique of NMR Spectroscopy one has to deal with intense magnetic fields requiring precisely controlled power supplies as well as frequencies. Moreover, most of the power required by the instrument is dissipated as heat and very little of it (in microwatts) is obtained as a signal which has to be amplified by a complex electronic system. A schematic diagram of an NMR instrument is given in Fig 2.28.

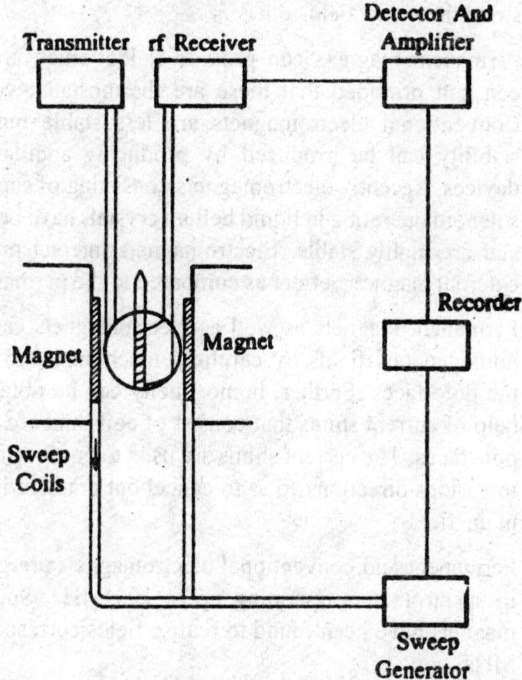

Fig. 2.28 : Schematic diagram of a NMR spectrometer.

The apparatus consists of the following essential components :

1. *Sample Holder :* Sample should be held in a holder which should be chemically inert, durable and transparent to rf radiation. Generally glass tubes are employed which are sturdy, practical

and cheap. They are generally about 7.5cm long and approximately 0.3 cm in diameter.

2. *Magnet :* Permanent magnet or electromagnet can be used in a nuclear magnetic resonance instrument. The important feature of the magnet is that it should give homogeneous magnetic field, *i.e.*, the strength and direction of the magnetic field should not change from point to point. Moreover, the strength of the field should be very high, *i.e.*, at least 20,000 gauss (G), because the chemical shifts are proportional to the field strength. Thus, the factors which are important in the design of the magnets for NMR Spectroscopy are homogeneity or uniformity of the field, the constancy of the field, strength and maximum obtainable strength of the field.

   Permanent magnets can provide fields which are sufficiently constant provided that these are thermostated very carefully. Conventional electromagnets are less stable but the desired stability can be produced by producing ancillary stabilising devices. Recently electromagnets consisting of superconducting solenoid operating in liquid helium crystals have been developed and are highly stable. Electromagnets interact much less with external magnetic effect as compared to the permanent magnets.

   Permanent magnets as well as electromagnets can give highly homogeneous fields by carefully machining and alignment of the pole faces. Further, homogeneity can be obtained with the help of current shims that consist of coils and are located at the pole faces. The current shims are used to produce field gradients in various directions so as to cancel out gradient inherent in the main field.

   Permanent and conventional electromagnets are generally used in spectrometers operating up to 100 MHz. Superconducting magnets have been found to realise fields corresponding to 230 MHz.

3. *Sweep Generator :* In order for a nucleus to resonate, the precession frequency of the nucleus should become equal to the frequency of the applied of radiation. If the applied magnetic field H_0 is kept constant, the precession frequency is fixed. In order to bring about resonance, the frequency of field should be changed so that it becomes equal to the resonance frequency

(frequency sweep method). On the other hand, if the rf radiation is kept constant, the resonance frequency of the nucleus must be changed by varying H_0 (field sweep method). Generally, the field sweep method is regarded as better because it is easier to vary H_0 than the rf radiation so as to bring about resonance in nuclei.

Practically, it is not easy to vary the magnetic field of a large stable magnet. This technical problem is solved by superimposing a small variable magnetic field on the main field. This can be done by fixing a pair of Helmholtz coils in the pole faces of the main magnet. The magnetic field induced by these coils can be varied by varying the current flowing through them. The small magnetic field generated by Helmholtz coil is in the same direction as the mainfield and therefore, it is added to the main field. Thus, the sample is exposed to both fields and they appear as one field to the molecule.

4. *Radio Frequency Generator :* In order to generate radio frequency radiation, radio frequency oscillator is used. To achieve the maximum interaction of the rf radiation with the sample, the coil of oscillator is wound around the sample container. The oscillator irradiates the sample with a rf radiation.

   The oscillator coil is wound perpendicular to the applied magnetic field. This is done so that the applied rf field should not change the effective magnetic field in the process of irradiation.

5. *Radi Frequency Receiver :* When the radio frequency radiation is passed through the magnetised sample, two phenomena, namely, absorption and dispersion, may occur. The line shapes associated with absorption and dispersion are shown in Figs. 4 and 5. The observation of either dispersion or absorption will enable the resonance frequency to be determined. It is found that the interpretation of absorption is easier as compared to dispersion spectrum.

The detector should be capable of separating absorption signal from dispersion signal and from that of the rf oscillator. There are two main methods of detection. These are as follows:

(i) The first method uses a radio frequency bridge. Its network balances out the transmitter signal and allows the absorption and dispersion signals to appear as an out-of-e.m.f. across the bridge.

In this method, the coil used for surrounding the samples serves as both a transmitter and a receiver coil.

(ii) The second method employs a separate receiver coil. This method is sometimes called the crossed coil or nuclear induction method. If the two coils are fixed at right angles to each other as well as to the direction of static magnetic field, they will not be effectively coupled. In this way, the transmitter signal is separated from the absorption and dispersion signals.

The separation of absorption and dispersion signals is achieved by knowing the fact that they differ in phase by an angle of 90°. The general method in practice is to use a phase sensitive detector which helps the operator to select the phase of the signal to be detected.

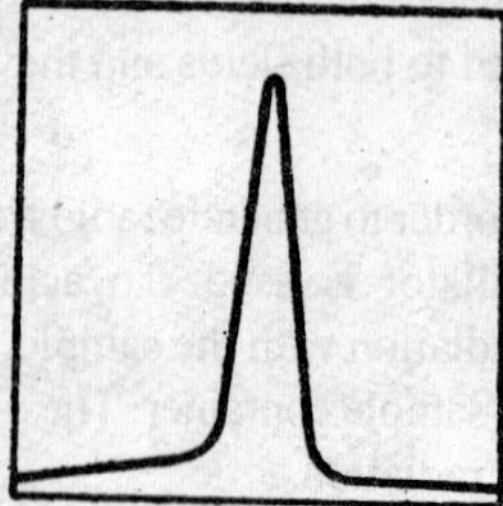

Fig. 2.29 : Absorption Signal **Fig. 2.30 : Dispersion Signal**

Readout System : The absorption signal received from radio frequency receiver is extremely weak. Therefore, it requires considerable amplification before it is fed to a chart recorder.

Chemical Shift

The frequency at which a nucleus comes into resonance in a magnetic field may be given by following equation

$$\nu = \frac{\gamma}{2\pi} H_0$$

From the above equation it is evident that the frequency can be calculated if the value of magnetic field at which resonance occurs could be derived from the NMR spectrum. But it is not easy to determine accurately the absolute value of the field corresponding to a peak and in practice the difference in peaks is observed. In order to solve this difficulty, the resonance frequencies of nuclei in a sample are measured relative to the resonance frequency of a nucleus in a reference compound and the frequencies are quoted relative to the reference frequency. The position of the peaks in an NMR spectrum relative to the reference peak is expressed in terms of the chemical shift, 8, which is defined as

$$\delta = \frac{H_0 \text{ (reference)} - H_0 \text{ (Sample)}}{H_0 \text{ (reference)}} \times 10^6 \text{ ppm} \qquad ...(8)$$

The value of H_0 for the reference is usually greater than H_0, for the sample, so subtraction in the direction indicated gives a positive δ. In terms of frequency unit δ takes the form

$$\delta = \frac{\text{v (sample)} - \text{v (reference)}}{\text{v (reference)}} \times 10^6 \text{ ppm} \qquad ...(9)$$

The chemical shift, δ, is dimensionless and expressed in parts per million (ppm) on account of the factor 10^6 in equations (8) and (9).

An alternative system which is generally used for defining the position of the resonance relative to the reference is assigned tau (τ) scale. On this scale, the reference is assigned the arbitrary position of 10 and the values of other resonances are given by τ = 10–δ where S has the same significance.

The normal reference compound used for ^{1}H nucluei in organic compounds is tetramethysilane (TMS). The protons in TMS are all equivalent and its NMR spectrum exhibits a single resonance line at a high applied field well beyond the protons in most of the organic compounds. In addition, TMS is chemically inert, relatively volatile (B.P. = 27°C) and soluble in most organic solvents. As TMS is not soluble in aqueous solutions, it is not suitable for aqueous solutions. For water soluble substances, DSS (2, 2-dimethyl-2 silapentane-5-sulphonate) may be used as a reference. The protons in the methyl groups of DSS give a strong line. The methylene protons of DSS gives a series of small peaks which can be ignored.

The chemical shift, or τ value of a resonance band in NMR Spectroscopy can be compared with the wave number value of an absorption band in electronic, vibration and rotational Spectroscopy.

Causes of Chemical Shift : In other branches of Spectroscopy, absorption bands are observed at different wave-number values and in NMR Spectroscopy bands are observed at different chemical shift 8 or v values. This difference may arise due to the simple reason that different nuclei in a molecule experience different magnetic fields as a result of the secondary magnetic fields associated with the molecule. The secondary fields arise due to the applied field H_0. The effect of secondary field may affect the chemical shift in one of the following ways :

(i) When the secondary fields produced by the circulating electrons oppose the applied field at a particular nucleus in the molecule, it means that effective field experienced by the nucleus is less than the applied field. This is known as positive shielding and the resonance position moves upfield in NMR spectrum because the value of applied field necessary to bring the nucleus into resonance will be greater than that when there were no secondary opposing held. This is shown in Fig. 2.31.

(ii) If the secondary field produced by the circulating electrons reinforces the applied field, the position of resonance moves downfield. This is known as negative shielding. This is shown in Fig. 2.31.

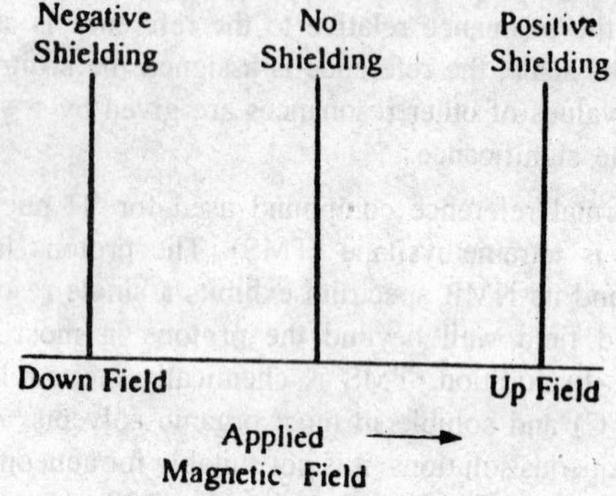

Fig. 2.31 : Effect of shielding and deshielding on position of resonance of nuclei.

There are two types of shielding phenomena which modify resonance peaks in an NMR spectrum. We will discuss these one by one.

(i) *Local Shielding* : The field experienced by a nucleus may be modified by fields due to induced circulation of electrons localised on that nucleus. This is known as local shielding. The extent of shielding resulting from local magnetic fields is related to the electron density in the immediate vicinity of the protons. The order of magnitude of δ can be predicted from the knowledge of electronegativity

$CH_3 - H$	$CH_3 - I$	$CH_3 - Ar$	$CH_3 - Cl$	$CH_3 - F$
δ(ppm)0.2	2.2	2.7	3.6	4.3

From the above example it is clear that the value of 8 increases as the value of electronegativity increases.

The local shielding is shown in Fig. 2.32.

(ii) *Low Range Shielding* : In aromatic-compounds the secondary fields set up by the induced circulation of re-electrons often influence the fields experienced by nuclei not directly associated with thc rclcctrons.

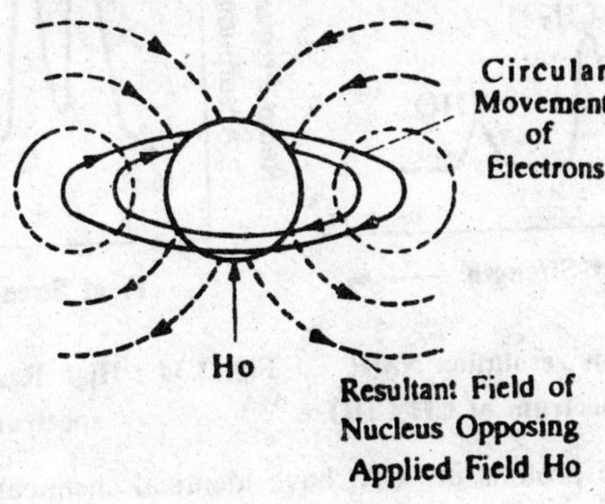

Fig. 2.32 : Local shielding

Due to this, additional shielding and deshielding effects may be included within the molecule.

The protons on an acetylenic linkage lie in a shielded region whereas those on a carbonyl group or ethylenic linkage lie in a deshielded region. This explains the observation that acetylenic protons are resonating at relatively high field (7 – 8τ) as compared with protons on a carbonyl group (0 – 0, 7τ) and ethylenic protons

The long range shielding effects are more pronounced in unsaturated groups than saturated groups.

Spin-Spin Coupling

The interaction between the spins of the neighbouring nuclei in a molecule may cause the splitting of the lines in the NMR spectrum. This is known as spin-spin coupling which occurs through bonds (not space) by means of a slight unpairing of the bonding electrons.

It should be kept in mind that the area under a broad NMR signal and the total area under the several split signals remain the same.

The NMR spectrum of CH_3CHO has the form shown schematically in Fig. 2.33.

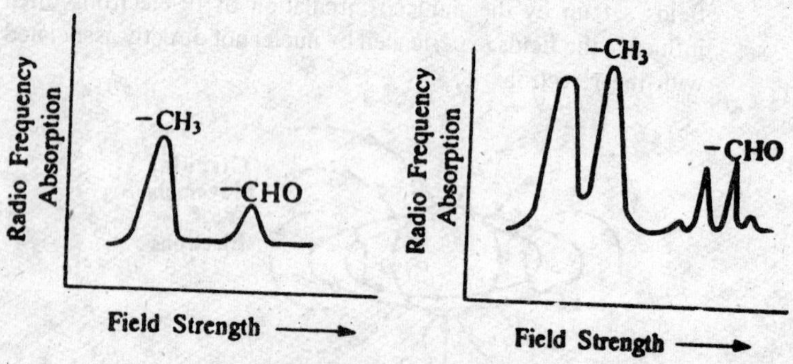

Fig. 2.33 : Low resolution NMR spectrum of CH_3CHO.

Fig. 2.34 : High Resolution NMR spectrum of CH_3CHO

The three protons in $–CH_3$ have identical chemical environments and so all absorb at the same energy. The proton in –CHO is in a different chemical environment and, therefore, it absorbs the radiation at a different applied field. The intensity of the $–CH_3$ peak is three times the intensity of the –CHO peak.

Further magnetic effects can be caused by other nuclei which have spin in the molecule and these give rise to a hyperfine structure. Thus, the protons in the $–CH_3$ group can experience two possible fields due to the magnetic moment of the –CHO proton because m_I can take the value of $\pm 1/2$.

It means that $–CH_3$ protons in CH_3CHO undergo spin-spin coupling and, therefore, the peak due to $–CH_3$ protons may undergo splitting to give rise to two peaks with equal intensities in the NMR spectrum of CH_3CHO Fig. 2.34.

The –CHO proton can experience four different fields depending on the values of m_I on each of the $–CH_3$ protons.

Table 2.7

m_I	m_I	m_I	M_I *(Total)*
$+\frac{1}{2}$	$+\frac{1}{2}$	$+\frac{1}{2}$	$-\frac{3}{2}$
$+\frac{1}{2}$	$+\frac{1}{2}$	$-\frac{1}{2}$	$+\frac{1}{2}$
$+\frac{1}{2}$	$-\frac{1}{2}$	$+\frac{1}{2}$	$+\frac{1}{2}$
$-\frac{1}{2}$	$+\frac{1}{2}$	$+\frac{1}{2}$	$+\frac{1}{2}$
$-\frac{1}{2}$	$-\frac{1}{2}$	$+\frac{1}{2}$	$+\frac{1}{2}$
$-\frac{1}{2}$	$+\frac{1}{2}$	$-\frac{1}{2}$	$-\frac{1}{2}$
$+\frac{1}{2}$	$-\frac{1}{2}$	$-\frac{1}{2}$	$-\frac{1}{2}$
$-\frac{1}{2}$	$-\frac{1}{2}$	$-\frac{1}{2}$	$-\frac{3}{2}$

As there are four different values of $M_1 \left(\frac{3}{2}, \frac{1}{2}, -\frac{1}{2}, -\frac{1}{2}\right)$ the –CHO peak is therefore split into four peaks with intensities in the ratio 1 : 3 : 3 : 1. The NMR spectrum of CH_3CHO under high resolution.

We can generalise the above example by stating that a proton with n equivalent protons on the neighbouring carbon atom will be split by the n protons into (n + 1) lines (a multiplet) with relative sub-areas given by the coefficients of the binomial expansion $(r + 1)^n$. In place of (n + 1), the multiplicity of a given group may also be given by 2nI + 1 where n is the number of protons on adjacent atoms and I is the nuclear spin quantum number of proton which is equal to 1/2.

The multiplicity of NMR signal caused by adjacent methylene group protons will be (2 + 1) or (2 × 2 × 1/2 + 1) or 3. Also, the multiplicity caused by the methyl group protons will be (3 + 1) or (2 × 3 × 1/2 + 1) or 4.

The relative intensities of the triplet caused by methylene protons can be evaluated from the coefficients of the terms of $(r + 1)^2$ or r^2 +

2r + l, *i.e.*, 1 : 2 : 1. Similarly, the relative intensities of the four fine peaks of $-CH_3$ protons will be $(r + 1)^3$ or $r^3 + 3r^2 + 3r + 1$ or 1 : 3 : 3 : 1.

Coupling through more than three bonds is very inefficient. It is normally not observed. Equivalent nuclei do not interact with each other to cause spin-spin spliting. Spin interactions are independent of the strength of the applied field.

Coupling constant J : The spacing of adjacent lines in the multiplets is a direct measure of the spin-spin coupling of the protons and is known as the spin-spin coupling constant.

The coupling constant is generally denoted by J. It is usually expressed in cycles per second. The magnitude of J in cps does not depend upon the magnetic field. However, it depends on the structural relationships between the coupled protons.

Coupling constants remain same in both multiplets of a pair which are interacting. Coupling constants rarely exceed 20 cps. On the other hand, chemical shifts vary over 1000 cps. The value of J decreases with distance. In Fig. 2.35, NMR spectrum of $CH_3Br-CHBr_3$ is shown. The values of J have been calculated for splitting of CH and CH_3.

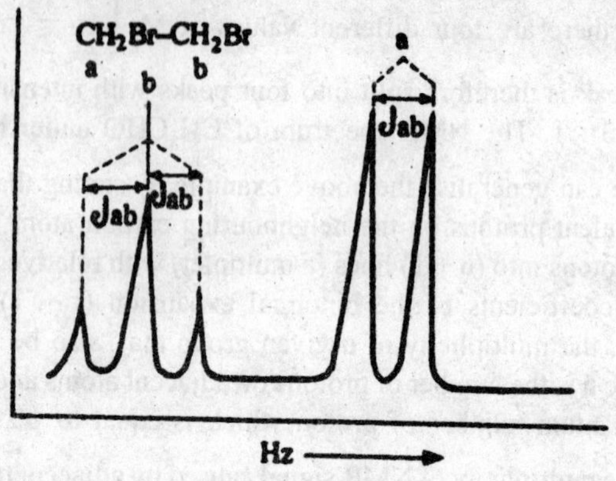

Fig. 2.35 : Spin-spin coupling

Applications of NMR Spectroscopy

The primary applications of NMR are in the field of structure determination and delinearation. There is literature of NMR spectra comparable to those for optical absorption spectra and X-ray diffraction spectra. Comparison of the NMR spectrum of an unknown compound with these in the literature may lead directly to identification of the unknown. Even if this is not the case, the NMR spectrum can help to distinguish between two possible isomeric structures and in many cases can furnish the critical information that the analyst needs to delineate the exact structure of the complicated compound.

Integration of the NMR spectrum furnishes information that can be used for quantitative determination of one or more components present in a mixture.

1. *Structural Diagnosis by NMR* : A large number of principles are known which will decide about the structure of an unknown compound from its NMR spectrum. Some of these are outlined as follows :
   (i) The number of main NMR signals should be equal to the number of equivalent protons in the unknown compound.
   (ii) The chemical shift indicates that what type of hydrogen atoms are present, *e.g.*, methylene, methyl groups, olefins, ethers etc.
   (iii) The spin-spin splitting or multiplicity reveals about the possible arrangement of groups in the molecule.
   (iv) From the area of peaks, one can draw the conclusion about the number of hydrogen nuclei present in each group- For example, the relative areas of methyl (CH,) and methylene (CH_2) peaks in $CH_3 - CH_2 - CH_3$ would be 6 : 2. In butane, it would be 6 : 4.
2. *Qaaotatire Analysis* : NMR Spectroscopy has been used to determine the molar ratio of the components in a mixture.

If the NMR spectrum of a mixture of compounds is recorded, it may exhibit a resolved band due to one of the components and if the number of protons in the molecule giving rise to that band is known, then the integrated area of the band will give a direct measure of the relative concentration of that component in the mixture. This can be illustrated by taking the NMR spectrum of a mixture of benzene and cyclohexane which exhibits bands at 2.73 and 8.56 τ from benzene and cyclohexane respectively, and the integrated area of the former band will be proportional to six protons whereas that of the latter band will be proportional to twelve protons. Obviously if the integrated areas of the two bands in the spectrum of the mixture are equal, then the molecular ratio of benzene to cyclohexane in the mixture must be 2:1.

In certain cases the analysis of a mixture becomes more complicated if the bands in the NMR spectrum overlap. However, the complete analysis is possible. If the spectrum of a mixture of n components exhibits a resolved band for each of the (n – 1) components, then the contribution. A' to the total integrated area, A" of an overlapped band from components which exhibit resolved bands may be given by

$$A' = A_1 \frac{x_1}{y_1} + A_2 \frac{x_2}{y_2} + ... + A_{n-1} \frac{x_n - 1}{y_n - 1}$$

where A_1, A_2, ... A_{n-1} denote the integrated areas of resolved bands due to components 1, 2, ..., (n – 1), y_2 y_2 ...., y_{n-1} denote the number of components 1, 2, ..., (n – 1) giving rise to the resolved bands, and x_1, x_2, ..., x_{n-1} represent the number of protons in components 1, 2, ... (n – 1) contributing to the area of the overlapped band.

The contribution of the nth component to the integrated area of the overlapped band will be given by A" – A' and if the number of the protons giving rise to this integrated area is known, the relative concentration of nth component can be evaluated.

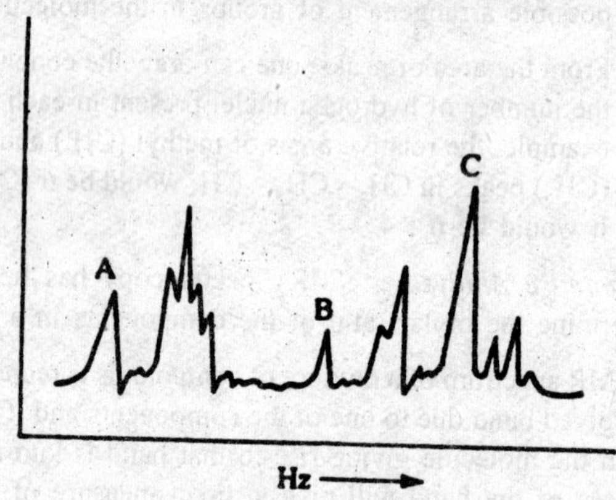

Fig. 2.36 : Analysis of a three component system by NMR

In some cases, peak heights rather than areas can be correlated with composition because in these cases, the signals chosen are extremely narrow and their widths depend upon the uniformity of the magnetic field. This can be illustrated by recording an NMR spectrum of a three component system having equal quantities of benzaldehyde, ethyl alcohol and toluene (Fig. 2.36). In the Fig 2.36, the heights of the peaks of aldehyde, hydroxyl and aromatic methyl groups are marked by A, B and C respectively. The relative heights of the peaks A, B, and C are recorded in the Table 2.8. From the relative heights, the composition can be calculated.

Table 2.8

Composition	C_6H_5CHO	C_2H_5OH	$C_6H_5CH_3$
Measured relative heights	1	1.9	3.4
Calculated relative heights	1	2.02	3.33
Composition from NMR data	33.3	32.3	34.4

NMR Spectroscopy is very successful in analysing a mixture of isomeric substituted compounds like ortho- and para-xylenes.

3. *Hydrogen Bonding :* The NMR can be used to study the hydrogen bonding in metal chelates as well as in organic compounds. Proton signal is shifted towards low field in the case of hydrogen bonding. This reveals that hydrogen bond formation results in the decrease in the electron shielding of the proton.

If there occurs an upfield shift of the signal, this may be ascribed to the breaking of intermolecular hydrogen bonds. This can be seen in the case of ethanol on increasing temperature, or on diluting the ethanol with carbon tetrachloride.

Schneider, Bernstein and Pople have considered that the shifts associated with hydrogen bonding arise from three terms :

(i) The electrostatic effect of the donor atom or group.

(ii) The diamagnetic anisotropy of the donor atom or group.

(iii) Changes in long range shielding of the acceptor group.

In the case of strong hydrogen bonds, the factor (i) is dominant. However, the factor (ii) may become dominant in the case of weak hydrogen bonds. The factor (iii) possesses little experimental foundation. It is only operating in the cases of intermolecular hydrogen bonding. This factor (iii) is important in explaining the gas-to-solution shifts for acetylene and hydrogen bond.

Some workers have attempted to correlate the proton shifts of chloroform in several solvents with the strengths of the corresponding hydrogen bonds. But no quantitative relationships between magnitudes of low field shifts and hydrogen bond strengths could be established. However, this study has provided quantitative indication of bond strengths.

Table 2.9 lists 8 values (low field shifts caused by hydrogen bonding) observed for various types of acidic protons.

Table 2.9 : δ Values of Acidic Protons

Compounds	*δ(ppm)*	*Compounds*	*δ(ppm)*
ArSH	3.4	ROH	0.5-4.5
ArOH	4.5-6.5	RSH	1-2
$ArNH_2$	3.6	RCOOH	9.13
ArNHR	3.6	$RCONH_2$	5-13
RNH_2	1.5	R CONHR	5-12
RNHR	1.5	>C = N–OH	9-12

4. *Intramolecular Conversion :* From the electron diffraction study of PF_5, it is clear that the compound is a trigonal bipyramid and contains two non-equivalent kinds of fluorine atoms. This clearly shows that the NMR spectrum of PF_5 should give different NMR signals. When the NMR spectrum (^{19}F) of PF_5 is recorded, it shows a 1 : 1 doublet whose separation does not depend upon field strength. This reveals that five fluorines are equivalent and their signal is splitted to a doublet by coupling with the ^{31}P nucleus (spin 1/2). This rather anomalous result has been explained by suggesting that in PF_5 there occurs intramolecular rearrangements which are rapid enough to prevent the appearance of separate signals for atoms lying at structurally non-equivalent sites in a molecule. Berry called this type of behaviour as "pseudorotation" and gave a schematic representation for this (Fig. 2.37). The pseudorotation is a vibrational process and is not involving any bond breaking process. Thus, this could explain the equivalence of the five fluorines as well as the spin-spin coupling between the ^{31}P and ^{19}F in PF_5.

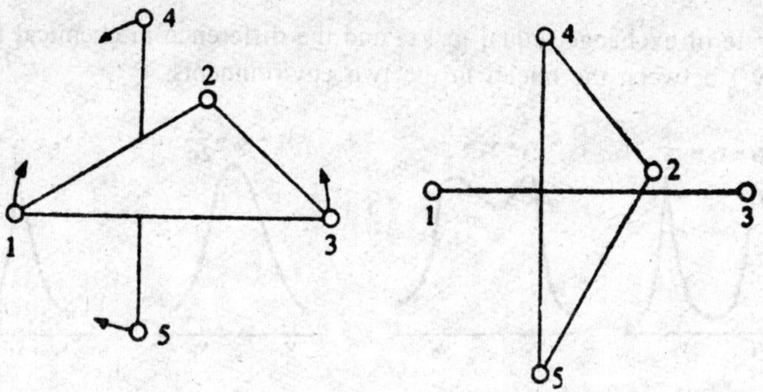

Fig. 2.37 : Pseudorotation or intramolecular conversion in a trigonal bipyramidal molecule.

5. *Exchange Effects :* The width of an absorption band in an NMR spectrum depends upon the physical state of the sample, and also upon the type of the nucleus, in a particular environment. In the case of liquids, the width of absorption bands in their NMR spectra is very small *i.e.*, of the order of 2-3 Hz. However, in some liquids, broad bands are observed in their NMR spectra which can normally be accounted for in terms of exchange effects. Let us illustrate the exchange effects in relation to the equilibrium between pyruvic acid and 2, 2. dihydoxypropionic acid :

$$\underset{a}{CH_3COCOOH} + H_2O \rightleftharpoons \underset{b}{CH_3C(OH)_2\ COOH}$$

As the chemical environment of the methyl protons is continually changing from a to b and if the mean life-time, T, of the nucleus in each environment is long as compared to the transition between the lower and upper magnetic energy states (~10^{-3} sec), then the spectrum for the methyl protons should consist of two sharp bands (Fig. 2.38). On the other hand, if mean life time T, is short as compared with the time required for the transition, between the lower and upper magnetic energy states, one will expect a single sharp band in the NMR spectrum at a position corresponding to the average environment of the nucleus. However, at intermediate life-times either one or two broad resonances may be observed (Fig. 2.38).

From the above discussion, one may draw an important conclusion that the appearance of the bands in NMR spectrum is dependent upon the rate of exchange (equal to l/τ) and the difference in chemical shift ($v_a - v_b$) between the nuclei in the two environments.

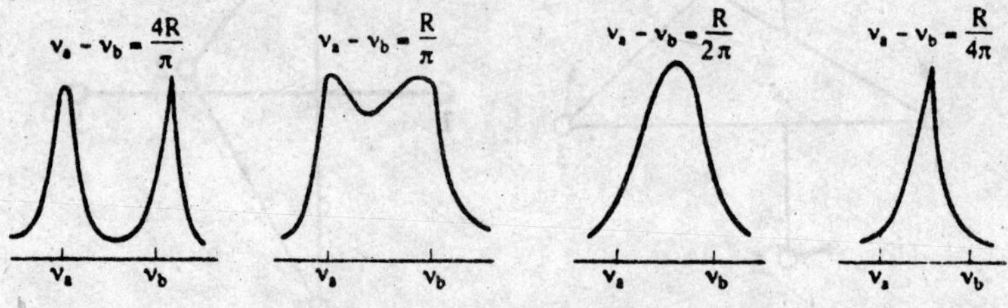

Fig. 2.38 : Change in appearance of the spectrum for two exchanging nuclei as the rate of exchange R is raised.

From the Heisenberg's uncertainty principle, one can deduce the relationship between the width of individual bands in the spectrum and the mean life time, τ, of the protons in each environment. According to Heisenberg's uncertainty principle, it follows that

$$\Delta E \times \Delta t = h/2\pi \qquad ...(1)$$

where Δ = the life time of the state.

ΔE = the uncertainty in the measurement of the energy of state, and

Δ = the Planck's constant.

In the case of equilibrium reactions involving exchange effects, the life time of the state, Δt is approximately equal to the mean life time, τ, of the protons in a particular environment. Thus, equation (1) becomes as

$$\Delta E.\ \tau = h/2\pi \qquad ...(2)$$

The uncertainty in the measurement of the energy of a magnetic state means that there will be a similar uncertainty in the measurement of the position of a line which arises from the absorption of energy by this state. Thus, a band instead of a line is observed in an NMR spectrum. Dividing equation (2) by $A\tau$ and using $E = Av$ yieids the following equation:

$$\frac{\Delta E}{h} = \Delta v = \frac{1}{2}\pi\tau \qquad ...(3)$$

where Δv denotes the uncertainty in the measurement of the frequency corresponding to the position of the line. We know that the uncertainty is generally measured as the width of the band at half the band height. Thus :

$$\Delta v = \frac{\Delta\omega}{2\pi} \qquad ...(4)$$

where $\Delta\omega$ denotes the uncertainty in the angular frequency of the processing nuclei. On comparing equations (3) and (4), we get

$$\Delta\omega = I/\tau \qquad ...(5)$$

In the case of a system having no exchange processes, each band in an NMR spectrum possesses natural width. If in a system exchange processes involving certain nuclei are occurring, the bands arising from these nuclei are broadened and the extent of broadening. $\Delta\omega$, is proportional to the reciprocal of τ, *i.e.*, the mean life time of the protons in a particular environment. This can be proved that the expression for the half-width of the broadened band arising due to exchange processes is as follows :

$$\frac{1}{T_2'} = \frac{1}{T_2} + \frac{1}{\tau} \qquad ...(6)$$

where $1/T_2$ = the half-width of the band in the absence of exchange processes, and

$1/T_2'$ = the half-width of the exchange broadened band.

From equation (6) one can calculate τ of nuclei in a particular environment if the half-widths of the appropriate bands in the NMR spectrum in the presence and absence of exchange processes are known.

In the case of example considered above, the forward and backward reactions are acid catalysed. As the concentration of acid present is increased, the bands due to the methyl protons will broaden and eventually merge into a single sharp band.

The reciprocal of the mean life time is equal to the specific rate of the reaction which in turn is equal to $k_0 + k_0^+ [H^+]$ where ky is the spontaneous rate constant, k_H^+ is the catalytic rate constant and (H^+) is the hydrogen ion concentration. Therefore, equation (6) can be written as :

$$1/T_2' = \text{constant} + k_H^+ \ [H^+] \qquad ...(7)$$

If, $1/T_2'$ is plotted against $[H^+]$, a straight line is obtained whose slope is equal to the value of the catalytic-rate constant. Thus, by measuring the band width at half-height of the methyl resonances in the spectra of pyruvic acid and 2, 2,-dihydroxypropanoic acid at different acid concentrations, the catalytic-rate constants for the forward and back reactions can be calculated. The equilibrium constant can also be calculated by measuring the integrated areas of the methyl resonances.

A rather interesting feature of this kind of kinetic study is that all measurements are carried out without disturbing the chemical equilibria.

Intermolecular Exchange Reactions

In discussing the high resolution spectrum of ethyl alcohol it should be kept in mind that spin-spin interaction between the hydroxyl group proton and the – CH_2 groups is to be taken into consideration. Such a spectrum arises due to the presence of acid or base because of the following reactions:

$$C_2H_5OH + H_3O^+ \underset{}{\overset{k_1}{\rightleftharpoons}} C_2H_5OH_2^+ + H_2O$$

$$C_2H_5OH' + C_2H_5OH_2^+ \overset{k_2}{\rightleftharpoons} C_2H_5OHH' + C_2H_5OH$$

$$C_2H_5OH + HO^- \overset{k_3}{\rightleftharpoons} C_2H_5O^- + H_2O$$

$$C_2H_5OH + C_2H_5O^- \overset{k_4}{\rightleftharpoons} C_2H_5O^- + C_2H_5OH$$

If a nucleus undergoes exchange between two chemically and hence magnetically different sites, the resonance of the nucleus in these two sites will be broadened if the frequency of exchange is of comparable magnitude to the difference in chemical shifts. From the Heisenberg's uncertainty principle it follows that :

$$\Delta E \,.\, \Delta t \sim \frac{h}{2\pi}$$

$$\Delta E = h\Delta v$$

$$\therefore \quad \Delta v = \frac{1}{\Delta T}$$

Thus, the uncertainty in frequency is approximately equal to the reciprocal of the uncertainty in time for which the nucleus has a given energy, which is actually mean life time τ , *i.e.*, $\Delta T = \tau$.

The broadening of the hydroxyl lines in ethanol proceeds roughly as shown below :

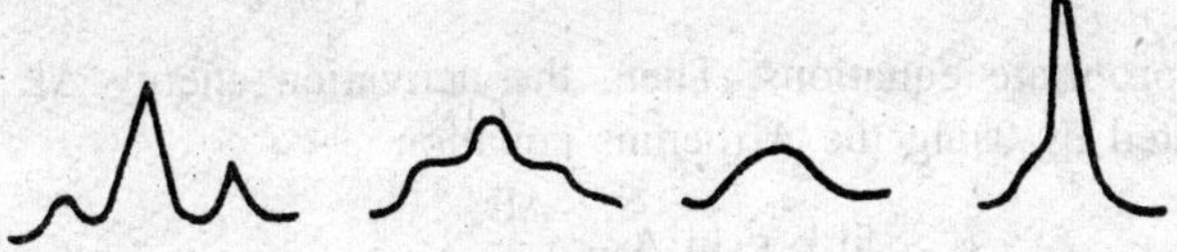

Fig. 2.39

As the rate of exchange increases the –OH triplet because of the spin-spin splitting by the $-CH_2$ group begins to broaden and finally coalesces into a single sharp line. The $-CH_2$ lines are similarly affected by the exchange process. The evaluation of rate constants from line broadening measurements is fairly easy for molecules having simple N.M.R. spectra. For ethyl alcohol, we get

$$k_1 = 2.8 \times 10^5 \text{ l mole}^{-1} \text{ sec}^{-1},$$

$$k_2 = 1.1 \times 10^6 \text{ l mole}^{-1} \text{ sec}^{-1},$$

$$k_3 = 2.8 \times 10^6 \text{ l mole}^{-1} \text{ sec}^{-1}{}_{v},$$

and $$k_4 = 1.4 \times 10^6 \text{ l mole}^{-1} \text{ sec}^{-1}.$$

Determination of Activation Energy

This application can be illustrated by taking a well studied example, *i.e.*, N, N' –dimethyl-acetamide (DMA)

$$\overset{CH_3}{\underset{O}{}}\!> C - N < \begin{matrix} CH_3 \\ CH_3 \end{matrix}$$

Because of the large separations, there is, no spin-spin coupling between protons on different methyl groups. When the NMR spectrum of DMA is recorded at –30°C, the spectrum shows a single peak and a doublet, indicating that the two methyl groups attached to the nitrogen are not equivalent. Magnetic equivalence of these methyl groups can be realised by increasing the rate the rotation, *i.e.*, by heating the compound. Thus, as the temperature of DMA is increased, the peaks of the doublet begin to broaden and ultimately convert into a single line at about 90°C. The rate of rotation may be given by

$$k = 1/2\pi$$

where k is the rate of rotation and π is the mean life tome of the particular configuration which is related to the linewidth and line separation of the methyl peaks. The value of π can be measured at any temperature using the appropriate equations. Then, the activation energy ΔE may be calculated by using the Arrhenius equation

$$\ln k = \ln A - \frac{\Delta E}{RT}$$

where A is a constant which is characteristic of the reaction. For DMA, the value of Δ E comes out to be 12 L cal $mole^{-1}$.

Limitations of NMR Spectroscopy

The various limitations of NMR spectroscopy are as follows :

(i) One of the serious problems with NMR is its lack of sensitivity. The minimum sample size is about 0.1 ml having minimum concentration of about 1%.

(ii) In some compounds two different types of hydrogen atoms resonate at similar resonance frequencies. This results in an overlap of spectra and makes such spectra difficult to interpret.

(iii) While characterizing the organic compounds, no information about molecular weight is given but the relative number of different protons present are only known.

(iv) In most of the cases, only liquids can be studied by NMR spectroscopy, although polymers, when preheated with various solvents, frequently becomes fluids which can be treated as liquids.

In spite of the above mentioned limitations, NMR is one of the most useful analytical tools yet developed for the elucidation of many inorganic and organic structural problems.

ELECTRONIC SPECTRA

Introduction

Molecular spectra, which appear in the ultraviolet and visible region, arise due to transitions from one electronic state to another. If the excitation energy given to a molecule is sufficiently large, the electron in the molecule may undergo transitions from one energy state to higher energy state. During such excitations, known as electronic excitations, vibrational and rotational transitions also occur. Because of this, *each band in the electronic spectra is made up of a number of the lines due to the simultaneous changes in rotational and vibrational energies.*

As in atomic system, the energies involved are generally large, the electronic spectra will, therefore, occur in ultraviolet region, the wavelength is shorter, and it is inversely related to the energy absorbed. Therefore, the energy absorbed will be greater in this region.

As explained earlier, pure rotational spectrum is given by those molecules which have permanent dipole moment while the rotational spectra require a change of dipole moment during the motion but electronic spectra are given by all molecules. This means that molecules like H_2 or N_2, which show no vibrational or rotational spectrum, give an electronic spectrum.

Consider E_e, E_r and E_v be the electronic, rotational and vibrational energies of a molecule before transition. E_e', E_r', and E_v' are corresponding values after the transition. Then, total energy E of the molecule initially will be :

$$E_{total} = E_e + E_r + E_v$$

E', the energy in the final state will be :

$$E'_{total} = E_e' + E_r' + E_v'$$

Then, ΔE_{total}, the change in energy involved during electronic transitions is given as

$$E_{total} = (E_e' - E) + (E_r' - E_r) + (E_v' - E_v) = \Delta E_e + \Delta E_r + \Delta E_v$$

The frequency of radiation emitted is given by the following expression :

$$\bar{v} = \frac{\Delta E_{total}}{hc} = \frac{\Delta E_e + \Delta E_r + \Delta E_v}{hc} \text{cm}^{-1} \quad ...(1)$$

As ΔE_r and ΔE_v may have a number of values which depend upon their rotational and vibrational quantum numbers. But the number of possible lines for a given change in the electronic level is large, therefore, the electronic band spectra are generally very-complex.

Equation (1) can be put as

$$\bar{v} = \frac{\Delta E_{total}}{hc} = \frac{\Delta E_e}{hc} + \frac{\Delta E_r}{hc} + \frac{\Delta E_v}{hc} \text{ cm}^{-1}$$

These approximate orders of magnitude of those changes are as follows :

$$\frac{\Delta E_e}{hc} \approx \frac{\Delta E_v}{hc} \times 10^3 \sim \frac{\Delta E_r}{hc} \times 10^6 \text{ cm}^{-1} \quad ...(2)$$

From equation (2) it follows that vibrational changes will produce a coarse structure and rotational energy changes fine structure on the spectra of electron transitions.

Each electronic level has its characteristic potential energy diagram. The forms of such curves may be completely unrelated. AS in some spectra the (0, 0) transitions is the strongest, in others the intensity increases to a maximum at some value of V (vibrational quantum number), while in others only a few vibrational lines are noted with large value of V. All these types of spectra can be easily explained in terms of Franck-Concon Principle.

The Franck-Concon Principle : The different cases of intensity distribution are explained in an easily visualised manner by the Franck-Condon principle. Franck's main idea, which was developed mathematically and later given a wave mechanical basis by Condon, is the following :

"The electron jump in a molecule takes place so rapidly in comparison to the vibration that immediately afterwards the nuclei still have nearly the same relative position and velocity as before the transitions."

In order to apply this principle, let us consider Fig. 2.40(a), (b), and (c), in which are drawn potential curves that were assumed by Franck and Condon for the upper and lower electronic states in the three typical cases of intensity distribution.

(i) In Fig. 2.40(a), the potential energy curves of the two electronic states have been so chosen that their minima lie very nearly one and above the other (equal internuclear distance). In absorption, the molecule is initially at the minimum of the lower potential curve, if we discard the position of the zero-point vibration. It can be seen that for a transition to the minimum of the upper potential curve (0 – 0) band, the requirement of the Franck-Condon principle is that the change of position and momentum be small. On the other hand, a transition into a high vibrational state [CD in Fig. 2.40(a)] would be possible only when at the moment of the electronic jump either the position (transition from A to C or the velocity) (transition from A to E) or both alter to an appreciable extent. At the point, E, of a course, the molecule has the same amount of kinetic energy as at B. Only at the turning point C or D the velocity and kinetic energy are zero, as in the initial state at A.

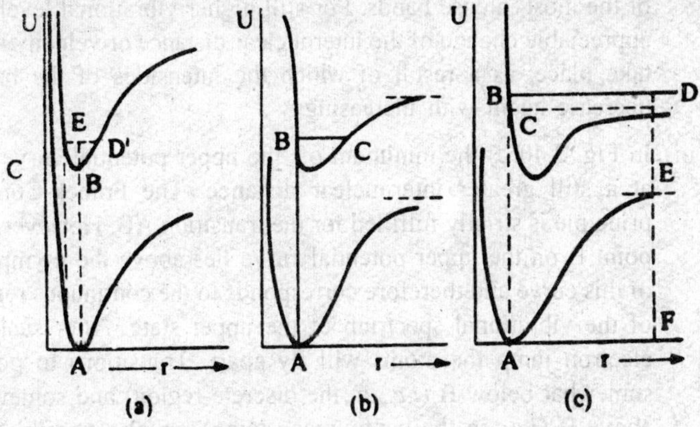

Fig. 2.40 : Potential energy curves exhibiting the intensity distribution according to Franck-Condon principle

From the above it follows that (on the basis of Franck-Condon principle) a transition from V" = 0 to such a high vibrational level is forbidded or atleast highly improbable. For the level V' = 1, the necessary alternation of the position or of the velocity during the elecfron jump is comparatively small. Therefore, the 1 – 0, band will still appear, though with much smaller intensity than the 0 – 0 bond. For the 2 – 0, 3 – 0, ... bands the necessary alternation of position and velocity increases and consequently, a rapidly decreasing intensity is to be expected.

(ii) In Fig. 2.40(b), the minimum of the upper potential curve lies as a somewhat greater r value than that at the lower. Therefore, the transition from minimum to minimum (0 –0), band is no longer the most probable, since the internuclear distance must alter somewhat in such a transition. The most probable transitions for from A to B in Fig. 2.40(b) (vertically upward). For this transition there is no change in the internuclear distance at moment of the jump and no change of the velocity. Thus, immediately after the electron jump the two nuclei still have their old distance from each other and zero relative velocity. Since, however, the equilibrium internuclear distance has a different value in the new electronic state, the nuclei start to vibrate between B and C. The vibrational levels whose left turning points lie in the neighborhood of B are the upper levels of the most intense bands. For still higher vibrational levels an appreciable change of the internuclear distance or velocity must take place, as a result of which the intensities of the bands decrease again with increasing V'.

(iii) In Fig. 2.40(c) the minimum of the upper potential curve lies at a still greater internuclear distance. The Franck-Condon principle is strictly fulfilled for the transition AB. However, the point B on the upper potential curve lies above the asymptote of this curve and therefore corresponds to the continuous region of the vibrational spectrum of the upper state. After such an electron jump the atoms will fly apart. Transitions to points somewhat below B (*i.e.*, in the discrete region) and somewhat above B (*i.e.*, in the continuous region) are also possible.

Summarising we can say that in absorption the most intense transition from V" = 0 is always that corresponding to a transition from the minimum of the lower potential curve vertically upward. In all cases in which a fine structure analysis has been carried out the internuclear distances, r actually have been found to be in agreement with the values assumed according to the above discussion for an explanation of the intensity distances,

Vibrational Coarse Structure of Electronic Spectra

As we have already described that electronic energy change is always accompanied by vibrational and rotational energy changes, we may therefore write

$$E' \text{ total} - E'' \text{ total} = (E_e' - E_e'') + (E_v' - E_v'') + (E_r' - E_r'') \quad ...(1)$$

However, we will ignore rotational energy changes. Thus, equation (1) becomes as

$$E' \text{ total} - E'' \text{ total} = (E_e' - E_e'') + (E_v' - E_v'')$$

The frequency of the spectrum arising due to electronic transition will be

$$\bar{v}_0 = \frac{E' \text{ total} - E'' \text{ total}}{hc} = \frac{(E_e' - E_e'') + (E_v' - E_v'')}{hc}$$

$$= \frac{E_e' - E_e''}{hc} + \frac{E_v' - E_v''}{hc}$$

$$= \bar{v}_0 + \frac{1}{hc}\left[\left\{\left(V' + \frac{1}{2}\right) hc\,\bar{w}' - \left(V' + \frac{1}{2}\right)^2 hcx'\bar{w}'\right\} - \left\{V'' + \frac{1}{2}\right\} hc\bar{w}''\right.$$

$$\left. - (V'' + 1/2)\, hcx''\, \bar{w}''\right\}\Big] \qquad ...(2)$$

where V' and $\bar{w}'$ refer to the vibrational quantum number and equilibrium vibrational frequency of the molecule in the initial state, V", $\bar{w}''$ are the corresponding quantities in the final state.

As we have taken rotational energy to be zero, equation (2) yields the frequencies v_0 of the centres (or origins) of the series of vibrational bands building the electronic spectrum.

When a molecule is undergoing an electronic transition, *i.e.*, ΔV may be positive, negative or zero, there is no selection rule for V. A large number of bands will be observed in the band system of any molecule because every transition (V' → V") possesses some probability. A series of bands possessing a constant value of V' → V" is termed as a sequence.

There is another set of bands possessing a different value of either V' or V" while the other varies regularly, for example, V' = 0, while V" = 0, 1, 2, 3, etc . This is termed as a progression.

When absorption occurs, the electronic spectrum generally contains a single progression only when V" becomes equal to zero. For this, the reason is that the most of the molecules will exist in the lowest vibrational state (V" = 0) at room temperature ad therefore intense transitions will be those which occur between V" = 0 and V' = 0, 1, 2, ... etc.

Rotational Fine Structure of Electronic Vibration Transition

The electronic spectrum is made up of one or more series of convergent lines constituting the vibrational coarse structure electronic transition. Generally, each of these lines is broad which when examined under a high resolution is seen as a cluster of closely spaced lines which is, of course, rotational line structure.

Energy Levels

If we regard that the electronic, vibrational and rotational energies of a molecule are independent of each other, we may therefore write the total energy (excluding the kinetic energy of transition) of a diatomic molecule as :

$$E_{total} = E_e + E_v + E_r \quad ...(1)$$

We have already proved that

$$E_r = hc\,BJ\,(J + 1) \quad ...(2)$$

In equation (2), we have injected D, the centrifugal distortion constant. Substitution of equation (2) in (1), we get

$$E_{total} = E_e + E_v + hc\,BJ\,(J + 1) \quad ...(3)$$

In terms of energy changes, equation (3) may be put as

$$\Delta E_{total} = E_e + E_v + hc\,\Delta\,\{BJ\,(J + 1)\} \quad ...(4)$$

The frequency of radiation which arises as a result of electronic

$$\bar{\nu} = \frac{1}{hc}\,\Delta E_{total} \quad ...(5)$$

Substituting equation (4) in (5), we get

$$\bar{\nu} = \frac{1}{hc}\left[\Delta E_e + \Delta E_r + hc\,\Delta\left\{BJ\,(J + 1)\right\}\right]$$

$$= \frac{\left[\Delta E_e + \Delta E_v\right]}{hc} + \Delta\left\{BJ\,(J + 1)\right\}$$

$$= \bar{\nu}_0 + \Delta\left\{BJ\,(J + 1)\right\} \quad ...(6)$$

where $\bar{\nu}_0$ refers to the frequency in wave numbers of the origin of particular band.

Selection Rule

The rule for J depends upon the type of electronic transition which the molecule is undergoing. If $^1\Sigma$ represents both the upper and lower electronic states in which there occurs no electronic angular momentum about the internuclear axis, the selection rule is

$$\Delta J = \pm 1 \text{ for } {}^1\Sigma \rightarrow {}^1\Sigma \text{ transitions}$$

From the above rule, it follows that for such transitions only P and R branches will occur. No Q branches will occur.

For all other transitions, which have either the upper or lower, or both states having angular momentum about the internuclear axis, the selection rule becomes as

$$\Delta J = 0 \text{ or } \pm 1$$

From the above rule it follow that P and R branches along Q branches ($\Delta J = 0$) will occur.

For both these cases, there is a restriction which means that a transition between the states which corresponds to $J = 0$ cannot occur, *i.e.*,

$$J = 0 \leftarrow\!/\!\rightarrow J = 0$$

Spectrum

If we represent the upper electronic state by B_1 and J_1 and lower electronic state by B_2 and J_2, we may put equation (6) in the form

$$\bar{v} = \bar{v}_o + B_1 J_1 (J_1 + 1) - B_2 J_2 (J_2 + 1) \qquad ...(7)$$

where B_1 and B_2 refer to the upper and lower electronic states respectively as, generally, to different vibrational states and so they are very much different from one another. As B_1 and B_2 are different from each other and also from B, one cannot therefore take $B_1 = B_2 = B$. Let us take P, R and Q branches in turn. Thus, we have

(1) P Branch

As $\Delta J = -1$, $J_2 = J_1 + 1$, it means that

$$\bar{v}(P) = \bar{v}_o - (B_1 + B_2)(J_1 + 1)(B_1 - B_2)(J_2 + 1)^2 \qquad ...(8)$$

where $J_1 = 0, 1, 2, ...,$ etc.

(2) R Branch

As $\Delta J = +1$, $J_1 = J_2 + 1$, it means that

$$\bar{v}(R) = \bar{v}_o + (B_1 + B_2)(J_2 + 1)(B_1 - B_2)(J_2 + 1)^2 \qquad ...(9)$$

where $J_2 = 0, 1, 2, ...,$ etc. By combining equations (8) and (9), we get the combined equation in the form

$$\bar{v}(PR) = \bar{v}_o + (B_1 + B_2)m + (B_1 - B_2)m^2 \qquad ...(10)$$

where $m = \pm 1, \pm 2, ...$ In equation (10), positive values result R branch ($\Delta J = +1$) whereas negative values gives rise to P branch ($\Delta J = -1$). It is important to remark here that m cannot be zero *i.e.*, $m \neq 0$. Because in P branch zero value of m ($m = 0$) corresponds to $J_1 = -1$ which is impossible. Same is true for R branch. Thus, we conclude that no line from P and R branches would appear at the band origin v_o

(3) Q Branch

As $\Delta J = 0$, $J_1 = J_2$, so that

$$\bar{\nu}(Q) = \bar{\nu}_o + (B_1 - B_2)J_2 + (B_1 - B_2)J_2^{\ 2} \quad ...(11)$$

where J_2 = 1, 2, 3, ... As transition from J = 0 to J = 0 is not possible, it means that J_1 and J cannot be equal to zero, *i.e.*, $J_1 = J_2 \neq 0$. In other words, it means that no line will appear a the band origin.

Let us now discuss the appearance of all these branches, *i.e.*, R, P, Q, with reference to the difference $(B_1 - B_2)$ which may be positive or negative.

Case I.

When $B_1 < B_2$, draw Fig. 2.41 in which R, P and branches are drawn separately represented as a, b and c respectively considering the difference between uppér and lower values of B to be about 10 per cent and choosing $B_1 < B_2$.

From Fig. 2.41, we draw the following important generalisations :

(i) The spacing between P branch lines increases with m, and P branch lines are on the low wave number side of the band origin.

(ii) R branch lines are on high wave number side of the band origin. The spacing between the R branch lines decreases rapidly with m. In R branch lines, there appears a point at which separation decreases at zero. This point is termed as band head.

(iii) Q branch lines are close to low wave number origin and their spacing increases with the value of J_2.

(iv) On the high wave number side of the origin a band head appears in R branch only which is said to be degraded (or shaded) towards the red. This shaded region is towards the red end (low frequency) of the spectrum.

Case II.

When $B_1 > B_2$, all cases (i) to (iv) discussed as in case I are reversed, *i.e.*,

(i) In P branch lines, there appears to be a crowding of lines to form a band head at the red end of the band. Such a band is said to be shaded towards the violet.

(ii) R branch lines still appear on highwave number side of the band origin. In the R branch lines, separation increases.

(iii) Q branch lines spread to high wave number.

Advantages of electronic spectra

Electronic spectra have the advantages that they occur in ultraviolet or in visible regions which can be photographed and therefore, examined in greater details and with greater precisions.

Applications of Electronic Spectra

These spectra provided information with regard to

(i) vibrational frequencies and moment of inertia

(ii) interatomic distances

(iii) dissociation energy

(iv) force constants.

THE FORTRAT DIAGRAM

Equation (10) for the P and R branch lines and equation (11) for Q lines can be expressed terms of variable parameters P and Q respectively :

$$\bar{v}\,(PR) = \bar{v}_o + (B_1 + B_2)p + (B_1 - B_2)p^2 \qquad ...(12)$$

$$\bar{v}\,(Q) = \bar{v}_o + (B_1 - B_2)q + (B_1 - B_2)q^2 \qquad ...(13)$$

Both equations (12) and (13) represent parabola provided if we take both positive and negative values for p while positive values for q only. The sketches of the parabola obtained from equation (12) and (13) are called Fortrat parabola (Fig. 2.42) in which we have taken 10% difference between the upper and lower B values and also taken $B_1 < B_2$. Fig. (2.42) regions of positive p values are labelled as $\bar{v}$ (R) whereas those of negative p values as $\bar{v}$(P). Both p and q may take only integral values but not zero values. The allowed values of p and q have been indicated by drawing circles round the allowed points on the Fortrat's diagram. This fact helps us the read $\bar{v}$ values of spectral line directly from the graph.

A useful property of Fortrat diagram is that the band head lies at the vertex of P and R parabola. The position of vertex can be obtained by differentiating equation (12) which is

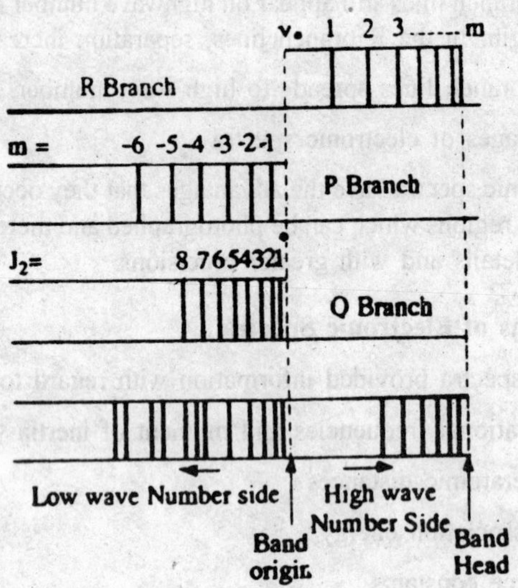

Fig. 2.41 : Rotational fine structure of electronic vibration of a diatomic molecule.

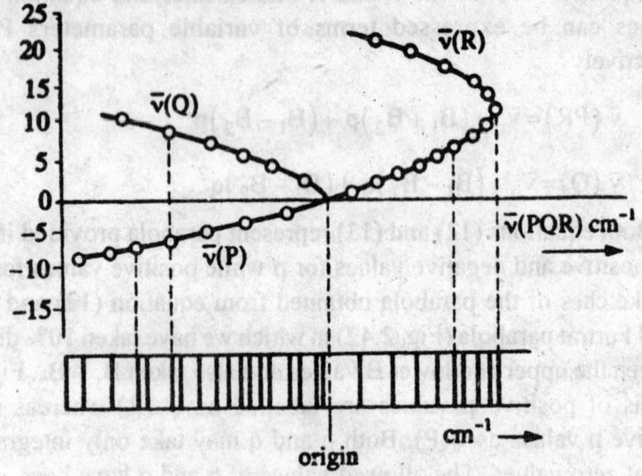

Fig. 2.42 : Fortrat diagram

$$\frac{d\overline{v}(PR)}{dp} = B_1 + B_2 + 2(B_1 - B_2)p \qquad \text{...(14)}$$

At the vertex, $\frac{d\overline{v}(PR)}{dp} = 0$, therefore, equation (14) can be put as

$$p = \frac{B_1 + B_2}{2(B_1 - B_2)} \quad \text{...(15)}$$

From equation (15) two conclusions can be drawn :

(i) If $B_1 > B_2$, band head will occur at positive p values in R branch.

(ii) If $B_1 < B_2$, band head will occur at negative p values in P branch.

THE CHARGE TRANSFER SPECTRA

The type of spectra is very common an mainly occurring in the ultraviolet rather than the visible region. This type of spectra arises due to the transfer of an electron from one atom to another atom. Alkali halides are the examples of the charge transfer spectra. In Fig. 2.43, there are absorption spectra of gaseous sodium halides and of crystalline potassium and rubidium halides. In Fig. 2.44, charge transfer transitions in gaseous sodium halides have been shown.

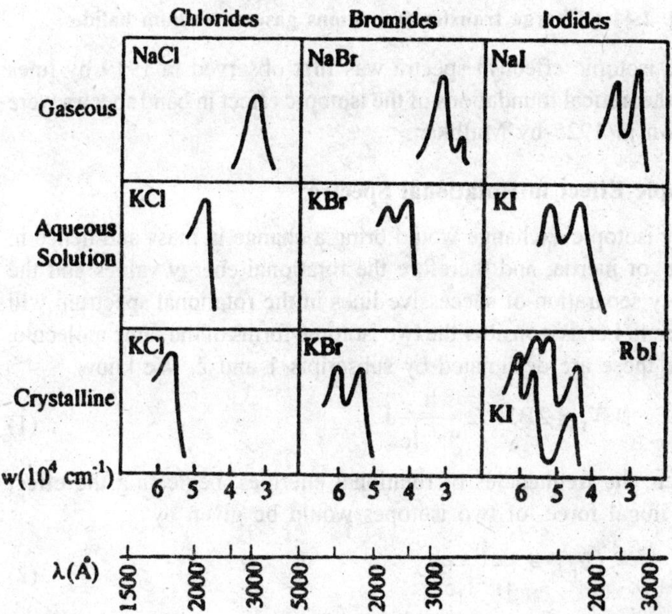

Fig. 2.43 : Absorption spectra of gaseous sodium halides, of potassium halides in aqueous solution, and of crystalline potassium and rubidium halides.

ISOTOPIC EFFECT IN MOLECULAR SPECTRA

Molecules having different isotopes of the same element, such as DCl and HCl, show different spectra. This is expected because the masses of the atoms of different isopotes are different and hence reduced masses and also the frequencies of vibration and of rotation would be different. The atomic distance and force constant which depend upon the extra nuclear electrons mainly are assumed to remain constant.

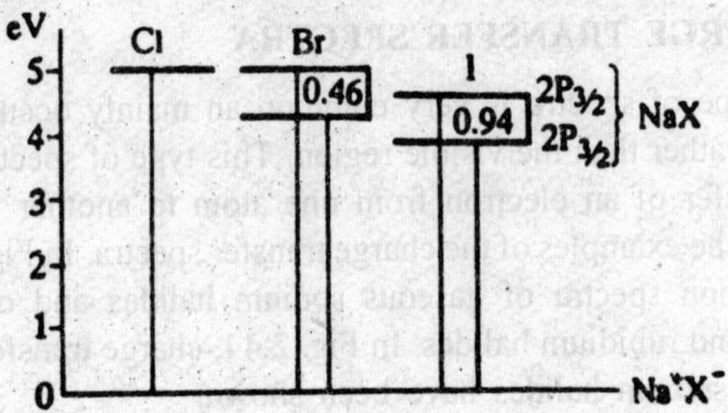

Fig. 2.44 : Charge transfer transitions gaseous sodium halides.

The isotopic effect in spectra was first observed in 1919 by Imes and the theoretical foundations of the isotopic effect in band spectra were laid down in 1925 by Mullikan.

1. Isotopic Effect in Rotational Spectra

Any isotopic exchange would bring a change in mass and hence in moment of inertia, and therefore the rotational energy values and the frequency separation of successive lines in the rotational spectrum will be different. Let us consider the two isotopic forms of the same molecule. Suppose these are designated by subscripts 1 and 2. We know

$$\bar{v}_r = 2BJ = 2\frac{h}{8\pi^2 Ic}J \quad \text{...(1)}$$

Then, the frequencies of rotational energies, neglecting the effect of centrifugal force, of two isotopes would be given by

$$(\bar{v}_r)_1 = \frac{h}{4\pi^2 I_1 c}J \quad \text{...(2)}$$

and

$$(\bar{v}_r)_2 = \frac{h}{4\pi^2 I_2 c}J \quad \text{...(3)}$$

In equations (2) and (3), I_1 and I_2 denote the moment of inertia of the isotopic species, 1 and 2. Now the shift in frequency due to isotopic change will be given by

$$\Delta\bar{v}_i = (\bar{v}_r)_1 (\bar{v}_r)_2$$

$$= \frac{h}{4\pi^2 c} J \left(\frac{1}{I_2} - \frac{1}{I_2} \right) = \frac{h}{4\pi^2 c I_1} J \left(1 - \frac{I_1}{I_2} \right)$$

$$= (\bar{v}_r)_1 \left(1 - \frac{I_1}{I_2} \right) \quad \text{[Using eq. (2)]} \quad \text{...(4)}$$

As internuclear distances do not change, the ratio of the moment of inertia will be equal to the ratio of reduced masses, *i.e.*,

$$\frac{I_1}{I_2} = \frac{\mu_1}{\mu_2} = \rho^2 \quad \text{...(5)}$$

where ρ^2 denotes the constant value of the ratio. On substituting equation (3) in (4), we get

$$\Delta\overline{v_i} = \overline{(v_r)} \left(1 - \rho^2\right) \quad \text{...(6)}$$

Within the limit of small error, $\bar{v}_r$ can be 2BJ [see equation (1)] and, thus, equation (6) can be put as

$$\Delta\overline{v_i} = 2BJ \left(1 - \rho^2\right) \quad \text{...(7)}$$

From equation (7) it follows that the isotopic shift increases with value of J. As the shift cf the lines in rotational spectrum accompanying an isotopic change could not be observed because of experimental difficulties. Such an isotopic effect in rotation is inferred from the study of electronic and vibration-rotation bands.

Isotopic Effects in Vibrational Bands

Although the two isotopic forms of the same molecule possess different reduced masses, yet they have the same force constants.

We know, $$K = 4\pi^2 \omega_e^2 \mu c^2 \quad \text{...(8)}$$

where K is the force constant. Suppose ω_1 and ω_2 are the values of ω_e for two isotopic forms μ_1 and μ_2 be their respective reduced masses. Thus we get

$$K = 4\pi^2 \omega_1^2 \mu_1 c^2 \quad \text{...(9)}$$

$$K = 4\pi^2\ \omega_2{}^2\mu^2c^2 \qquad ...(10)$$

Dividing Eq. (10) by (9), we get

$$\left(\frac{\omega_2}{\omega_1}\right)^2 = \frac{\mu_1}{\mu_2} \text{ or } \left(\frac{\omega_2}{\omega_1}\right) = \left(\frac{\mu_1}{\mu_2}\right)^{1/2} = \rho$$

or $$\omega^2 = \rho\ \omega_1 \qquad ...(11)$$

From theoretical conditions, the value of an harmonicity constant, x, is proportional to the equilibrium frequency and thus, equation (11) becomes as

$$x_2 = r\ x_1 \qquad ...(12)$$

We have proved earlier, the frequency of the centre of any band involving lower vibrational level V – 0 will be given by

$$\overline{v}_v \to_o = V_{\omega e} - \{V_1(V+1)\}\ x_{\omega e} \qquad ...(13)$$

With the aid of two equations (11) and (12), equation (13) may be written for two isotopic species :

$$(\overline{v}_v \to_o)_1 = V_{\omega 1} - \{V(V+1)\}\ x_{1\omega 1}$$

$$(\overline{v}_v \to_o)_2 = V_{\omega 2} - \{V(V+1)\}\ x_{2\omega 2}$$

$$= V\ \rho_1\omega_1 - \{\ V\ (\ V + 1)\}\ \rho^2\ x_1\omega_1 \text{ [Using eqs. (11) and (12)]}$$

In the above equations the subscripts 1 and 2 refer to the two isotopic forms of the same molecule. Now the frequency difference of the centres of the two isotopic bands, called isotopic shift, $\Delta\overline{v_i}$ will be given by

$$\Delta\overline{v_i} = (v_v \to_o)_1 (v_v \to_o)_2 \qquad ...(14)$$

$$\Delta\overline{v_i} = (1-\rho)\{1-(V+1)_\rho\ x_1\}Vx_1 \qquad ...(15)$$

From relation (15), the values of isotopic shifts may be calculated, *i.e.*,

Fundamental band (V = 1 to V = 0)

$$\Delta\overline{v_i} = (1-\rho)(1-2\rho\ x_1)\ \omega_1 \qquad ...(16)$$

(b) First overtone band (V = 2 to V = 0)

$$\Delta\overline{v_i} = (1-\rho)(1-3\rho\ x_1)\ 2\omega_1 \qquad ...(17)$$

(c) Second overtone band (V = 3 to V = 0)

$$\Delta\overline{v_i} = (1-\rho)(1-4\rho\, x_1)\, 3\omega_1 \qquad ...(14)$$

From equations (15), (16) and (17) it follows that it is possible to calculate the frequency shift corresponding to one isotopic form of the same molecule. The isotopic shift depends upon the factor $(1 - \rho)$. If $\rho > 1$, the isotopic shift Δv_i is in the direction of lower frequency whereas if $\rho < 1$, the shift is in the opposite direction.

The vibrational isotope effect, expected according to the theory, was first applied to the rotation vibration spectrum of HCl. As chlorine atom (Cl) possesses two isotopes Cl^{35} and Cl^{37}, two bands should appear, one corresponding to HCl^{35} and the other to HCl^{37}. In the case of HCl^{37}, the isotopic frequency shift will occur by a small amount towards the longer lengths with respect to the band belonging to HCl^{35}. This has been confirmed by experimental investigation of the rotation-vibration bands of HC1. The isotopic shifts obtained from experimental results were found to be in good agreement with those calculated theoretically as discussed earlier. This can be seen from the following Table 2.10.

Table 2.10. Observed and calculated vibrational isotope displacement for the infrared HC1–bands :

Table 2.10

Band	$\Delta \bar{v}_i$ ***(obs)***	$\Delta \bar{v}_i$ ***(cal)***
1.0	2.01	2.105
2.0	4.00	4.033
5.0	5.834	5.845

ELECTRON SPIN RESONANCE SPECTROSCOPY

Introduction

Electron spin resonance is a branch of absorption Spectroscopy in which radiation having frequency in the *microwave region* is absorbed by *paramagnetic substances* to induce transitions between magnetic energy levels of electrons with *unpaired spins*. The magnetic energy splitting is done by applying a static magnetic field. The electron spin resonance phenomenon is shown by atoms having an *odd number* of electrons, ions having partly filled inner electron shells and other mo ecules that carry angular momentum of electronic origin. The main interest of electron spin resonance lies in the study of free radicals having unpaired electrons. These electrons remain after the homolytic fission of a covalent bond which is brought about by the irradiation of the sample with the ultraviolet or gamma radiation. The electron spin resonance (ESR) is also called by other names such as *"electron paramagnetic resonance"* (EPR) and *"electron magnetic resonance."* All these names carry the same meaning and simply describe the different aspects of the same phenomenon.

Electron spin resonance was invented by *Zavoiskii* in 1944. For some time this technique remained of interest only to solid state physicists until after the initial applications of molecular orbital theory in the mid fifties by *Stevens Owen*, and *McGarvey* that it became of general interest to Chemists The technique of electron spin resonance may be regarded as a fascinating extension of the Stern-Gerlach experiment.

Theory of ESR

Electron Behaviour : In ESR, the energy levels are produced by the interaction of the magnetic moment of an unpaired electron in a molecule ion with an applied magnetic field. The ESR spectrum results in due to the transitions between these energy levels by absorbing radiations of microwave frequency.

For an electron of spins = 1/2 the spin angular momentum quantum number will have values of $m_s = \pm 1/2$. In the absence of magnetic field, the two values of m_s, *i.e.*, +1/2 and –1/2 will give rise to a doubly degenerate spin energy state. If a magnetic field is applied, this degeneracy is removed and this leads to two non-degenerate energy levels. The low energy state will have the spin magnetic moment aligned with the field and corresponds to the quantum number, $m_s = -1/2$. On the other hand, the high energy state will have the spin magnetic moment opposed to the field and corresponds to the quantum number, $m_s = +1/2$ These energy states are illustrated in Fig. 1. These two states will possess energies that are split up from original state with no applied magnetic field by the amount $+\mu_e H$ and $\mu_e H$ for the low energy and high energy states respectively ; here H is the magnetic field acting on the unpaired electron and μ_e is the magnetic moment of the spinning electron. In ESR, a transition between the two different energy levels takes place by absorbing a quantum of radiation of frequency in the microwave region. Thus, the ESR spectrum of a free electron would consist of a single peak corresponding to a transition between these levels. When the absorption occurs, the following relation holds good. *i.e.*,

$$2\mu_{e\ H} = h\nu \qquad ...(1)$$

where ν is the frequency of the absorbed radiation in cycles per second. As the relation (1) holds good for a free electron, the energy (ΔE) of transition is more accurately given by the following relation:

$$\Delta E = h\nu = g\beta H \qquad ...(2)$$

In equation (2)

(i) h is the Planck's constant,

(ii) H is the applied magnetic field,

(iii) β is the Bohr's magneton which is a factor for converting angular momentum into magnetic moment. The value of p is defined as follows :

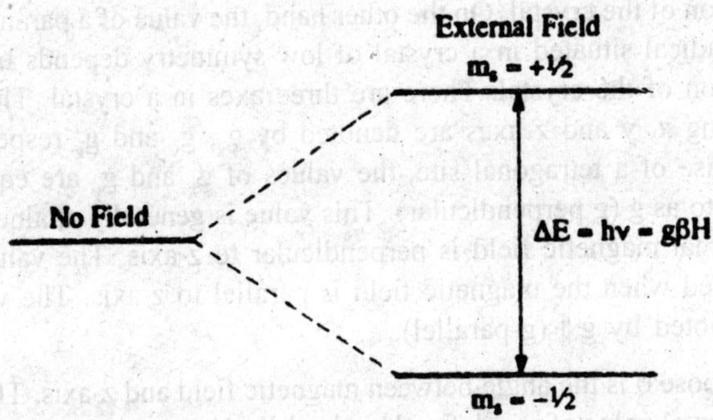

Fig. 2.45

$$\beta = \frac{eh}{4\pi mc}\ 0.9723 \times 10^{-20} \text{ erg/gauss}$$

where e is the electronic charge, m the mass of electron and c and the velocity of light.

(iv) g is the proportionality factor which is a function of the electron's environment. It is sometimes called the *spectroscopic splitting factor or Lande's splitting factor*. The value of g is not constant but it is a tensor quantity. For a free electron its value is 2.0023 but this value is slightly modified for electrons in molecules. Virtually all free radicals and some ionic crystals have almost same value of g as for free electrons, but this value is modified by +0.003. The reason for this is essentially that in free radicals, the electron behaves in very much the same way as an electron in free space.

In some ionic crystals, the values of g vary from 0.2 to 0.8. The reason for this is that the unpaired electron is localised in a particular orbital about the atom and the orbital angular momentum couples with the spin angular momentum giving rise to a low value of g in ionic crystals.

The values of g depend upon the orientation of the molecule having the unpaired electron with respect to applied magnetic field. In the case of a gas or solution, the molecules have free motion and the value of g is averaged over all orientations. But in the case of crystals, the movement of electrons is not free. However, in the case of a paramagnetic ion or radical situated in a perfectly cubic crystal site, the value of g is same in all directions, *i.e.*, the value of g does not depend upon the orientation of the crystal. On the other hand, the value of a paramagnetic ion or radical situated in a crystal of low symmetry depends upon the orientation of the crystal. There are three axes in a crystal. The value of g along x, y and z axes are denoted by g_x, g_y and g_z respectively. In the case of a tetragonal site, the values of g_x and g_y are equal and referred to as g (g perpendicular). This value is generally attained when the external magnetic field is perpendicular to z-axis. The value of g_z is obtained when the magnetic field is parallel to z-axis. The value of g, is denoted by g || (g-parallel).

Suppose θ is the angle between magnetic field and z-axis. Then, the experimental value of g is defined by the following equation for a system possessing axial symmetry.

$$g^2 = g\,\|^2 \cos^2\theta + G \perp \sin^2\theta$$

We will now drop the discussion about g and continue it in a later section.

Instrumentation

The energies required to bring about a transition are different in ESR and NMR. In ESR, transitions occur at frequencies in the microwave region whereas in NMR, transitions occur in the radio frequency region. This reveals that the instrumentation for the NMR and ESR must be different.

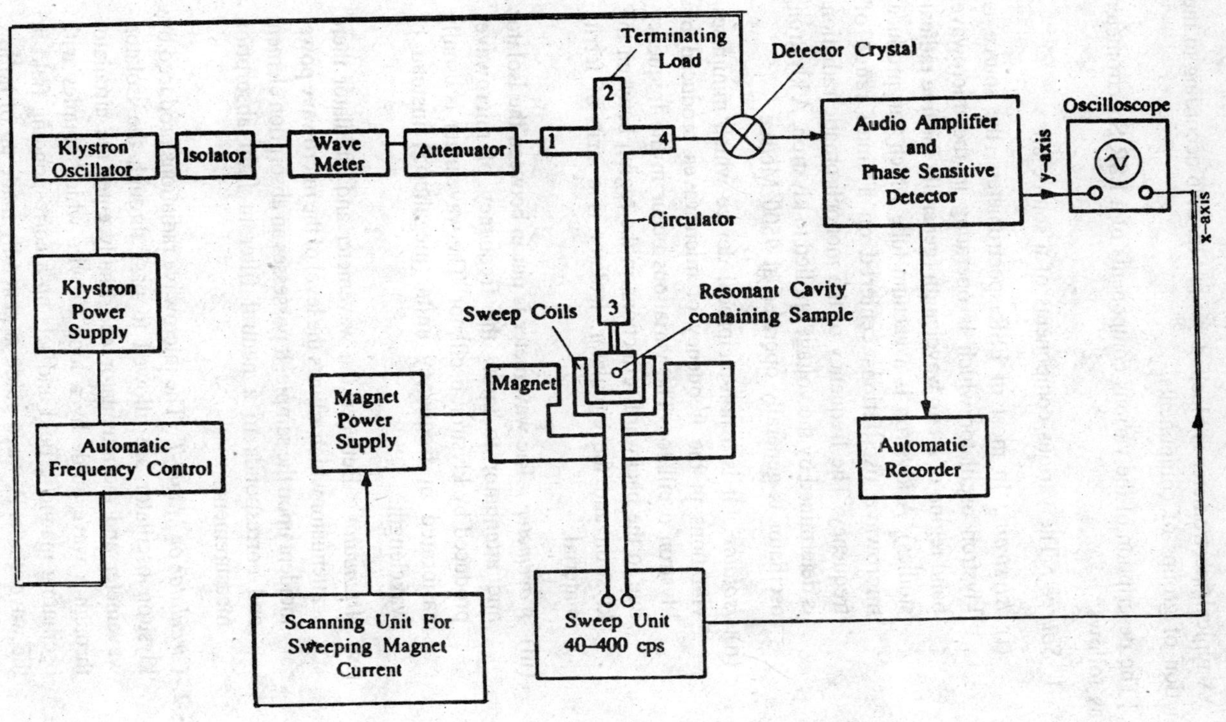

Fig. 2.46 : Block diagram of a typical ESR spectrometer.

A typical spectrometer is blocked out in Fig. 2.46 according to the function of groups of components.

The description of the various components of a ESR Spectrometer is as follows:

1. *Source* : The various components of a source are :

   (i) *Klystron* : In most of ESR spectrometers, the source is klystron oscillator which is operating in the microwave band region of 3 cm. wavelength (generally for free radical studies). A klystron is a vacuum tube which can produce microwave oscillations centered on a small range of frequency. The frequency of the monochromatic radiation is determined by the voltage applied to klystron. A klystron oscillator is generally operated at 9500 Me/see.

   (ii) *Isolator* : It is a non-reciprocal device which minimises vibrations in the frequency of microwaves produced by klystron oscillator. The variations occur in the frequency due to the backward reflections in the region between the klystron and the circulator. Isolator is a strip of ferrite material.

   (iii) *Wavemeter* : The wavemeter is put in between the isolator and attenuator to know the frequency of microwaves produced by klystron oscillator. The wavemeter is usually calibrated in frequency units (megahertz) instead of wavelength.

   (iv) *Attenuator* : Between the wavemeter and circulator there is attenuator which adjusts the level of the microwave power incident upon the sample. It possesses an absorption element and corresponds to a neutral filter in light absorption measurements.

2. *Circulator or Magic-T* : The microwave radiations produced by klystron oscillator are allowed to pass through the isolator, wavemeter and the attenuator and finally enter the circulator through a wave guide by a loop of wire which couples with oscillating magnetic field and sets up a corresponding field in the wave guide. A wave guide is generally made up of hollow rectangular copper or brass tubing (2.2cm × 10cm) having silver or gold plating inside to produce a highly conducing flat surface.

The operation of a four-port circulator is indicated in Fig. 2.47. The microwave radiations enter the arm 1. Arm 2 is connected to resonant cavity and sample. Arm 3, generally having a terminating load, seems to absorb any power which might be reflected from the detector arm. Arm 4 is attached to the detector.

3. *Sample Cavity :* The heart of an ESR spectrometer is the resonant cavity containing the sample. The cavity system is constructed in such a way to maximize the applied magnetic field along the sample dimension. A sample volume of 0.15 to 05 ml can be used with samples which do not possess a high dielectric constant. Flat cells with a thickness of 0.25 mm and sample volume of 0.05 ml are generally used for such samples which possess high dielectric constant. For studying anisotropic effects in single crystals and in solid samples, rotatable cavities are generally utilized.

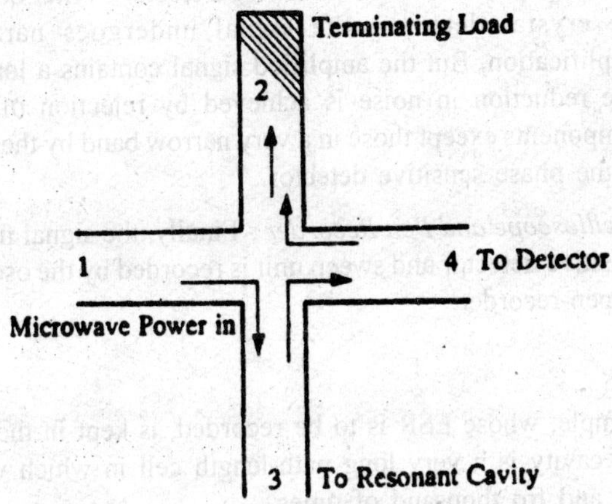

Fig. 2.47 : A four-port microwave circulator showing the directions of microwave transmission among the several arms.

In most of the ESR spectrometers, dual sample cavities are generally used. This is done for simultaneous observation of a sample and a reference material. By the use of a reference material, the sources of error are compensated by comparing relative signal heights.

4. *Magnet System :* The resonant cavity is placed between the pole pieces of an electromagnet. This provides a homogeneous

magnetic field and can be varied from zero to 500 gauss. The field should be stable and uniform over the sample volume. The stability of field is achieved by energizing the magnet with a highly regulated power supply. The stability of 1 part in 10^6 is satisfactory for resolution of ESR spectra of such samples whose g-factor ranges from 1.5 to 6. On the other hand, the stability might be as low as 1 part in 10^3 for paramagnetic ions and for free radicals in solid matrices.

In order to sweep the magnetic field over a small range, the provision is made by varying the current in a pair of sweep coils.

5. *Crystal Detectors* : The most commonly used detector is a silicon crystal which acts as a microwave rectifier. This converts microwave power into a direct-current output.

6. *Autoamplifier and Phase Sensitive Detector* : After detection by the crystal detector, the signal undergoes narrow-band amplification. But the amplified signal contains a lot of noise. The reduction in noise is achieved by rejection of all noise components except those in a very narrow band by the operation of the phase sensitive detector.

7. *Oscilloscope and Pen Recorder* : Finally, the signal from phase sensitive detector and sweep unit is recorded by the oscilloscope or pen-recorder.

Working

The sample, whose ESR is to be recorded, is kept in the resonant cavity. The cavity is a very long path-length cell in which waves are reflected to and fro thousand of times.

The klystron oscillator is set to produce microwaves which after passing through the isolator, wavemeter and attenuator are received by circulator through arm. The microwave power entering arm 1 will divide between arms 2 and 3. The arm 2 is generally having a balancing load. If the impedances of arms 2 and 3 are same, microwave power is absorbed completely and no power will be received by the detector through arm 4. If the impedance of arm 3 (the sample cavity) changes because of some ESR resonance absorption by a sample in it, some microwave power will enter into detector through arm 4. The detector acts as a rectifier converting the microwave power into direct current. If the magnetic field around the resonant cavity having the sample is changed to the value required for resonance, the recorder will show an absorption peak (Fig. 2.48). Further, if the main magnetic field is swept slowly over a period of several minutes, the recorder will show the derivative of the microwave absorption spectrum against magnetic field (Fig. 2.49). Due to instrumental considerations associated with the signal-to-noise ratio, the ESR spectra are generally recorded as first derivative spectra.

In case of low-frequency modulation (400 Hz or less) the coils are fixed outside and in some ESR spectrometers, they may be mounted on the magnet pole pieces. But in the case of higher modulation frequencies, the coils are mounted inside the sample cavity; the main reason for this is that higher modulation frequencies cannot penetrate metal effectively.

Sensitivity of ESR Spectrometer : The sensitivity of an ESR spectrometer may be expressed as

$$N_{min} = 1 \times 10^{11} \frac{\Delta H}{\sqrt{\tau}} \qquad \text{...(4)}$$

where N_{min} = minimum number of detectable spins per gauss,

ΔH = width between deflection points on the derivative absorption curve, and

τ = the time constant of the detecting system ; this is inversely proportional to the bandwidth of the detection circuit.

Working at higher magnetic field strength increases the sensitivity of ESR spectrometers. For example, the sensitivity of an ESR spectrometer working at 35,000 MHz is twenty times greater than that working at $500 MHz.

Presentation of the ESR Spectrum

The ESR spectrum, similar to NMR spectrum, may be obtained by plotting intensity against the strength of a magnetic field. However, the better way is to represent ESR spectrum as a derivative curve in which the first derivative (the slope) of the absorption curve is plotted against the strength of magnetic field. In Fig. 2.48 there is a single absorption peak having no fine structure. In Fig. 2.49, there is a derivative curve that corresponds to Fig. 2.49.

The results represented by derivative curves in ESR can be readily interpreted. Each negative slope in the derivative curve represents a peak or shoulder in the absorption spectrum. Every crossing of the derivative axis with a negative slope indicates a true maximum whereas a crossing with positive slope indicates a minimum. It means that the number of peaks or shoulders in the absorption curve can be determined from the number of minima or maxima in the derivative curve.

The peak height of the absorption curve or derivative curve does not provide much information. However, the total area covered by either the absorption or derivative curve is proportional to the number of unpaired electrons in the sample. In order to calculate the number of electrons in an unknown sample, comparison is made with a standard sample having a known number of unpaired electrons and possessing the same line shape as the unknown Gaussian or Lorentzian). The most widely used standard is 1, 1, diphenyl-2-picryl hydrazyl free radical (DPPH) whose structure is as follows :

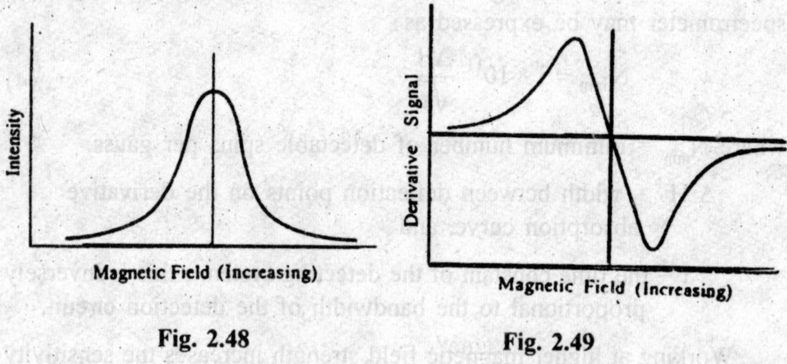

Fig. 2.48

Fig. 2.49

DPPH is a chemically stable material having the splitting factor g = 2.0036. DPPH contains 1.53 × 10^{21} unpaired electrons per gram.

NO_2

N—Ṅ— —NO_2

NO_2

Fig. 2.50

DPPH cannot be employed as an internal standard for other free radicals because there is only slight variation in the g-values and the unknown substance cannot be distinguished from the standard substance.

In the case of free radicals, an internal standard is a trace of Cr (III) entrapped in a tiny chip of ruby crystal cemented permanently to the sample cell. This standard shows a strong resonance and its g value is 1.4.

The quantitative measurements must be made on the integral of the ESR curve.

Hyperfine Splitting

There is no phenomenon which parallels the NMR chemical shift in ESR studies. However, the ESR spectrum exhibits hyperfine splitting which is caused by the interactions between the spinning electrons and adjacent spinning magnetic nuclei. When a single electron interacts with one nucleus, the number of splittings will be equal to 21 + 1, where I is the spin quantum number of the nucleus. In general, if a single electron interacts magnetically with n equivalent nuclei, the electron signal is split up into a (2nI + 1) multiplet.

Let us illustrate the hyperfine splitting by considering an example of a hydrogen atom having one proton and one electron (I = 1/2 for the proton). In the absence of a magnetic field, the single electron of spin (s = 1/2) gives rise to a doubly degenerate spin energy state. When a magnetic field is applied, the degeneracy is removed and two energy levels, one corresponding to $m_s = -1/2$ aligned with the field and the other corresponding to $m_s = +1/2$ aligned opposing the field, will be obtained Fig. 2.51. The spectrum of a free electron would consist of a single peak corresponding to a transition between these levels. This is shown in Fig. 2.52.

When the interaction between the two energy states and the nuclear spin due to proton is considered, each energy state is further split up into two energy levels corresponding to $m_I = +1/2$ and $m_I = -1/2$. where m is the nuclear spin angular momentum quantum number. Thus, corresponding to two energy states, four different energy levels are obtained Fig. 2.51. It means that ESR spectrum of hydrogen would consist of two peaks Fig. 2.51 corresponding to two transitions shown by two arrows in Fig. 2.51,

The energies of four energy levels Fig. 2.51 are as follows :

$$E_{1/2, 1/2} = 1/2\, g\, \beta\, H + 1/4\, A \quad ...(5)$$

$$E_{1/2, -1/2} = 1/2\, g\, \beta\, H - 1/4\, A \quad ...(6)$$

$$E_{1/2',-1/2} = -1/2\ g\ \beta\ H + 1/4\ A \qquad ...(7)$$

$$E_{1/2',1/2} = -1/2\ g\ \beta\ H - 1/4\ A \qquad ...(8)$$

The general expression for all the four energy levels is as follows:

$$E = h\beta Hm_s + Am_sm_I \qquad ...(9)$$

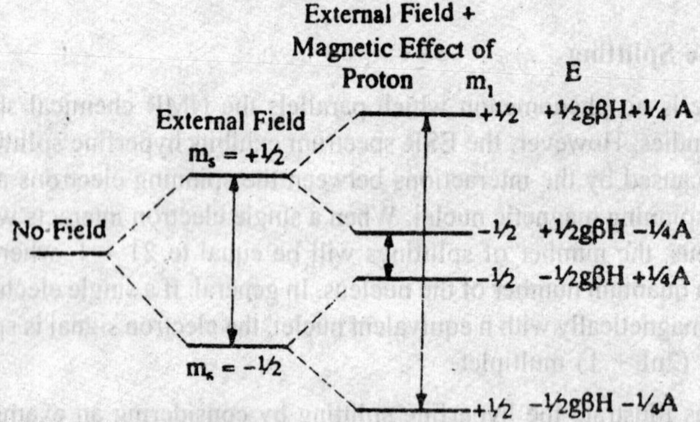

(a) Effect of external field on the energy states for an electron. **(b) Effect of a nuclear spin 1/2 of proton on these states.**

Fig. 2.51

where A is termed as the hyperfine coupling constant. When the proper values of m_s and m_I are substituted in equation (9), the values of energies indicated in Fig. 2.51 are obtained. The selection rules in ESR are

$$\Delta w_I = 0 \text{ and } \Delta m_s = \pm 1$$

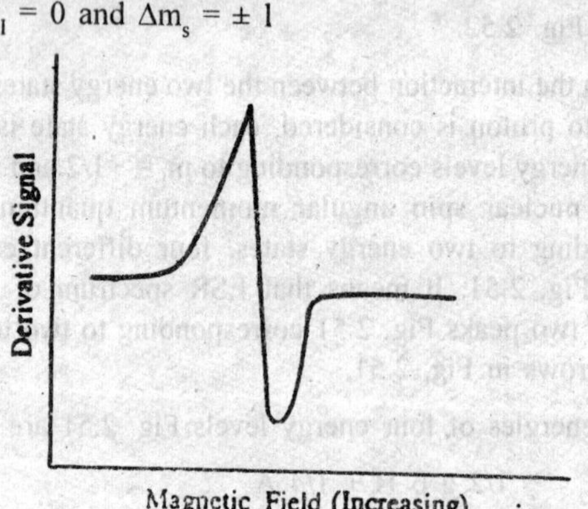

Fig. 2.52

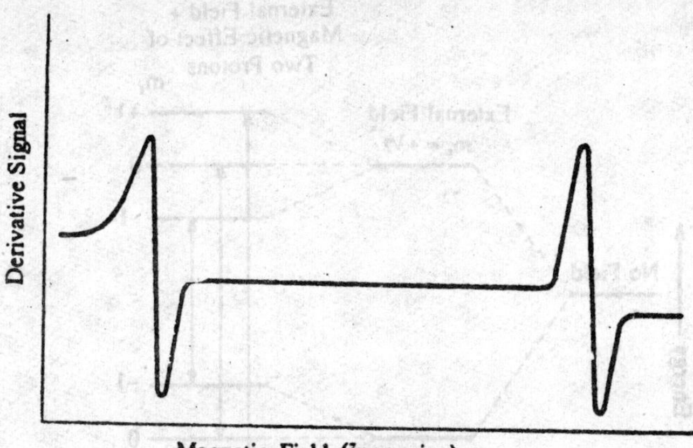

Fig. 2.53

The deuterium ($_1^2H$) is a simple example of a system with I = 1. Energy levels are computed in a similar manner as done in the case of hydrogen atom. Corresponding to m_s = +1/2. there are three values of m_I, *i.e.*, m_I = 1, 0 and –1. Similarly corresponding to m_s = –1/2. there are also the three values of m_I, *i.e.*, m_I = –1. 0 and + 1. The energy of all the six energy levels are as follows :

$E_{1/2,1} = 1/2\ g\ \beta\ H + 1/2\ hA$ $\quad E_{-1/2,-1} = -1/2\ g\ \beta\ H + 1/2\ hA$

$E_{1/2,0} = 1/2\ g\ \beta\ H$ $\quad E_{-1/2,0} = 1/2\ g\ \beta\ H$

$E_{1/2,-1} = 1/2\ g\ \beta\ H - 1/2\ hA$ $\quad E_{-1/2,1} = 1/2\ g\ \beta\ H - 1/2\ hA$

By virtue of the selection rules $\Delta m_s = \pm 1$ and $\Delta m_I = 0$, there are three allowed transitions and thus three lines corresponding to these transitions in the ESR spectrum of deuterium. These lines will be of equal intensity, since there is no coincidence of states, *i.e.*, the states are all non-degenerate. The various transitions are shown in Fig. 2.54. Corresponding to these transitions, a typical derivative spectrum of deuterium in an increasing magnetic field is shown in Fig. 2.55.

Let us now consider the case for s = 1/2 and I = 3/2. An example of this is the methyl radical. When the interaction is considered between the single unpaired electron on the carbon atom and three hydrogen nuclei, there are four values corresponding to m_s = +1/2. Similarly, there are four values corresponding to m_s = –1/2, *i.e.*,

For m_s = +1/2 m_I = + 1/2, –1/2 –3/2

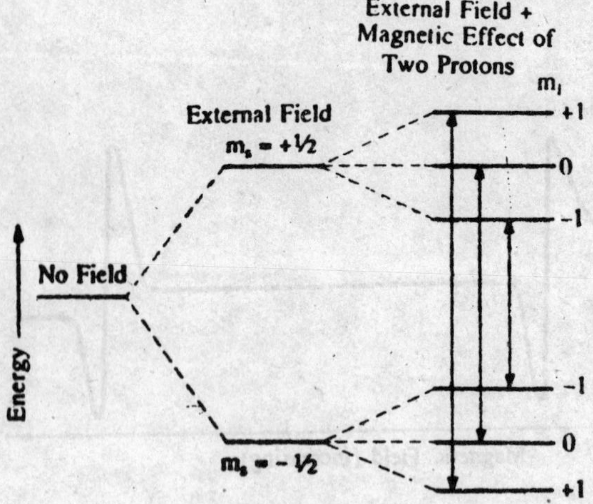

Fig. 2.54

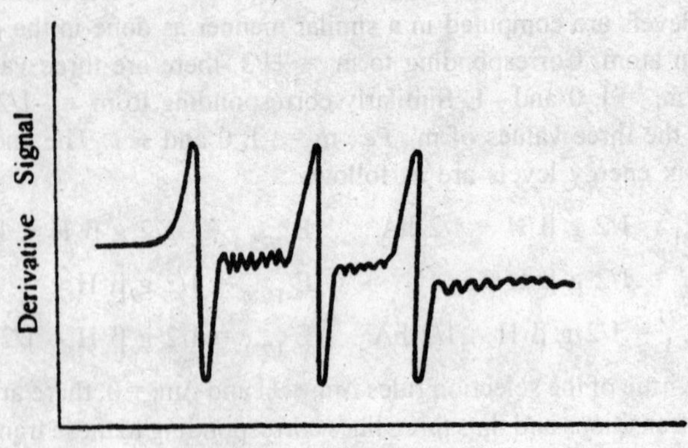

Fig. 2.55

and for $m_s = -1/2$, $m_I = -3/2, -1/2, +1/2, +3/2$

On applying the selection rule, $\Delta m_I = 0$ and $\Delta m_s = \pm 1$, four transitions are possible, resulting four lines in the ESR spectrum of methyl radical. The observed relative intensities for the four lines are in the ratio 1 : 3 : 3 : 1. The four transitions that occur in the ESR spectrum of methyl radical are shown in Fig. 2.56. The derivative spectrum of methyl radical is shown in Fig. 2.57.

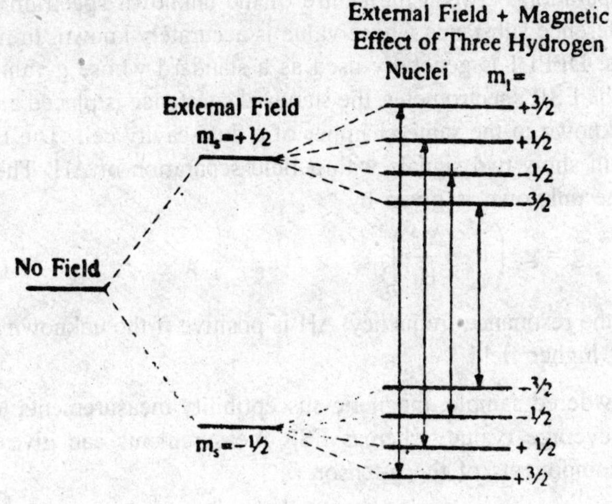

Fig. 2.56

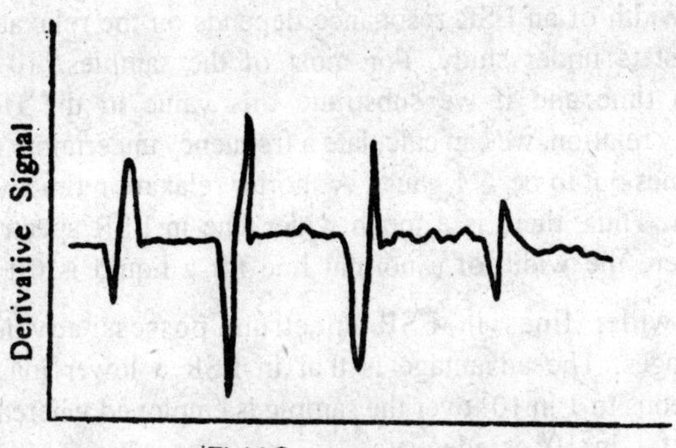

Fig. 2.57

In general, when the absorption spectrum of an unpaired electron undergoes interaction with n equivalent nuclei of equal spin I, the number of lines obtained in ESR spectrum will be given by (In I + 1). When the interaction involves a set of n equivalent nuclei-of spin I, and a set of equivalent nuclei m of spin I_j, the number of lines will be given by $(2n\ I_i + 1)\ (2m\ I_j + 1)$.

Determination of g-Value

In order to measure the value of g, the bust method is to measure the field separation between the centre of the unknown spectrum and that of a reference substance whose value is accurately known. In most of the cases, DPPI I is generally used as a standard whose g-value is 2.0036. In the ESR spectrometer, the standard substance is placed along with the unknown in the same chamber of a dual-cavity cell. The ESR spectrum will show two signals with a field separation of ΔH. The g-value for the unknown is given by

$$g = g_s \left(1 - \frac{\Delta H}{H}\right) \quad ...(10)$$

where H is the resonance frequency. ΔH is positive if the unknown has its centre at higher field.

In a powdered sample, magnetic susceptibility measurements lead only to an average rvalue, whereas ESR measurements can give the individual components of the g-tensor.

Line Width

The width of an ESR resonance depends on the relaxation time of the spin state under study. For most of the samples, 10^{-7} sec. is a relaxation time and if we substitute this value in the Heisenberg's uncertainty relation, we can calculate a frequency uncertainty (line width) which comes out to be $\approx$ 1 gauss. A shorter relaxation time will increase this width. Thus, there is a much wider line in ESR spectrum than in NMR where the width of a normal line for a liquid is 0.1 gauss.

The wider lines in ESR spectrum possess advantages and disadvantages. The advantage is that in ESR a lower magnetic field homogeneous to 1 in 10^5 over the sample is employed whereas for NMR a figure of 1 in 10^8 is adequate.

The major disadvantage of the wider lines is that these are more difficult to observe and measure than sharp lines. For this simple reason, ESR spectrometers are generally operated in the derivative mode.

The increase in width of spectral lines in ESR spectrum is termed as the broadening of the lines. There are often peak-broadening phenomena which can only be overcome by working at very low temperatures. The broadening becomes too much in those cases where the ground state has another orbital level above it, not having very different energy. For this reason, it is very difficult to observe spectrum of titanium (III) (d^1).

The .broadening of spectral lines in ESR may also occur if the concentration of the paramagnetic species is very high. For this reason, the ESR spectrum is obtained for such a substance whose trace quantity is incorporated as an impurity on a diamagnetic compound, isomorphous with that under study.

The Heisenberg uncertainty- principle can be utilised to estimate line width from relaxation times. It states that the product of uncertainty in energy and the uncertainty in time is constant and equal to $h/2\pi$.

$$\Delta E \times \Delta t = h/2\pi$$

Applications of ESR Spectroscopy

By careful interpretation, an ESR spectrum provides the following types of information.

(i) It decides about the site of unpaired electron(s).

(ii) The number of line components decide about the number and type of nuclei present in the neighbourhood of the old electron.

(iii) The relative intensities of the spectrum lines in an ESR spectrum confirm the type of nuclei which are responsible for the splitting pattern. Summation of the intensities can be utilised to evaluate the total number of free electrons in the sample.

(iv) From the ESR spectrum, the value of g can be measured by comparing the position of the line with that of a standard substance of known g value, *e.g.*, DPPH powder ($g = 2.0036$) or $K_2(SO_3)$, NO ($g = 2.0057$).

(v) If the electric field is not spherical, the ESR spectrum is anisotropic, *i.e.*, the rotation of the sample shifts the ESR spectrum,

In simple substances the interpretation of an ESR spectrum is very easy, provided spectrum is fully resolved and absorption lines are narrow. In complex substances, the ESR spectrum will not contain all the lines expected because the g-factor and coupling values are such that two lines may overlap and, thus, the lines are not resolved. In some other substances, many equivalent nuclei may undergo interaction to give rise to peaks of relative heights, causing the difficulty in seeing the smallest peaks which are very important in the correct interpretation of these complicated molecules. The best method to interpret the extremely complicated ESR spectrum is to introduce the approximate coupling constants in an electronic computer and then comparison of the computed spectrum is made with the experimental one. If the two are same, one can be sure about the structure of a complex substance whose ESR spectrum is being studied.

The various applications of ESR Spectroscopy are as follows :

1. *Study of Free Kadicals* : Free radicals can be readily studied by ESR, even in very low concentrations. For example, a signal for DPPH radical can be detected even if there is 10^{-12} gm of material in the spectrometer,

   The ESR spectrum of a quinone-hydroquinone redox system is shown in Fig. 2.58. This proves conclusively that a semiquinone free radical anion exists as an intermediate. The first derivative spectrum of this system contains fiveline spectrum which arises due to the magnetic spin interaction between the odd electron and the four hydrogen atoms on the ring. The intensities of five lines are in the ratios 1:4:6:4:1.

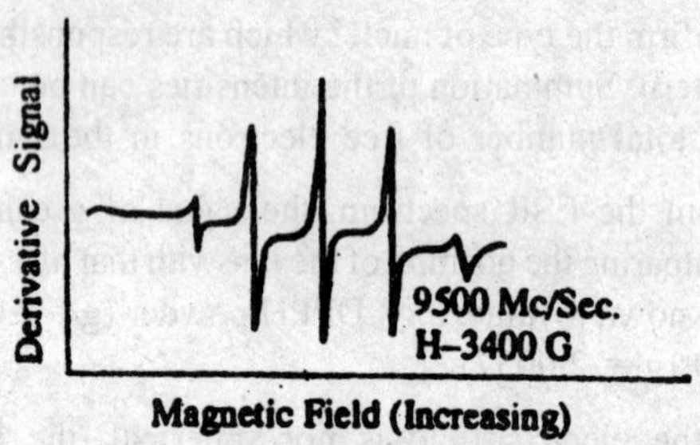

Fig. 2.58

2. *Reaction Velocities and Mechanisms* : A large number of organic reactions are known which proceed by a radical mechanism. Most of the radicals formed during organic reactions are not stable but are very reactive. In order to maintain a high and enough steady concentration for ESR studies, the rapid-flow system is generally utilised. For this, the special sample cell is designed to permit the compounds (whose interaction yields the paramagnetic intermediates) to mix together and immediately

flow through the observation area. In this way a steady concentration of the radicals can be maintained. With this method, radicals of lifetimes of about 0.01 second have been characterised. An interesting example of this method is the characterisation of CH_2OH radical which contains one set of two equivalent protons. This radical may be produced in a flow system by mixing a solution of T_t^{3+} and CH_3OH with H_2O_2 solution just outside the microwave cavity. If the pH is sufficiently low, the OH proton contributes no detectable hyperfine splitting, since this proton is rapidly exchanged. Hence, a simple 1:2:1 triplet arises from thc CH_2 protons (Fig. 2.59). The hyperfine splitting is 17.2 G.

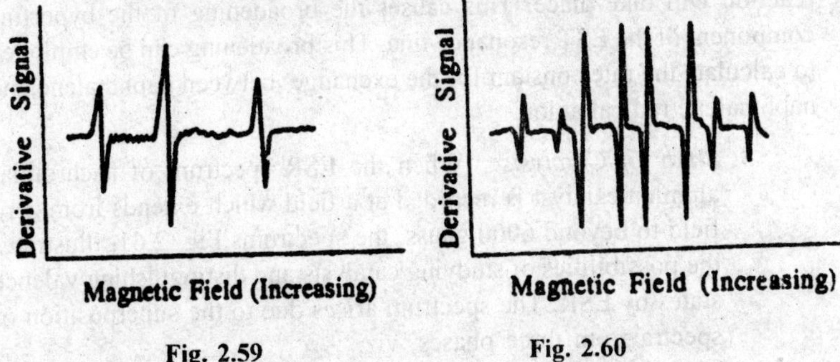

Fig. 2.59 **Fig. 2.60**

The following reactions are known to take place to produce CH_2OH radical :

$$Ti^{3+} + H_2O_2 \rightarrow Ti^{4+} + OH + OH-$$

$$Ti^{3+} + HO \rightarrow Ti^{4+} + OH-$$

$$OH + H_2O_2 \rightarrow H_2O + O_2H$$

$$CH_3OH + OH \rightarrow CH_2OH + H_2O$$

Similarly, this method can be used to detect CH_3CHOH radicals which are generated by mixing an acidified solution having C_2H_5OH and hydrogen peroxide. Fig. 2.60 shows the first derivative spectrum of CH_3CHOH.

In ESR spectroscopy special cells have been used in which radicals are produced *in situ* by irradiation with ultraviolet, gamma or X-rays or by electrolytic redox reactions. For example ethyl radicals re produced when ethyl alcohol is irradiated with X-ray radiation. When ESR of the ethyl radicals is taken, its spectrum shows five lines which confirms the formation of ethyl radicals. Similarly, polyethylene, after irradiation with g-rays, yields a seven-fold resonance pattern which has been shown to be due to the formation of

$$
\begin{array}{c}
-CH - C - CH_2 - \\
| \\
CH_2 \\
|
\end{array}
$$

The ESR spectroscopy can be used to study very rapid electron exchange reactions. An interesting example is that when naphthalene is added to a solution of naphthalene radical anion, an electron exchange reaction will take place. This causes the broadening of the hyperfine component of the EST resonance line. This broadening can be employed to calculate the rate constant for the exchange between naphthalene and naphthalene radical anion.

3. *Study of Catalysts* : When the ESR spectrum of a chromia-alumina catalyst is recorded at a field which extends from zero field to beyond 6000 gauss, the spectrums Fig. 2.61, illustrates the possibilities of studying catalysts and distinguishing valence states by ESR. The spectrum arises due to the superposition of spectra from three phases, viz.

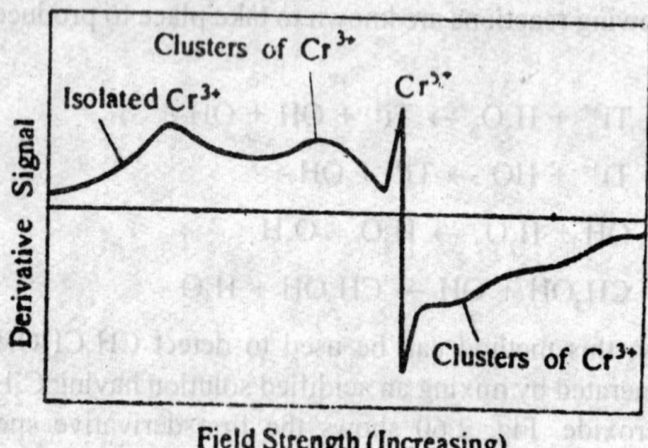

Fig. 2.61

(i) A broad line centred at g = 2 arises from clumped or clustered Cr^{3+} ions.

(ii) A low field line appears at ~ 1600 gauss. This arises due to isolated Cr^{3+} ions.

(iii) A sharp singlet near g = 2 is attributed to Cr^{5+} ions.

ATOMIC SPECTRUM

Introduction : Whent the light coming directly from a source is allowed to enter a prism or grating spectroscope, and '*emission spectrum*' of the source is observed in the spectroscope. On the other hand, when light from a source showing a continuous emission spectrum is passed through an absorbing material and then into the spectroscope, and 'absorption spectrum' of the material is observed. A beam of white light is passed through a prism (or grating), it breaks up into the beams of constituent colours. The different coloured beams, when focussed on a screen by a converging lens, from an array of colours on the screen. This is called the 'spectrum' of white light.

The are two main types of spectra :

(i) 'Emission spectra' and (ii) 'Absorption spectra'.

Emission of Spectrum

This type of spectrum may be obtained, when the light coming after passing through a prism or a grating, is examined directly with a spectroscope. the hydrogen atom gets sufficient energy from outside by some means, the electron from an inner orbit of lower energy goes up to an outer orbit of higher energy. This excited state of the atom lasts for jumping down, it emits the difference in energy between the two orbits as electromagentic radiation. If the electron jumps from an orbit ni to an orbit n_f, the wavelength of the emitted radiation will be

$$\frac{1}{\lambda} = R\left(\frac{1}{n_f^2} - \frac{1}{n_i^2}\right).$$

It is found that for,

$n_f = 1$,	$n_i = 2, 3, 4, \ldots$	we obtain	Lyman series,
$n_f = 2$,	$n_i = 3, 4, 5, \ldots$	we obtain	Balmer series,
$n_f = 3$,	$n_i = 4, 5, 6$	we obtain	Paschen series,
$n_f = 4$,	$n_i = 5, 6, 7, \ldots$	we obtain	Brackett series,
$n_f = 5$,	$n_i = 6, 7, 8, \ldots$	we obtain	Pfund series,

The corresponding energy level diagram is shown in Fig. 2.62. The top horizontal line represents zero energy, that is, energy of the electron outside the atom (n = ∞). The other horizontal lines represent energies of different orbits given by the formula,

$$E = -\frac{me^4}{8\varepsilon_0^2 h^2}\left(\frac{1}{n^2}\right).$$

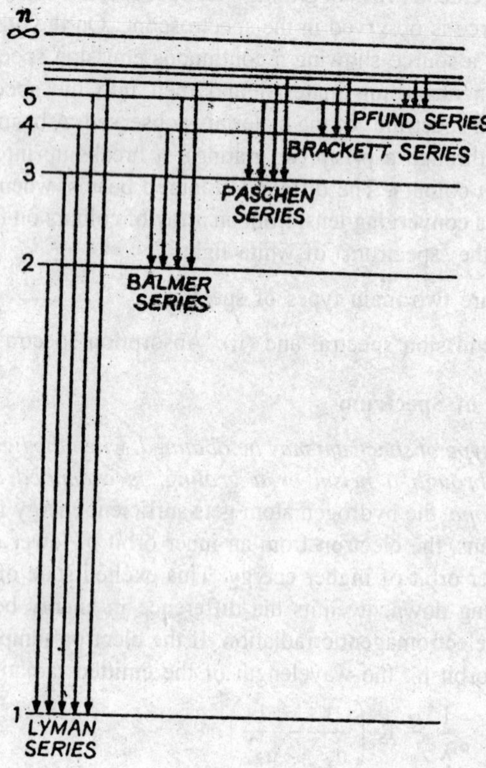

Fig. 2.62

The arrows ending at the lines n = 1, 2, 3, 4, and 5 represent the transitions responsible for the Lyman, Balmer, Paschen, Brackett and Pfund series respectively.

Absorption Spectrum

Absorption spectrum are obtained when the absorbing substance is placed between a source emitting a continuous spectrum and the slit of the spectroscope. In such cases, certain colours (wavelengths) are absorbed by the substance. Hence, the spectrum is found to consist of dark lines or bands against a bright background.

An example of such spectra is sun's spectrum. It is a line absorption spectrum. It was studied in detail by Fraunhoffer offer who named the absorption lines as A, B, C, D. These lines inform us about the elements which are present around the sun. Similarly, when an intense beam of continuous white light is passed through sodium vapour and then sent into a spectroscope, we obtain two dark lines on a continuous background in the same positions as the yellow D_1 and D_2 lines in the sodium, emission spectrum. If the sodium vapour is replaced by Iodine vapour (I2), an absorption band spectrum of I2 molecule is obtained.

Series Relationship in Atomic Spectrum

Atom spectrum consist of a large number of lines. A quantitative experimental study of these spectral lines was made in the second half of the 19th century when several regularities were observed in the spacings of the lines. For example, in 1870. Lveing and Dewar noticed that the spectral lines of various elements could be grouped into distinct 'series'. *In each series the spacing and intensity of lines decrease regularly towards shorter wavelengths,* until it becomes impossible to distinguish the individual lines. The point at which the lines of the series finally converge is called the 'series limit'. The various series in complicated spectra overlap.

The simplest atomic spectrum is that of hydrogen. Its visible part consists of a single series which was first observed by Balmer in 1885. This is called the 'Balmer series' of hydrogen. Its first line having longest wavelength of 6563 Å, is named Ha, the next Hb, and so on, the series limit reaching at 3646 Å. Besides this, there is a series of lines in the ultraviolet part of the hydrogen spectrum which is known as 'Lyman series, and three series in the infra-red part which are known as 'Paschan series', 'brackett series' and 'Pfund series'.

Balmer discovered a formula for the wavelengths of all the lines of the Balmer series. His formula is

$$\lambda = 3646 \frac{n^2}{n^2 - 4} \text{ Å}, \; n = 3, 4, 5, \ldots$$

n = 3 gives the wavelength of H_α line, n = 4 gives H_β line, ... and n = ∞ gives the series limit.

Subsequently, Rydberg found that Balmer's formula was a special case of a more general formula which is as follows.

$$\bar{\nu} = \frac{1}{\lambda} = R_H\left(\frac{1}{m^2} - \frac{1}{n^2}\right),$$

where $\bar{\nu}$ is wave number (reciprocal of wavelength), m and n are positive integers (n > m) and RH is the Rydberg constant for hydrogen. Its value is 1.097 × 107 m^{-1}. To obtain formula for Lyman series we set m = 1 and n = 2, 3, 4, ..., for Balmer series m = 2 and m = 3, 4, 5, ... and so on. Thus.

$$\bar{\nu} = R_H\left(\frac{1}{1^2} - \frac{1}{n^2}\right), \; n = 2, 3, 4, \ldots \qquad \text{(Lyman)}$$

$$\bar{\nu} = R_H\left(\frac{1}{2^2} - \frac{1}{n^2}\right), \; n = 3, 4, 5, \ldots \qquad \text{(Balmer)}$$

$$\bar{\nu} = R_H\left(\frac{1}{3^2} - \frac{1}{n^2}\right), \; n = 4, 5, 6, \ldots \qquad \text{(Paschen)}$$

$$\bar{\nu} = R_H\left(\frac{1}{4^2} - \frac{1}{n^2}\right), \; n = 5, 6, 7, \ldots \qquad \text{(Brackett)}$$

$$\bar{\nu} = R_H\left(\frac{1}{5^2} - \frac{1}{n^2}\right), \; n = 6, 7, 8 \ldots \qquad \text{(Pfund)}$$

We see that the wave numbers of hydrogen lines can be expressed as differences of two terms of the form $\frac{R_H}{n^2}$.

The next simplest spectra are of the 'monovalent' atoms of alkali metals Li, Na, K, etc. The lines in the spectrum of an alkali atom can be grouped unto four distinct series, a principal series' of intense lines a 'sharp series' of fine lines, a diffuse series' of comparatively broader lines and a 'fundamental series' which lies in the infra-red region. The sharp and diffuse series line in the visible part and converge to a common limit. Rydberg represented the lines of a particular series by the formula.

$$\bar{\nu} = Z^2R_A\left[\frac{1}{(m-\Delta_1)} - \frac{1}{(n-\Delta_2)^2}\right],$$

where R_A is the Rydberg constant for a particular element A, Z is the atomic number, m and n are positive integers, and Δ_1 and Δ_2 are constants for the particular series. Actually each line of an alkali spectrum is a close doublet. Again, we find that the wave numbers of the lines can be expressed as differences of two terms like $\frac{Z^2 R_A}{(n - \Delta)^2}$.

After alkali spectra, next in complexity are the spectra of 'divalent' atoms of alkaline-earths Be, Mg, Ca. In a typical alkaline-earth spectrum, we can distinguish two distinct systems of lines, a system of singlets and a system of distinguish two distinct systems of lines, a system of singlets and a system of triplets. Each system has a principal series, a sharp series, a diffuse series and a fundamental series.

As we proceed to atoms having several valence electrons, the spectra becomes more complex and the groupings of lines into series becomes less pronounced. Still regularities can be observed in complex spectra, and it is possible to express the wave number of any spectral line as the difference of two terms.

Emission and absorption spectra are further classified according to their appearance. There are three classes.

(i) Line Spectrum

If the light source is a low-pressure gas (as in a discharge tube), flame, arc or spark, the spectrum is discontinuous, showing a number of sharp bright coloured lines. These lines are the images of the slit of the spectroscope formed by lights of different colours. The entire series of images is called a 'line spectrum'. The different lines differ in intensity and nature. Some are sharp, some are sharp on one side and diffused on the other, and others are diffused on both sides.

The line spectrum is the characteristic of the atom or the ion. It means that a particular atom or ion always gives the characteristic set of spectral lines, and no two atoms or ions can give the same spectral line. For example, sodium atom gives two intense yellow lines called D_1 and D_2 lines.

(ii) Band Spectrum

Such spectrum are obtained by the radiation from gas molecules such as oxygen (O_2), nitrogen (N_2), cyanogen (CN), etc. They consist of illuminated regions, called 'bands', separated by dark spaces. With a high-resolving instrument each band is seen to consist of very fine lines which become closer and closer on one side of the band until they coincide. This side thus has a sharp and bright edge called the 'head' of the band.

(iii) Continuous Spectrum

When the emitting source is an incandescent solid, or liquid, such as a lamp filament or a gas at high pressure, the spectrum is continuous containing all colours (wavelengths) from red to violet. Its appearance is like an unbroken luminous band of light in which the colour changes gradually from point to point but without any sharp boundary. In this spectrum, the intensity is maximum at a certain point and decreases on both sides of it. The point of maximum intensity shifts towards the violet end of the spectrum as the temperature of the source increases.

Rydberg-Ritz Combination Principle

The principle states *the wave numbers of spectral lines can be expressed by the difference of spectroscopic terms in such a way that other difference of those terms give also the wave numbers of lines in the same spectrum.*

Suppose in a spectrum the wave numbers of two lines are given as

$$\bar{v}_a = T_2 - T_3 \text{ and } \bar{v}_d = T_1 - T_4,$$

Then lines of the following wave numbers are also expected in the same spectrum,

$$\bar{v}_b = T_2 - T4 \text{ and } \bar{v}_c = T_1 - T_3.$$

This means that constant differences exist between the wave numbers.

$$\bar{v}_b - \bar{v}_a = \bar{v}_d - \bar{v}_c$$

$$\bar{v}_c - \bar{v}_a = \bar{v}_d - \bar{v}_b.$$

Displacement Law

According to this law *the spectrum of any neutral atom of atomic number Z closely resembles the spectrum of the singly ionised atom of atomic number Z + 1.* For example, the spectrum of H^+ (Z = 2) closely resembles the spectrum of H (Z = 1) and can be represented by similar formula,

$$\bar{v} = 4\, R_{He} \left[\frac{1}{m^2} - \frac{1}{n^2}\right].$$

Other elements deprived of all but one electron, also produce hydrogen-like spectra which an be represented by the general formula

$$\bar{v} = Z^2R_A \left[\frac{1}{m^2} - \frac{1}{n^2}\right].$$

In a similar way, the spectrum of a singly ionised alkaline-earth atom resembles the spectrum of an alkali atom. From this we conclude that it is the number of valence electrons in an atom which determines the qualitative character of the spectrum of that atom.

Hydrogen Spectrum

The spectrum of hydrogen atom consists of a number of lines. These lines have been grouped into a number of 'series'. The lines in each series are such that *their separation and intensity decrease regularly towards shorter wavelengths,* converging to a limit called the 'series 'limit. The wavelengths in each series can be given by a simple empirical formula. The first such spectral series was observed by balmer in 1885 and is called the *Balmer series of hydrogen.* The first line with the longest wavelength (6563 Å) is named Ha. The next Hb, and so on. The series limit lies at 3646 Å, beyond which is a faint continuous spectrum. Balmer's formula for the wavelengths of the series is

$$\frac{1}{\lambda} = R\left(\frac{1}{2^2} - \frac{1}{n^2}\right), \ n = 3, 4, 5, \ldots \qquad \text{(Balmer)}$$

The quantity R is called the 'Rydberg constant' and has the value

$$R = 1.097 \times 10^7 \text{ meter}^{-1}.$$

The Ha line corresponds to n = 3, the H_β line to n = 4, and so on. The series limit experiment.

The Balmer series contains only those spectral lines which fall in the visible part of the hydrogen spectrum. The lines falling in the ultraviolet and infrared parts form other series. The lines in the ultraviolet form the Lyman series whose wavelengths are give by

$$\frac{1}{\lambda} = R\left(\frac{1}{1^2} - \frac{1}{n^2}\right), \ n = 2, 3, 4, \ldots \qquad \text{(Lyman)}$$

In the infrared, three spectral series have been observed whose lines have the wavelengths given by the formulas

$$\frac{1}{\lambda} = R\left(\frac{1}{3^2} - \frac{1}{n^2}\right), \ n = 3, 5, 6, \ldots \qquad \text{(Paschen)}$$

$$\frac{1}{\lambda} = R\left(\frac{1}{4^2} - \frac{1}{n^2}\right), n = 5, 6, 7, ... \quad \text{(Brackett)}$$

$$\frac{1}{\lambda} = R\left(\frac{1}{5^2} - \frac{1}{n^2}\right), n = 6, 7, 8, ... \quad \text{(Pfund)}$$

Bohr's Theory of Hydrogen Spectrum

The existence of sharp spectral lines cannot be explained from the classical electro-magnetic theory. Bohr explained it by applying Planck's quantum hypothesis to the Rutherford's atomic model.

The Rutherfod's atom consists of a central massive nucleus containing the positive charge of the atom, and the electrons move round the nucleus in circular planetary orbits. The centripetal force required for the orbital motion is provided by the electrostatic attraction between the positively-charge nucleus and the negatively-charged electron.

Bohr proposed two postulates :

(i) An electron can move only in those orbits for which the angular momentum L of the electron is an integral multiple of $h/2\pi$ where h is Planck's constant. (Thus, Bohr quantised toe angular momentum of the electron.). The electron moving in any of the permitted orbits does not radiated energy in spite of its acceleration towards the centre of the orbit. The atom, therefore, is said to exist in a *stationary* state.

(ii) The emission (or absorption) of radiation by the atom takes place when an electron jumps for one permitted orbit to another. The radiation is emitted (or absorbed) as a single quantum (photon) whose energy hv is equal to the difference in energies of the electron in the two orbits involved. Thus, if E_i be the energy of the initial orbit of the electron and Z_f that of the final orbit then we have

$$hv = E_i - E_f$$

where v is the frequency of the emitted (or absorbed) radiation.

Let e, m and v be the charge, mass and velocity of the electron (measured in column, kg and meter/sec respectively) and r the radius of the orbit measured in meter. The positive charge on the nucleus is Z e, where Z is the atomic number (Fig. 2.63) In case of hydrogen

Z = 1, so that positive charge on the nucleus is e. As the centripetal force is provided by the electrostatic attraction, we have

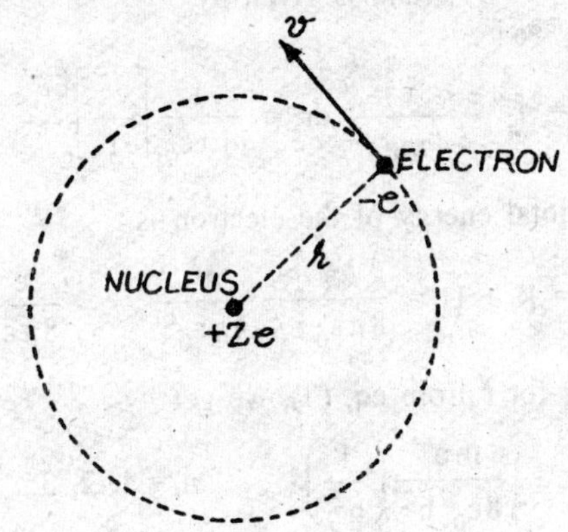

Fig. 2.63

$$\frac{mv^2}{r} = \frac{1}{4\pi\varepsilon_0}\frac{e^2}{r^2}$$

or $$m\,v^2 = \frac{e^2}{4\pi\varepsilon_0 r}. \qquad ...(1)$$

From the first postulate, the angular momentum of the electron is given by

$$L = m\,v\,r = n\,\frac{h}{2\pi}, \qquad ...(2)$$

where n is called as 'quantum number' having values 1, 2, 3, ...

Squaring eq. (2) and dividing by eq. (1) we get

$$r = n^2\,\frac{h^2\varepsilon_0}{\pi m e^2},\ n = 1, 2, 3, ... \qquad ...(3)$$

This is the expression for the radius of the permitted orbits.

Now, the energy E of the electron in an orbit is the sum of kinetic and potential energies. The kinetic energy of the electron is

$$K = \frac{1}{2}\,mv^2 = \frac{e^2}{8\pi\varepsilon_0 r}. \qquad \text{[from eq. (i)]}$$

The potential energy at a distance r to infinity against the electrostatic attraction $\left(-\dfrac{e^2}{4\pi\varepsilon_0 r^2}\right)$, and is given by

$$U = \int_r^\infty -\frac{e^2}{4\pi\varepsilon_0 r^2}\,dr = \frac{1}{4\pi\varepsilon_0}\left[\frac{e^2}{r}\right]_r^\infty = -\frac{e^2}{4\pi\varepsilon_0 r}.$$

Hence the total energy of the electron is

$$E = K + U = \frac{e^2}{8\pi\varepsilon_0 r} - \frac{e^2}{4\pi\varepsilon_0 r} = -\frac{e^2}{8\pi\varepsilon_0 r}.$$

Substituting for r from eq. (3), we get

$$E = -\frac{me^4}{8\varepsilon_0^2 h^2}\left(\frac{1}{n^2}\right) \qquad n = 1, 2, 3... \qquad ...(4)$$

This is the expression for the energy of the electron in the n th orbit. We see that it in negative.

Let E_i and E_f be the energies of the electron corresponding to the initial (higher) and final (lower) orbits of the excited atom. Then, we have

$$E_i = \frac{me^4}{8\varepsilon_0^2 h^2}\left(\frac{1}{n_i^2}\right)$$

and $$E_f = -\frac{me^4}{8\varepsilon_0^2 h^2}\left(\frac{1}{n_f^2}\right),$$

where n_i and n_f are the corresponding quantum numbers. The energy difference between these states is

$$E_i - E_f = \frac{me^4}{8\varepsilon_0^2 h^2}\left(\frac{1}{n_f^2} - \frac{1}{n_i^2}\right).$$

Hence, from Bohr's second postulate, the frequency ν of the emitted photon is

$$\nu = \frac{E_i - E_f}{h}.$$

$$= \frac{me^4}{8\varepsilon_0^2 h^2}\left(\frac{1}{n_f^2} - \frac{1}{n_i^2}\right)$$

The corresponding wavelength l is given by

$$\frac{1}{\lambda} = \frac{v}{c} = \frac{me^4}{8\varepsilon_0{}^2ch^3}\left(\frac{1}{n_f{}^2} - \frac{1}{n_i{}^2}\right)$$

This equation indicates that, *Since ni and nf can take only integral values, the radiation emitted by excited hydrogen atoms should contain certain discarte wavelengths only.*

The value of the constant *term* $\frac{me^4}{8\varepsilon_0{}^2ch^3}$ comes out to be the same as the Rydberg constant R in the Balmer's empirical formula. Thus we have

$$\frac{1}{\lambda} = R\left(\frac{1}{n_f{}^2} - \frac{1}{n_i{}^2}\right).$$

Shortcomings of Bohr's Theory

Bohr's theory although very successful l in explaining the hydrogen spectrum and giving valuable information about atomic structure, has the following shortcomings.

(i) An individual line of hydrogen spectrum, when examined under a high resolving spectroscope, is found to be accompanied by a number of theory as such.

It can, however, be explained when the relativistic variation in the mass of the electron and the electron 'spin' are taken into account.

(ii) Bohr's theory cannot explain the variation in intensity of the spectral lines of an element. The intensity can be explained by quantum mechanics.

(iii) The theory is only applicable to one-electron atoms such as hydrogen isotopes, singly-ionized helium, doubly-ionised lithium, etc. It does not explain the spectra of complex atoms.

(iv) The success of Bohr's theory in explaining the effect of magnetic field on spectral lines is only partial. The theory cannot explain the 'anomalous' Zeeman effect.

(v) The theory does not satisfactorily explain the distribution of electrons in atoms.

No Balmer Lines in Absorption Spectrum of Hydrogen

The Bohr theory also explains the absorption line spectrum of hydrogen. When a beam of - continuous light (containing all wavelengths) is passed through hydrogen and then sent into a spectrograph, a set of dark lines is obtained. In terms of quantum theory the incident light is a beam of quanta (photons) of all sorts of energies. Now according to Bohr theory the hydrogen atoms absorb only those quanta whose energies correspond to transitions between its discrete energy levels.

The resulting excited hydrogen atoms re-radiate the absorbed energy almost atones but these photons come off in random directions with only a few in the same direction as the original beam of continuous light. The dark lines in the absorption spectrum are therefore never completely black. Obviously the absorption lines will have exactly the same frequencies as the emission lines.

Now it is found that all the emission lines of hydrogen spectrum do not appear in the absorption spectrum. The reason is that normally the atom is always in the ground state $n = 1$. Therefore, absorption transitions can only occur from $n = 1$ to $n > 1$. Hence lines of only the Lyman series can appear in absorption spectrum.

To obtain Balmer series in absorption, the atom must initially be in the state $n = 2$, because Balmer lines require transitions from $n = 2$ to $n > 2$. Since atoms are usually in the ground state. Balmer lines are not obtained in absorption.

DIFFERENT SPECTRAL LINES OF HYDROGEN

Every atom when excited emits radiations. The radiations form a line spectrum which is the characteristic of the emitter. Each atom has its own particular line spectrum which is regarded as the characteristic of that element to which the atom belongs. Like other elements, hydrogen possesses its own characteristic line spectrum. Hydrogen spectrum consists of a number of lines. These have been grouped into five series which are named after their discoverers, Fig 2.64.

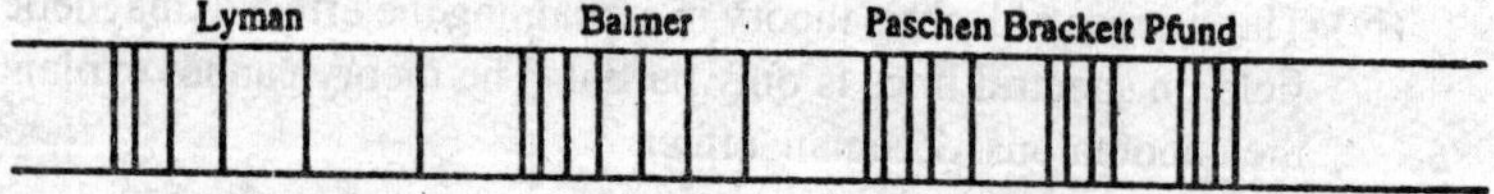

Fig. 2.4

Many attempts were made by various workers to find a rule for underlying relationship which governed the wave lengths of these lines. We will discuss these by one.

Calculation of Rydberg's Constant : The energy of the hydrogen atom when the electron is in the n_1th orbit.

$$E n_1 = -\frac{me^4}{8\varepsilon_0^2 h^2} \cdot \frac{1}{n_1^2} Z^2$$

and atom the energy of the atom when the electron is in the n_2th orbit

$$E n_2 = -\frac{me^4}{8\varepsilon_0^2 h^2} \cdot \frac{1}{n_2^2} Z^2$$

and the frequency of the photon emitted, when an electron jumps from n_2th to n_2th orbit is given by Bohr's third postulate, *i.e.*,

$$h\nu = E n_1 - E n_2 = -\frac{me^4}{8\varepsilon_0^2 h^2}\left(\frac{1}{n_1^2} - \frac{1}{n_2^2}\right) Z^3$$

or

$$\nu = \frac{me^4}{8\varepsilon_0^2 h^2}\left(\frac{1}{n_2^2} - \frac{1}{n_1^2}\right) Z^2.$$

Since the velocity of light $c = \nu\lambda$, we have

$$\frac{\nu}{c} = \frac{1}{\lambda} = \frac{me^4}{8\varepsilon_0^2 ch^2}\left(\frac{1}{n_2^2} - \frac{1}{n_1^2}\right) Z^2 \quad ...(9)$$

where

$$R = \frac{me^4}{8\varepsilon_0^2 ch^2}.$$

On substituting the values, we get $R = 109737.302\ cm^{-1}$. Here ε_0 is a constant whose numerical value is equal to $8.854. \times 10^{-12}$. This value is found to be in agreement with the value obtained from the spectroscopic data of the Balmer's series which is $109677.67\ cm^{-1}$.

Success of Bohr's Theory : This theory explains the following facts about atomic spectra.

FAILURES OF BOHR'S THEORY

(i) It failed to explain *Stark Effect.* It is similar to Zeeman effect and is produced in the presence of external electrostatic field (Fig. 2.65).

(ii) It failed to explain Zeeman effect. When a substance emitting a line space trum is placed in a magnetic field, its lines would split up into a number of closely spaced lines. This is known as *Zeeman effect* (Fig. 2.65).

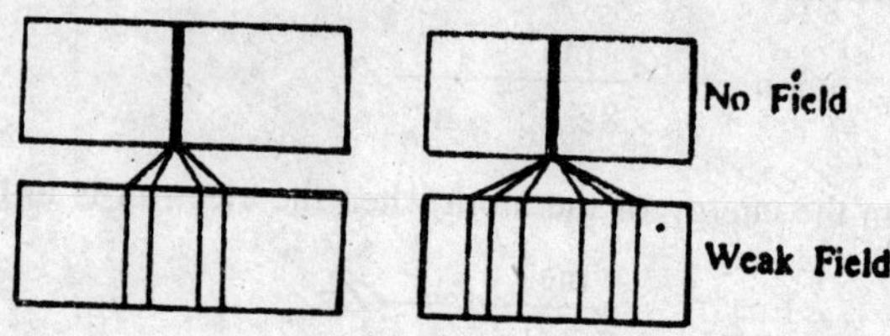

Fig. 2.65

(iii) When spectral lines of hydrogen are observed very closely, each line is further made up of much closely spaced lines.

(iv) It failed to explain spectra of atoms other than hydrogen.

(v) Another fundamental objection against Bohr's theory is that is used two theories which are opposed to each other, *i.e., quantum theory* was used to account for the existence of stationary orbits and for frequencies of radiations emitted while motion of electron in its orbit obeyed *the law of classical mechanics.*

(vi) In the light of the Heisenberg's uncertainty principle, both the velocity and the position of the electron cannot be specified at a given time as Bohr did. Thus, it is another weakness of Bohr's theory.

(vii) Another weakness of Bohr's theory is that it did not throw light on the distribution and arrangement of electrons in atoms.

(viii) Bohr assumed that the nucleus is stationary and only electrons are revolving around it. Detailed facts have revealed that both the nucleus and the electrons move in closed orbits around their centre of mass. Therefore, it is major weakness of Bohr's theory.

THE COMPTON EFFECT

From the classical theory X-ray scattering, one is aware of the following conclusions.

(i) The scattered X-ray should posses the same wavelength as the incident one.

(ii) The scattering constant (denoted by s) should have a constant value 0.2 for all types X-ray scatterings. It means that the scattering constant should be independent of the wavelength of the incident radiation.

(iii) The distribution of intensity f X-rays should be symmetrical in the scattered X-rays.

When the experimental results from the research of scattering of hard X-rays and γ-rays were analysed, the following points of discrepancy from classical expectation were brought out.

(i) The distribution of scattered radiation was not symmetrical. Scattering was taking place in the same direction as that of the incident radiation.

(ii) The scattering coefficient s was found to very with the wavelength of the incident radiation and its value is diminished as the wavelength of the incident radiation is decreased.

(iii) The scattering radiation was found to have a greater wavelength that of the incident radiation.

In order to explain the above experimental facts about the scattering of X-rays, Compton in 1925 proposed that the phenomenon of scattering might be regarded as an elastic collision of two particles, the photon of X-rays and the electron of the scatterer. When a photon of energy hv collides with an electron, it will lose a part of its energy in the form of kinetic energy to the electron.

Thus, the scattered photon will have a smaller energy hv, and, in consequence, a lower frequency or greater wavelength than that of the incident photon. *This observed change in radiation frequency or the wavelength of the scattered is known as Compton effect.*

In such a case an electron is also ejected from the scatter with an energy depending upon its direction. This ejected electron is known as *Compton recoil electron.*

Theory : Compton calculated mathematically the increase in wavelength by making use of quantum theory. Suppose the incident X-rays consist of photons each of energy hv, where v is the frequency.

This energy (hv) is much greater than that required to eject a free electron from the target or scatterer.

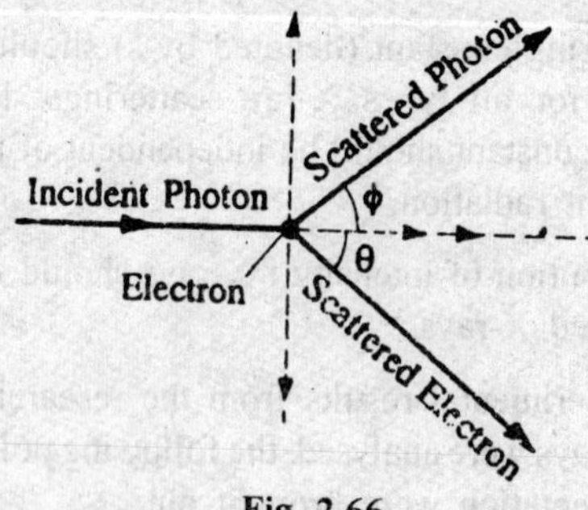

Fig. 2.66

Let the photon collide with an electron at rest in the target. A part of its energy is imparted to the electron which is ejected with a velocity u in a direction making an angle 0 with that of the incident X-rays photon. The remaining energy is associated with the X-rays ϕ photon, having lower frequency v′ and moving in direction making an angle f with that of the incident photon. If m_0 is the rest mass of the electron, its mass, when moving with a velocity u, will be given by the theory of relativity as.

$$m = \frac{m_0}{\sqrt{\left(1-\frac{u^2}{c^2}\right)}} \quad ...(1)$$

where c is the velocity of light.

On applying the principle of conservation of energy, we get

$$hv + m_0c^2 = hv' + mc^2$$

$$\Rightarrow \quad mc^2 = h(v - v') + m_0c^2$$

Squaring the above equation we get

$$m^2c^4 = h^2(v^2 - 2vv' + v'^2) - 2h(v - v')\, m_0c_2 + m_0^2c^4 \quad ...(2)$$

On applying the principle of conservation of momentum in the direction incident photon, we get

$$\frac{hv}{c} = \frac{hv'}{c}\cos\phi + mu\cos\theta \quad ...(3)$$

and in a direction perpendicular to the direction of incident photon

$$0 = \frac{hv'}{c}\sin\phi - mu\sin\theta \quad ...(4)$$

Equation (3) and (4) on simplification yield′

$$muc\cos\theta = h(v - v'\cos\phi) \quad ...(5)$$

and $$muc\sin 0 = hv'\sin '. \quad ...(6)$$

Squaring and adding equations (5) and (6), we get

$$m^2u^2c^2(\cos^2\theta + \sin^2\theta) = h^2(v - v'\cos\phi)^2 + h^2v'^2\sin^2$$

$$\Rightarrow \quad m^2u^2c^2 = h^2(v^2 + v'^2\cos^2\phi - 2vv'\cos\phi + v'^2\sin^{2'}) \quad ...(7)$$

$$= h^2(v^2 - 2vv'\cos\phi + v'^2)$$

Subtracting equation (7) from (1), we get,

$$m_0{}^2c^2(c^2 - u^2) = 2 - 2vv'h^2(1 - \cos\phi) + 2h(v - v')m_0c^2 + m_0{}^2c^4$$

$$\Rightarrow \frac{m_0{}^2c^2}{1-\frac{u^2}{c^2}}(c^2 - u^2) = m_0{}^2c^4 + 2h(v - v')m_0c^2$$

$$- 2vv'h^2(1 - \cos\phi)\,2h(v - v')m^2c^2 = 2vv'h^2(1 - \cos\phi)$$

$$\Rightarrow \frac{v - v'}{vv'} = \frac{h(1-\cos f)}{m_0c^2} \text{ or } \frac{1}{v'} - \frac{1}{v} = \frac{h}{m_0c^2}(1 - \cos\phi)$$

or $$\frac{c}{v'} - \frac{c}{v} = \frac{h}{m_0c}(1 - \cos f) \quad ...(8)$$

If l and l¢ are the wavelengths corresponding to the frequencies v and v¢ respectively, then equation (8) may be put as

$$\lambda' - \lambda = \Delta\lambda = \frac{h}{m_0c}(1 - \cos\phi)$$

From the equation (9), it follows that the increase in wavelength $\Delta\lambda$ is independent of the wavelength of the incident radiation and the nature of the scattering substance, but depends upon the angle of scattering. The quantity $h/.m_0c$ is known as Compton wavelength.

But $$1 - \cos\phi = 2\sin^2\frac{f}{2}$$

Therefore equation (9) becomes

$$\lambda' - \lambda = \Delta\lambda = \frac{2h}{m_0c}\sin^2\frac{f}{2}$$

or $$\lambda' = l + \frac{2h}{m_0c}\sin^2\frac{f}{2} \quad ...(10)$$

The following three cases may arise.

(i) When the scattering angle is zero ϕ, is zero, and then

$$\sin\frac{f}{2} = \sin 0 = 0$$

Therefore, equation (10) becomes as

$$\lambda' = \lambda \qquad ...(11)$$

Thus, there is no scattering.

(ii) When $\theta = 90°$, $\sin^2\theta/2 = 1/2$, equation (10) becomes as

$$\lambda' - \lambda = h/m_0c \qquad ...(12)$$

As h, m_0 and c are constants, equation (12) becomes as

$$\lambda' - 1 = \text{constant} = 2h/m_0c$$

or $$\Delta\lambda = \text{constant} = 2h/m_0c \qquad ...(13)$$

(iii) When $\theta = 180°$, $\sin^2 180°/2 = 1$, equatior (10) becomes as

$$\lambda' - 1\ 2h/m_0c. \qquad ...(14)$$

VECTOR ATOM MODEL

We know that each electron has an orbital quantum number l and spin quantum numbers. Both these quantum numbers are vector quantities. The angular momenta and magnetic dipole moments which are also vector quantities are calculate by using the values of l and s.

If an orbit is having many electrons, the orbital quantum numbers of each electron are to be added vectorially to obtain the total orbital quantum number. Thus, the total orbital quantum number. L, may be defined as follows.

$$L = \sum_{l=1}^{m} l_l \, .$$

where m is the number of electrons.

Similarly, the total spin quantum numbers, S, may be defined as follows.

$$S = \sum_{J=1}^{m} S_i \, .$$

If $l_1 = 2$ and $l_2 = 1$.

Then $L = l_1 + l_2$ and the values of L can be 3, 2 and 1 (Fig. 2.67). The angular momenta will be $\sqrt{6}h/2\pi$ and $2h/2\pi$ and the resultant angular momentum will be as follows.

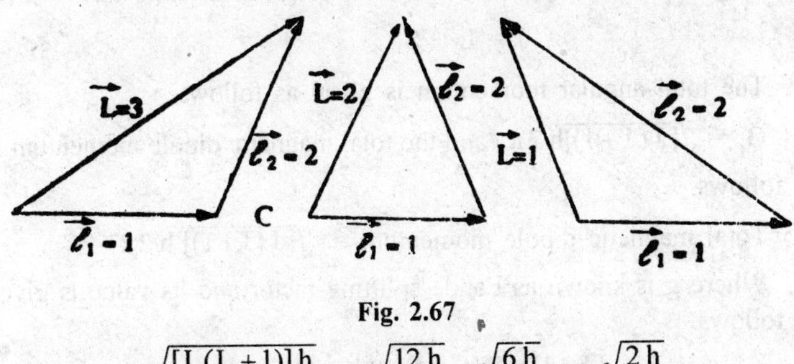

Fig. 2.67

$$G_L = \frac{\sqrt{[L(L+1)]}\,h}{2\pi} = \frac{\sqrt{12}\,h}{2\pi}, \frac{\sqrt{6}\,h}{2\pi} \text{ or } \frac{\sqrt{2}\,h}{2\pi}.$$

Thus, there are three possible ways of combining the two vectors to get the three different values for the resultant.

Whenever the spin quantum numbers of different electrons are to be combined, one has to take H and's rule into consideration. Thus, $S = s_1 + s_2$ will be either 0 or 1 ($s_1 = s_2 = 1/2$). The spin of each electron is a half integral value but the resultant spin of the two electrons $Gs = \sqrt{S(S+1)}\,\frac{h}{2\pi} = 0$ or $\sqrt{2h/2\pi}$, the angular momentum of each electron was $\frac{\sqrt{3}}{2}\frac{h}{2\pi}$. This is shown in Fig. 2.68.

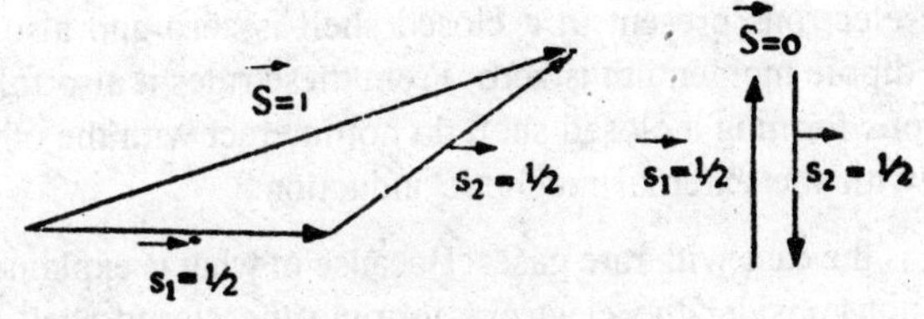

Fig. 2.68

Similarly the values of resultant magnetic dipole moments are calculated. The dipole momentum because of the orbital quantum number would be

$$\sqrt{[L(L+1)]}\mu_B = \sqrt{[L(L+1)]}.eh/4\pi m,$$

and that because of the spin quantum number would be

$$\sqrt{[S(S+1)]}_{\mu B} = 2\sqrt{[S(S+1)]}.eh/4\pi m$$

In general, total quantum number J is given as follows.

$$J = L + S,$$

The total angular momentum is given as follows.

$G_J = \sqrt{[J(J+1)]}h/2\pi$, and the total magnetic dipole momentum is as follows.

Total magnetic dipole momentum $= \sqrt{[J(J+1)]}\ h/2\pi$

Where g is known as Lande splitting factor and its value is given as follows.

$$g = 1 + \frac{J(J+1) + S(S+1) - (L+1)}{2J(J+1)}$$

We also have to remember the following rules.

(i) The vector sum of the orbital quantum numbers of all the electrons present in a closed shell is always zero, *i.e.*,

$$L = \Sigma\, l_t = 0.$$

(ii) The vector sum of the spin quantum number of all the electrons present in a closed shell is zero, *i.e.*,

$$S = \Sigma\, s_i = 0.$$

From the rules (i) and (ii), it means that the total angular momentum of all the electrons present in a closed shell is zero and also the total magnetic dipole momentum is zero. From these rules it also follows that the electrons forming a closed shell do not interact with the other atoms and also with the external magnetic induction.

Such is the case with rare gases. Because of what is explained above, we need not consider the electrons forming the closed shell when we are calculating the total angular momentum.

Only the electrons in the outermost shell need be considered. For example, when we consider the optical properties of lithium (2s electron), sodium, (3s electron) and potassium (4s electron), we will only consider the electron in the highest shell.

Similarly, we consider the optical properties of beryllium, magnesium, calcium, etc., we have to consider only the two s electrons in the outermost shell.

The orbital angular momentum vector l and spin angular momentum vector s combine in several ways to give the resultant vector which represents the atom in vector model. The commonly known two schemes of additions are L – S coupling and j – j coupling. These are the limiting cases of the whole range of permutation and combination of l and s for the electron of a complex atom.

1. L – S coupling : In this coupling scheme, all the angular momentum vectors l of the electrons combine to form the resultant L. Similarly and independently all their spin angular momentum vectors combine to form the result S. These two resultant vectors and vectorially to give the total angular momentum of the atom. This may be symbolically represented as follows.

$$(l_1 + l_2 + l_3 + ...) + (s_1 + s_2 + s_3...) = L + S = J.$$

The total quantum number J can take all the positive integral values from L + S to L – S or S – L depending upon whether L ³ S or S > L. *i.e.*, J can have the following values

$$J\ (L + S).\ (L + S - 1),\ (L + S - 2),\ ...\ (L - S + 1),\ (L - S),$$

The L – S coupling is possible when the interaction between individual orbital momenta on one hand and the individual spin angular momenta on the other hand are very strong. This coupling is also called normal coupling or Russell coupling.

2. j – j Coupling : In this coupling scheme the interaction between the orbital angular momentum and spin angular momentum in each electron is stronger than that of the orbital or the spin vectors of the other electrons in the atom. Hence the contribution to the total angular momentum due to several electrons of the same atom is individual contribution from each electron of the atom. If the total angular momentum of a single electron in a multi electron atom is.

$$j = l + s$$

then the total angular momentum of the atom due to several electrons may be given by

$$= (l_1 + s_1) + (l_2 + s_2) + (l_3 + s_3) + ... + (l_n + s_n).$$

$$= j_1 + j_2 + j_3 + ... + j_n = J.$$

This type of coupling is experienced very rarely and for majority of the cases L – S coupling is effective.

Notations : For a single electron the states are represented as

s, p, d, f, g, h,...

When l = 0 1, 2, 3, 4, 5, ... respectively. When several electron are present in the atom depending upon the total quantum number L the notations given are S, P, D, F, G, H, ... when L = 0, 1, 2, 3, 4, 5, ... respectively.

Further, the multiplicity of an energy level is given by (2S + 1). This number is given on the left side above as a subscript. On the right hand side after the letter denoting the value of L, the value of J is written below the level of the value L. On the left hand side of the whole is written the value of n, the principal quantum number.

Thus, the $3\,^2S_{1/2}$ means n = 3, L = 0, S = 1/2, multiplicity. (2S + 1) = 2 and J = 1/2, $3\,^2P_{3/2}$ means n 1/2, multiplicity (2S + 1) = 2 and J = 3/2. The 3^3D_5 means n = 3, L = 2, S = 1, multiplicity (2S + 1) = 3 and J = 3.

SPECTRA OF ALKALI METALS

Like the hydrogen atom, the spectra of the alkali metals consist of a series of lines with regularly decreasing separation and decreasing intensity. One important feature of the alkali-spectra is that the lines are mostly doublets though each and every line has not been resolved into two in the spectroscope

Quantum defect : The spectral lines obtained from alkali metals cannot be represented by a formula exactly analogous to Balmer's formula. As the lines converge to a limit, one is able to represent them by the difference of two spectral terms. Thus the values of the terms are not of Balmer's form *i.e.*, R/n^2 but of the form R/n^{*2}, where n is the principal quantum number and n* is the effective quantum number and its value is given by $(n + \alpha)$, where a is a small quantity and is known as the quantum defect.

The *Magnitude* of the *quantum defect* depends upon the penetration of the core by the orbit of the emitting electron. Increased penetration of the core depends upon the *shape of the orbit*. It is, therefore, expected that a series of different values of a will occur for different values of azimuthal quantum number.

There are *four* types of quantum defect (α) designated by S, P, D, and F. These words stand for the first letter of sharp, principal, diffuse and fundamental. The names of the series are attributed according to the order, as to which of them appears in the quantum defect of the running term. The principal series is most easily excited and is hence so calculate. The lines in the sharp series are well defined and those of the diffuse series are blurred.

In emission spectra, other series in addition to the principal series may be observed for the alkali metals. These series partly overlap each other. The last named series viz., the fundamental of Bergmann series, occur in the infra-red.

However, the frequencies of the lines of the four series may consequently be represented by the following general expression.

Sharp series $\bar{v}$ = aP – nS, Principal series $\bar{v}$ = aS – nP,

Diffuse series $\bar{v}$ = aP – nD, Fundamental series $\bar{v}$ = aD – nF,

where a is a constant, and n is the principal quantum number.

***Explanation*:**

As earlier stated, the volume of n* is given by (n + α). Ωηεν we calculate the values of n and n* for the S, P, D and F terms of the different alkali metals, we get the following values.

	Terms Series	***n*(= n + α)***	***n***	***l***
Lithium	S	1.59	2	0
	P	1.96	2	0
	D	3.00	3	2
	F	4.00	4·	3
	Terms Series	n* (= n + α)	*n*	*l*
Sodium	S	1.63	3	0
	P	2.12	3	1
	D	2.99	3	2
	F	4.00	4	3.

From the above table that the discrepancies between n* and n decrease in the order of S, P, D, and F of terms series. The discrepancies [n* – n (n + α) – n = α], also called quantum defect have been explained as follows.

If we compare the structure of an alkali metal with hydrogen, it may be regarded that its nucleus is a single unit and around which the additional electron rotates. The additional electron is known as *optical electron.*

In hydrogen, the electron remains in its fixed orbits and therefore, definite spectral lines will be observed. In the case of an alkali metal, the optical electron does not always rotate in its own orbits. It has a tendency to penetrate into the other inner orbits and this penetration depends upon the eccentricity of the orbit. The greater the eccentricity of the orbit, the more closely the electron may approach the nucleus in the course of its rotation. It is, therefore, more likely that greater the penetration of the inner levels, the greater will be divergence from hydrogen like behaviour.A simplified energy level diagram for sodium is shown in (Fig. 2.69), those for the other alkali metals are similar, except for the difference in the principal quantum number of the optical electrons. The integers indicate the various values of the principal quantum number n in each spectral state, represented by S, P, D and ZF. It may be noted that the S, P, D and F terms are actually doublets but this is not shown in the Fig. 2.68.

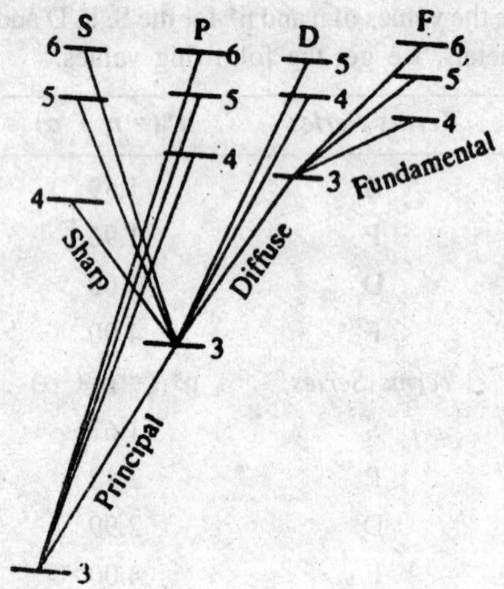

Fig. 2.69

EXPLANATION OF THE DOUBLET STRUCTURE

When we study the spectra of elements other than hydrogen, we observe that many lines are actually multiples consisting of two, three or more lines close together.

In order to explain Uhlenbeck and Goldsmit (9135) suggested that this multiplicity of spectral lines is due to the spin of the electron. The contribution due to the spin is quantised and is expressed in terms of spin quantum numbers. The value of this number is + 1/2 and – 1/2. The resultant of the spin and azimuthal quantum number is given by j = l + s, where 'j' is called the *inner quantum number.* As 's' can have + 1/2 and + 1/2, it follows that for every value of 'l' there are two values of 'j' viz.,

$$j_1 = l + 1/2$$

and $$j_2 = l - 1/2.$$

The above values hold good except when l is zero. In that case the two 'j' values are identical, *i.e.*, + 1/2, because it is not the + ve or – ve sign but is the numerical value only of 'j' which determines the momentum. It means that every 'l' level except l = 0 is consequently split into two levels with different energies. Let us now illustrate this discussion by applying to alkali metals.

(i) For an electron in an 'S' level, the value of l = 0. It means that the two values of j are numerically identical and hence it is a singlet level.

$$j_1 = l + \frac{1}{2} \text{ and } j_2 = l - \frac{1}{2}$$

$$j_1 = 0 + \frac{1}{2} \text{ and } j_2 = 0 - \frac{1}{2}$$

$$j_1 = + \frac{1}{2} \text{ and } j_2 = - \frac{1}{2}$$

(ii) In the P level, l = 1 and so the corresponding values of 'j' are 3/2 and 1/2.

$$j_1 = l + \frac{1}{2} \text{ and } j_2 = l - \frac{1}{2}$$

$$j_1 = 1 + \frac{1}{2} \text{ and } j_2 = 1 - \frac{1}{2}$$

$$j_1 = \frac{3}{2}\ j_2 = \frac{1}{2}.$$

Thus, each spectral line in 'P' level is doublet.

Similarly in the D and F levels, the values of 'j' are 5/2 and 3/2, 7/2 respectively. Again 'j' has two values and therefore each line will be a doublet. The doublet of the sharp series of spectra may be represented by the expressions

$$\bar{\nu} = aP_{1/2} - nS$$

and $$\bar{\nu} = aP_{3/2} - nS.$$

where S is a singlet and P is a doublet.

THE SOMMERFELD MODEL

Inspite of many success, Bohr's theory was found to be inadequate to explain certain details in the spectrum of hydrogen. For instance, the Hx line of the Balmer series was found to contain several components. This fine structure of spectral lines could not be explained by Bohr's theory which assumed that there was only one orbit for each quantum number, whereas the observed fine structure suggested that for any given quantum number n there might be several orbits of slightly different energies.

Sommerfeld, in 1915, guided by the above suggestion modified Bohr's theory by introducing the following modifications.

(a) Concept of elliptical orbits, and

(b) Relativistic variation of the mass of electron.

We will discuss these modifications one by one.

(a) *Elliptical Orbits :* In order to explain the multiplicity of spectral lines. Sommerfeld introduced the concept of elliptical orbits. He postulated that.

(i) Since the electron is moving around and under the influence of a massive nucleus, like a planet around the central massive sun, it might describe elliptical orbits as well.

(ii) While retaining the first circular orbit suggested by Bohr, Sommerfeld assumed one additional elliptical orbit in the case of Bohr's second orbit and added two additional elliptical orbits to Bohr's third orbit and so on (Fig. 2.70).

(iii) The nucleus is one of the foci for all these orbits.

(iv) In an elliptical orbit, the major and minoraxes will differ in lengths, but as the orbit broadens, they will approach each other and becomes equal when the orbit becomes circular. Circular orbit is only a special case of the elliptical orbit. The orbit is defined by two quantum numbers, *i.e.*, n and k which correspond to the major and minor axes of the ellipse. They are related as follows.

$$\frac{\text{Principal quantum number}}{\text{Azimuthal quantum number}} = \frac{n}{k} = \frac{\text{length of major axis}}{\text{length of minor axis}}.$$

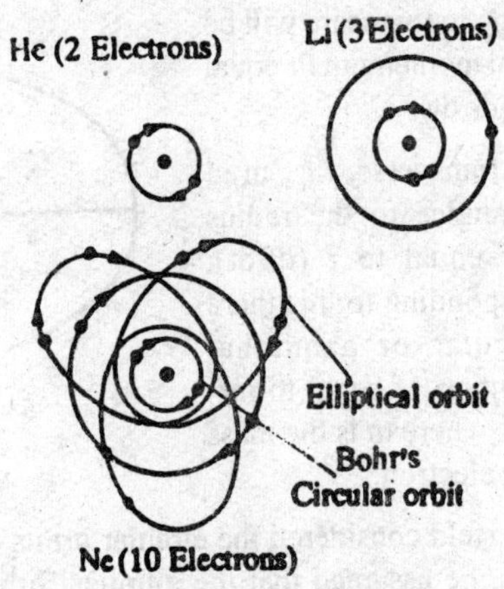

Fig. 2.70

It is clear from the above that for any given value of n, k cannot be zero as in that case the ellipse would degenerate into a straight line passing through the nucleus. Further, k cannot be more than n since b is always less than a. When n = k the path becomes circular. This is a limit to the number of different orbits that an electron may have in any energy level.

Such possible paths are referred to as sub-levels. For a given value of n, k can have only n different values which mean that there may be only n elliptical orbits or sub levels with different eccentricity. For elliptical orbits the values of k would be (n – 1), (n – 2) etc., down to k = 1. When k = 0, the ellipse would be a straight line. When n = 3, k = 3, (circular), 2, 1, two of sommerfeld model lines in its subdivision of the original Bohr stationary levels into various sub-levels of slightly differing energies as given by difference in orbit shapes.

Simple mathematical approach to Sommerfeld model : Let us consider the motion of an electron (– c) in an elliptical orbit as shown in (Fig. 2.71).

Its position at any instant can be fixed in terms of polar coordinates, r and f where r is the distance of the electron from the nucleus (+ e) at one of the foci of the ellipse and f is the angle which the radius vector makes with the major axis of the ellipse. The tangential velocity v of the moving electron at any instant can be resolved into two components.

(i) One radial, *i.e., dr/dt* along the radius vector. Corresponding to this there will be a radial momentum Pr equal to m (dr/dt).

(ii) Other transverse, *i.e.*, m at right angles to the radius vector equal to r (df/dt). Corresponding to this there is angular or azimuthal momentum of equal to mr^2 (df/dt), where m is the mass of the electron.

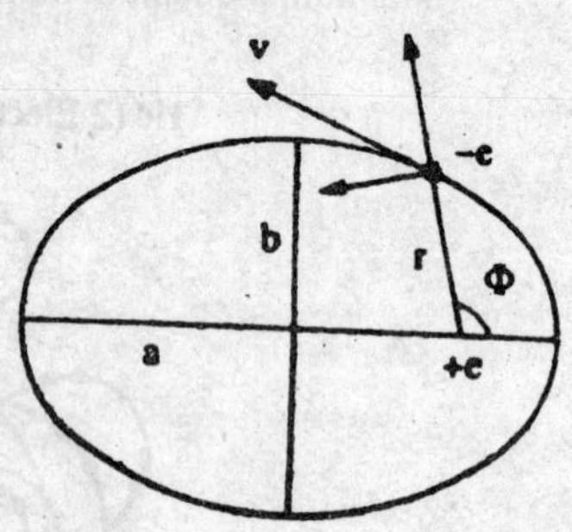

Fig. 2.71

As Sommerfeld considered the circular orbits to be special cases of elliptical orbits, he assumed that the elliptical orbits should satisfy the quantum condition of Bohr just as the circular orbits, *i.e.*,

$$\oint p_r\, dr = n_r\, h \qquad ...(1)$$

$$\oint p_f\, d\phi = n_\phi\, l_t \qquad ...(2)$$

Thus, the single n of Bohr's theory has been replaced by the two new quantum numbers n_r and n_ϕ, *i.e.*,

$$n = n_r + n_\phi \qquad ...(3)$$

where n_r is the radial quantum number and nf is the angular or azimuthal quantum number. The total energy E is given by

$$E = \text{P.E.} + \text{K.E.}$$

$$= \text{P.E.} + \text{Radial K.E.} + \text{Angular K.E.}$$

$$= -\frac{e^2}{r} + \frac{1}{2}m\left(\frac{dr}{dt}\right)^2 + \frac{1}{2}mr^2\left(\frac{d\phi}{dt}\right)^2 \qquad ...(4)$$

From equations (1), (2), (3) and (4), it can be shown that

$$1 - e^2 = \frac{b^2}{a^2} = \frac{n_\phi^{\ 2}}{(n_\phi + n_r)^2}$$

where e is the eccentricity of the ellipse whose semi-major and semi-minor axes are a and b respectively.

and $$E = -\frac{2\pi^2 mZ^2 e^4}{h^2}\left(\frac{1}{n_\phi + n_r}\right)^2 \qquad ...(6)$$

From equation (5), we get

$$(1 - e^2)^{1/2n} = \frac{b}{a} = \frac{n_\phi}{n_\phi + n_r} = \frac{n_\phi}{n}.$$

Thus, equation (6) can be written as

$$E = \frac{2\pi^2 mZ^2 e^4}{n^2 h^2} \qquad ...(7)$$

From equation (7), it follows that

(i) The elliptical orbits which have the same value of n, though of different eccentricities, have the same energy. It is not correct.

(ii) All orbits having the same values of the semi major axis possess the same energy, since the length for the semi-major axis is determined solely by the total quantum number n $(= n_f + n_r)$. Hence the energy of any of these permitted elliptical orbits is identical with that of a circular Bohr orbit, whose radius is equal to the semi-major axis of the ellipse.

From the above it follows that the theory of elliptical orbits in spite of two new quantising conditions involved, introduces not new energy levels other than those given by Bohr's theory of circular orbits. No new spectral lines, which would explain the fine structure, are there fore predicted. Thus, Sommerfeld has to modify his own model by suggesting the variation of mass of electron with velocity called relativistic variation.

Relativistic Variation of the Mass of Electron : The velocity if an electron moving in an ellitical orbit varies considerably at different parts of the orbit, being a maximum when the electron is nearest to the nucleus.

When allowance is made for the relativistic variation of electronic mass, as Sommerfeld did, the path of the electron is found to be no longer a simple ellipse, it is indeed, no longer a closed figure, but is transformed into a complicated curve known as rosette –a processing ellipse (Fig. 2.72). The total energy E of the system corrected by the relativistic variation of the mass of electron can be shown to be

$$E = -\frac{2\pi^2 mZ^2e^4}{h^2}\left[\frac{1}{n^2}+\frac{4\pi^2e^4Z^2}{c^2h^2}\left(\frac{n}{n\phi}-\frac{3}{4}\right)\frac{1}{n^4}+\ldots\right] \quad \ldots(8)$$

The relativistic correction, therefore, results in splitting up a given energy level E_α into n levels differing slightly from one another in energy. The splitting up of each energy levels gives rise to a fine structure of single spectral line, one application of the usual Bohr frequency conditions, When one explains the fine structure of lines, one should not take into account all theoretical possible transitions of the electron from one elliptical orbit to another that actually occur.

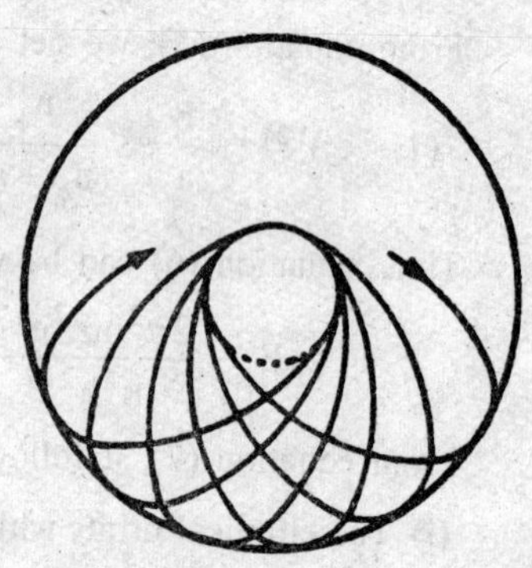

Fig. 2.72

According to a principal known as the selection rule, transition can take place only between orbits for which the azimuthal quantum number changes by + 1 or – 1, *i.e.*, nf = ± 1.

Limitations of Sommerfeld model : (i) Though Sommerfeld's theory is fairly well verified yet it leads only to three components for the structure of Hx line, while there should be really five. Thus the relativistic atom model has met with a partial success.

(i) The concept of elliptical orbits due to Sommerfeld gives the correct total (n) of possible azimuthal quantum number, but the actual values are not correct. The experimental studies as well as theoretical treatment based on wave mechanics show that azimuthal quantum number can be zero, so that the values can be 0, 1, 2,... –1, thus making a total of n possibilities. The correct and new azimuthal quantum number is denoted by 'l' to avoid the confusion. Thus, l is equal to K – 1.

(ii) It provides no idea about the number of electron which can be accommodated in a particular orbit.

(iii) It provides macerate values for angular momentum.

(iv) Sommerfeld model could not explain Zeeman and Stark effect.

Explanation of fine details in line spectrum by Sommerfeld model : It can be explained as follows.

When an electron moves in an elliptical path there will be a displacement each time in its elliptical path and thus resulting in a small difference in energy. Thus, the path of the moving electron is no longer a simple closed ellipse but a complicated curve, known as rosette, which is made up of different elliptical orbits of slightly different energies. This *difference in energies of elliptical orbits explains the existence of components of spectral lines in the spectrum of higher elements.*

Criticism of the Theory : (1) Suppose there are two orbits with principal quantum numbers, n = 2 and n = 3, (Fig. 2.73) respectively. Consider an electron which falls from third to the second orbit. There may be six possible transitions, each giving one fine line. But actual observations yield only five lines (Fig. 2.73).

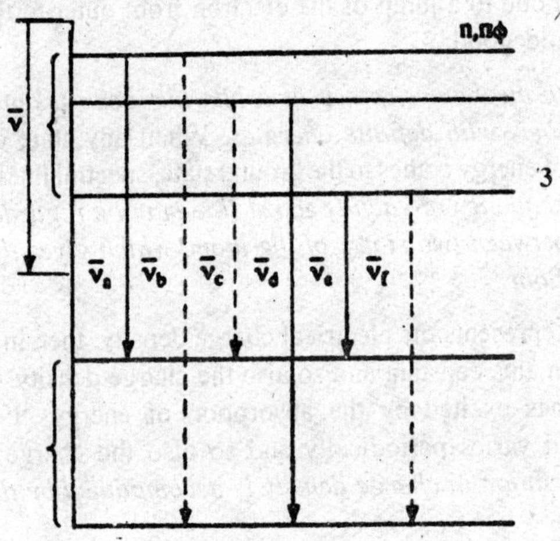

Fig. 2.73

Possibilities.

(i) $3K_3 \rightarrow 2K_2$ *i.e.*, $\Delta K = 3 - 2 = + 1$ (iv) $3K_2 \rightarrow 2K_1$ *i.e.*, $\Delta K = 2 - 1 = + 1$

(ii) $3K_2 \rightarrow 2K_1$ *i.e.*, $\Delta K = 3 - 1 = + 2$ (v) $3K_1 \rightarrow 3K_2$ *i.e.*, $\Delta K = 1 - 2 = - 1$

(iii) $3K_2 \rightarrow 2K_3$ *i.e.*, $\Delta K = 2 - 2 = 0$(vi)

$3K_1 \rightarrow 2K_1$ *i.e.*, $\Delta K = 1 - 1 = 0$,

where ΔK indicates the change in azimuthal quantum number. Sommerfeld applied *selection rule* which limited the number of transitions. According to such rule only those transitions are possible for which the quantum number K changes by – 1 or + 1 *i.e.*, $\Delta K = \pm 1$. Hence, transitions like ($3K_1 \rightarrow 2K_1$), ($3K_3 \rightarrow 2K_1$) and ($3K_2 \rightarrow 2K_2$) are forbidden. Therefore, according to this theory, the fine structure of H_α lines should made only of three lines. But actually it splits up into five lines. Sommerfeld theory fails to explain complicated systems.

(2) Sommerfeld's model could not explain Zeeman and Stark effect.

(3) The model gives no information regarding the relative intensities of lines, whose frequencies alone are predicted.

Wave mechanics and spectral lines : According to wave mechanics orbits do not exist in the atom. Thus, the interpretation of the emission of radiation due to a jump of the electron from outer to the inner orbits does not hold good.

In wave mechanics, place of orbits has been taken by stationary states of atoms with definite energies. When any state excited by the absorption of energy comes to the ground state, spectral lines are produced. Thus, *the frequency of each spectral line may be regarded as a 'beat' frequency between two states of the atom, which gives the same result as that of Bohr.*

As Ψ^2 represents the electrical charge density, then in the stationary state this remains constant and so also the charge density. But when any state becomes excited by the absorption of energy, Ψ is no longer constant and varies periodically and so also the charge density. *This periodic variation in charge density is accompanied by the emission of radiation.*

Excitation and Ionisation Potentials of an Atom : According to the Bohr's theory, in an atom there are certain discrete orbits only in which an electron can revolve without radiating energy.

Each of these orbits of a given atom is characterised by a certain (quantised) energy. Hence we say that there are certain discrete energy levels in an atom.

When an electron in an atom absorbs sufficient energy from an outside source, it rises from its present energy level to an higher level. The atom is than said to be 'excited'. This excited state lasts only for about 10^{-8} second after which the electron jumps back to the inner level, emitting the absorbed energy is emitted as electromagnetic radiations.

The energy required to excite or to ionise a given atom is perfectly definite.

For example, in case of hydrogen atom, the energy of the nth orbit is

$$E_n = -\frac{me^4}{8\varepsilon_0^2 h^2}\left(\frac{1}{n^2}\right).$$

Substituting the known values of m (= 9.1 × 10^{-31} kg), e (= 1.6 × 10^{-19}C), h (= 6.62 × 10^{-34} J-s) and ε_0 (= 8.85 × $10^{-12}C^2/N\ m^2$), we get

$$E_n = -2.17 \times 10^{-18}\left(\frac{1}{n^2}\right) \text{ joule (J)}$$

$$= -\frac{2.17\times10^{-18}}{1.6\times10^{-19}}\left(\frac{1}{n^2}\right)$$

$$= -\frac{13.6}{n^2} \text{ eV.} \qquad [1 \text{ eV} = 1.6 \times 10^{-19} \text{ J}]$$

Putting n = 1, 2, 3, ... ∞, the energies of the 1st, 2nd, 3rd, ... ∞ energy levels come out to be – 13.6, – 3.4 – 1.51, ... 0 electron volts respectively. Hence the energy to be supplied to the atom to raise an electron from the first to the second orbit is (13.6 – 3.4) = 10.2 electron-volts, from first to the third orbit is (13.6 – 1.51) = 12.09 electron volts,* and to the ionized state is (13.6 – 0) = 13.6 electron volts.

Now, the atoms are usually excited or ionised by bombarding them with electrons accelerated through a potential. But the excitation or ionisation of an atom can take place only when the bombarding electron has the required amount of energy.

The minimum accelerating potential which imparts to the bombarding electron an energy sufficient to excite a given atom is called the 'excitation potential' of the atom. Similarly, the minimum accelerating potential which imparts to the bombarding electron an energy sufficient to ionise the atom is called the 'ionisation potential' of the atom.

Franck-Hertz Experiment

Franck and Hertz, in 1914, performed a series of experiments to measure the excitation potentials of atoms of different elements. These experiments showed directly that in an atom discrete energy levels do exist.

Their apparatus is shown in Fig. 2.74. It consists of a glass tube in which are mounted a filament F, a grid G, and a plate P, as shown. The filament F is heated by a small battery B. An accelerating potential V is applied between F and G, and *a small fixed* retarding potential V_0 (about 0.5 volt) between G and P. The gas of the element whose atoms are to be studied is introduced in the tube at about 1 mm of mercury pressure.

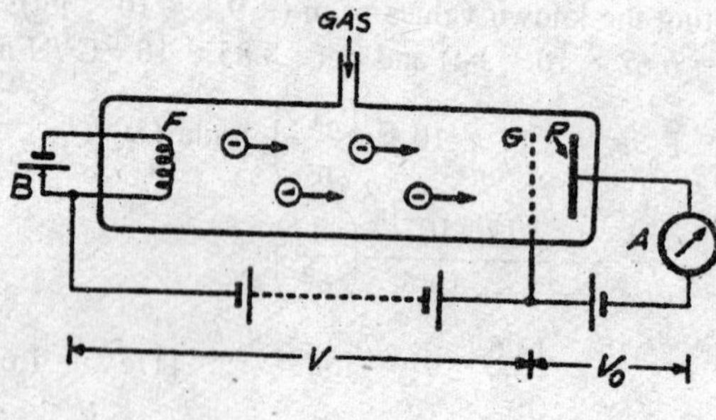

Fig. 2.74

The electrons emitted from the hot filament F are accelerated between F and G by the potential V, and retarded between G and P by the potential V_0. Thus, only electrons having energies greater than eV_0 at G are able to reach P. The current to P is recorded by an ammeter A.

The current is plotted against the *accelerating* potential V which is gradually increased from zero. The curve obtained shows a series of regularly spaced peaks (Fig. 2.75).

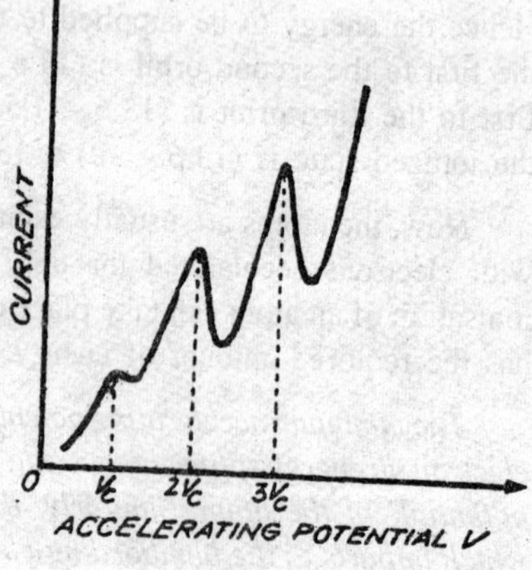

Fig. 2.75

Interpretation of the Curve

Electrons are emitted from F with a range of small energies. In the beginning, they acquire a small additional energy eV on reaching G. Those electrons whose energy is now greater than eV_0 reach P, and a current is obtained. As the accelerating potential V is increased, more and more of the electrons arrive at the plate and the current rises. Between F and G the electrons collide with the gas atoms. But since the electrons energies are not enough to excite the atoms, the electrons do not lose energy in these (elastic) collisions.

When the accelerating potential V becomes equal to the first excitation potential V_c of the as atoms, the electron energy at G is eV_c. The electrons can now suffer *inelastic* collisions with gas atoms near G and excite them to an energy level above their ground state. The electrons which do so, lose their energy and are unable to reach P against the retarding potential. Hence, the current drops sharply. Thus, the position of the first peak gives the excitation potential V_c of the gas atoms.

As the accelerating potential V is increased further, the electrons suffer inelastic collisions nearer and nearer to the filament, F, so that when they reach the grid G, once again they acquire enough energy to reach the plate P. The current thus begins to rise again. When V becomes equal to $2V_c$, a second inelastic collision occurs near the grid and the current drops again. Thus, the second peak gives $2V_c$. This process repeats as V is further increased.

The first peak occurs at a potential slightly less than V_c. This is because electrons are emitted from the filament with a finite velocity and hence with some energy. The true excitation potential V_c is obtained by measuring the difference between two successive peaks.

Limitations : Atoms have more than one excitation potentials and also an ionisation potential. Therefore, in an actual experiment the curve obtained is quite complicated and we cannot distinguish between which are excitation and which are ionisation potentials.

Demonstration of the Existence of Discrete Energy Levels : The experiment shows that electrons transfer energy to the atoms indiscrete amounts, and that they cannot excite atoms if their energy is less than eV_c. Franck and Hertz demonstrated it directly by observing the spectrum of the gas during electrons collisions. They showed that a particular spectral line does not appear until the electron energy reaches a *threshold* value. For example, in the case of mercury vapour, they found that minimum electron energy of 4.9 eV was required to obtain the 2536 Å line of mercury, and a photon of 2536 Å light has an energy of just 4.9 eV. This shows that discrete energy levels do exist in atom and the electrons of the atom can exist only in these levels.

Bohr's Correspondence Principle : The quantum theory gives results which are altogether different from those given by the classical theory in the microscopic world. Yet the two theories approach each other as the quantum number in question increases. This fact was pointed out by Bohr in 1932 who enunciated the following principle.

The predictions of the quantum theory for the behaviour of any physical system must coincide with the predictions of the classical theory in the limit in which the quantum numbers specifying the state of the system becomes very large. This is knows as 'Bohr's correspondence principle'. We establish it by means of an example.

According to classical theory, an electron moving in a circular orbit must emit radiation only of a particular frequency, namely, the frequency of revolution of the electron itself (or its integral multiples). According to quantum theory, on the other hand, radiation is emitted when the electron jumps from one orbit to the other and the frequency of radiation is determined by the difference in energy between the two orbits. Let us try to find a relation between the two frequencies.

The basic equations in the Bohr's theory of the hydrogen atom are the following.

$$\frac{mv^2}{r} = \frac{1}{4\pi\varepsilon_0}\frac{e^2}{r^2}$$

and $$mvr = n\frac{h}{2\pi}, \quad n = 1, 2, 3, ...$$

where v is the velocity of the electron of mass m in the orbit of radius r. These equations give

$$v = \frac{nh}{2\pi m r}$$

and $$r = \frac{n^2 h^2 \varepsilon_0}{\pi m e^2}.$$

Hence, the classical frequency of revolution of the electron is

$$f = \frac{\text{electron speed}}{\text{orbit circumference}} = \frac{v}{2\pi r}$$

$$= \frac{nh}{4\pi^2 mr^2} = \frac{nh}{4\pi^2 m\left(\frac{n^2h^2\varepsilon_0}{\pi me^2}\right)^2}$$

$$= \frac{me^2}{4\varepsilon_0{}^2 n^3h^3} = \frac{me^4}{8\varepsilon_0{}^2ch^3}\frac{2c}{n^3}$$

$$= \frac{2Rc}{n^3} \quad ...(i)$$

where $R\left(=\frac{me^4}{8\varepsilon_0{}^2ch^3}\right)$ is the Rydberg constant. This is the frequency which must be radiated by the moving electron classically.

Now, the frequency radiated on the basis of quantum theory is given by

$$v = \frac{E_i - E_f}{h} = \left(=\frac{me^4}{8\varepsilon_0{}^2h^3}\right)\left(\frac{1}{n_f{}^2} - \frac{1}{n_i{}^2}\right),$$

when the electron drops from the ni th orbit the nf the orbit. Again, introducing R, we have

$$v = Rc\left(\frac{1}{n_f{}^2} - \frac{1}{n_i{}^2}\right).$$

This may be written in the form

$$v = Rc\,\frac{(n_i - n_f)(n_i + n_f)}{n_f{}^2\,n_i{}^2} \quad ...(2)$$

Let us write $n_f = n$ and $n_i = n + 1$. Then the frequency of the emitted radiation for the transition $n + 1 \rightarrow$ (so that $\Delta n = 1$) is given by

$$v = Rc\,\frac{2n+1}{n^2(n+1)^2}.$$

When n is very large, we can write $\frac{2n+1}{n^2(n+1)^2} \simeq \frac{2}{n^3}$.

Under this condition, the emitted frequency is

$$v = \frac{2Rc}{n^3}. \qquad ...(iii)$$

Equations (i) and (iii) yield

$$v = f.$$

If we consider transitions Dn = 2, 3, 4, ... we shall have

$$v = 2f,\ 3f,\ 4f,...$$

Thus, for very large quantum numbers, the quantum theory frequency v of the radiation is identical with the classicai frequency of the revolution (or its harmonics) of the electron in the orbits. This is in accordance with Bohr's correspondence principle.

SOLVED EXAMPLES

Example 1:

Calculate the g value if the methyl radical shown ESR at 3290 G (0. 3290 T) in a spectrometer operating at 9230 MHz.

Solution:

From equation (2) we have

$$v = \frac{g\beta H}{h}$$

$$\text{or } g = \frac{hv}{\beta H} = \frac{\left(.627\times 10^{-34}\ \text{Js}\right)\left(9.230\times 10^{9}\ \text{s}^{-1}\right)}{\left(9.274\times 10^{-24}\ \text{JT}^{-1}\right)(0.3290\ \text{T})} 2.004729.$$

Example 2:

The ESR Spectrum of H atom studied in an ESR spectrometer operating at 9300 MHz shows two lines at 3570 G and 3044 G. Calculate hyperfine splitting constant (HSC) for H atom.

Solution:

Hyperfine splitting constant (HSC)

$$= (3570\ \text{G} - 3044\ \text{G})$$

$$= 526\ \text{G} = \frac{526}{10000} \times 1000 = 52.6\ \text{T}.$$

Example 3(a):

Calculate the radius of the second Bohr orbit of hydrogen atom and velocity of electron in this orbit. $\frac{1}{4\pi\varepsilon_0} = 9.10^9 \frac{N-m^2}{C^2}$, *m = 9.1 × 10 ³¹ kg, e = 1.6 × 10 ¹⁹ C and h = 6.63 × 10 ³⁴ J-s)* .

Solution:

2.1 Å, 1.1 × 10^6 m/s.

Example 3(b:

Calculate the time taken by the electron to traverse the first orbit in the hydrogen atom. Electron mass and charge are 9.1 × 10 ³¹ kg and 1.6 × 10 ⁻¹⁹ C and h = 6.63 × 10 ³⁴ J-s.

Solution:

The radius of the n the Bohr orbit is

$r_n = 4\pi\varepsilon_0 \frac{n^2 h^2}{4\pi^2 m e^2}$ **and the velocity of electron in the n th orbit is**

$$v_n = \frac{1}{4\pi\varepsilon_0} \frac{2\pi e^2}{nh}.$$

The time taken by the electron to traverse the n the orbit is therefore

$$T_n = \frac{2\pi r_n}{v_n} = 2\pi(4\pi\varepsilon_0)^2 \frac{n^2 h^2}{4\pi^2 m e^2} \frac{nh}{2\pi e^2}$$

$$= \frac{4\varepsilon_0^2 h^3 n^3}{me^4}.$$

For the first orbit, n = substituting the given values of h, m, e and ε_0 = 8.85 × 10^{-12} $C^2/N\text{-}m^2$, we get

$$T_1 = \frac{4.(8.85\times10^{-12})^2 (6.63\times10^{-34})^3}{(9.1\times10^{-31})(1.6\times10^{-19})^4} = 1.5 \times 10^{-16} \text{ s.}$$

Example 4:

Calculate the ESR frequency of an unpaired electron in a magnetic field of 3000 G (0. 3 T).

Solution:

From equation (2), we have

$$\Delta E = h\nu = g\,\beta H$$

were $g = 2$, $\beta = 9.274 \times 10^{-24}$ JT–1,

$H = 0.3$ T and $h = 6.627 \times 10^{-34}$ Js

$$\therefore\ \nu = \frac{g\beta H}{h} = \frac{2 \times 9.274 \times 10^{-24}\ \text{JT}^{-1} \times 0.3\,\text{T}}{6.627 \times 10^{-34}\ \text{Js}}$$

$$= 8.397 \times 10^{9}\ \text{Hz} = 8397\ \text{MHz} = 8.397\ \text{k MHz}.$$

Example 5:

Find an expression for the radius of the electron orbit in the hydrogen atom in its n th state. What will be the approximate quantum number n for an electron in an orbit of radius 1 mm ? (Take required values from problem 1).

Solution:

The basic equation in the Bohr's theory of hydrogen atom are.

$$\frac{mv^2}{r} = \frac{1}{4\pi\varepsilon_0}\frac{e^2}{r^2} \qquad \text{...(1)}$$

and

$$mvr = \frac{nh}{2\pi}. \qquad \text{...(2)}$$

Squaring (2) and dividing by (1), we get

$$r = 4\,\pi\,\varepsilon_0\,\frac{n^2h^2}{4\pi^2me^2}.$$

Substituting the given values, we get

$$r = \frac{1}{9 \times 10^9\ \text{Nm}^2/\text{C}^2} \times$$

$$\frac{n^2\,(6.63 \times 10^{-34}\ \text{Js})^2}{4 \times (3.14)^2 \times (9.1 \times 10^{-31}\ \text{kg}) \times (1.6 \times 10^{-19}\ \text{C})^2}$$

$= 0.53 \times 10^{-10}\ n^2$ meter

$= 0.53\ n^2$ Å.

For $r = 1$ mm $= 10^7$ Å, we have

$$10^7 = 0.53\ n^2.$$

$$\therefore \qquad n^2 = \frac{10^7}{0.53} = 18.87 \times 10^6$$

or $\qquad$ n; 4350.

Example 6:

Calculate the speed of electron in the n th orbit of the hydrogen atom. If relativistic is necessary.

(e = 1.6 × 10 19 C, h = 6.63 × 10 34 J-s, c 3.0 × 108 m s l

and $\frac{1}{4\pi\varepsilon_0} = 9 \times 10^9\ N\ m^2/C^2$*).*

Solution:

Let v be the speed of the electron (mass m, charge e) in the n th orbit whose radius is r, all in SI units. The charge on hydrogen nucleus is + e. As the centripetal force is provided by the electrostatic attraction, we have

$$\frac{mv^2}{r} = \frac{1}{4\pi\varepsilon_0}\frac{e^2}{r^2} \qquad \text{...(1)}$$

Further, from Bohr's postulate, the angular momentum of the electron is an integral multiple of h/2π, that is,

$$mvr = \frac{nh}{2\pi}. \quad n = 1, 2, 3, ... \qquad \text{...(2)}$$

Dividing eq. (1) by eq. we get

$$v = \frac{1}{4\pi\varepsilon_0}\frac{2\pi e^2}{nh}.$$

Substituting the given values, we get

$$v = (9 \times 109\ N\ m^2/C^2) \times \frac{2 \times 3.14 \times (1.6 \times 10^{-19}\ C)^2}{n\,(6.63 \times 10^{-34}\ Js)}$$

$$= \frac{1}{n}\ (2.18 \times 106)\ m\ s^{-1}.$$

This is the required value. Now

$$\frac{v}{c} = \frac{1}{n}\ \frac{2.18 \times 10^6}{3.0 \times 10^8} = \frac{0.0073}{n}.$$

Thus, for n = 1, 2, 3, ... we have

$$\frac{v}{c} = 0.0073,\ 0.0036,\ 0.0024, ...$$

Hence, the relativistic correction in necessary for n = 1 orbit only.

Example 7(a):

(i) How much energy is required to remove an electron from hydrogen atom in the ground state and also in a state with n = 8?

(ii) Calculate the corresponding energies of the singly-ionised helium atom (m = 9.11 × 10–31 kg, e = 1.6 × 10^{-19} C, h = 6.63 × 10^{-34} J s, e_0 = 8.85 × 10^{-12} $C^2/N\ m^2$ and 1 eV 1.6 × 10^{-19} J).

Solution:

(i) The energy required to remove an electron from the hydrogen atom in the ground state (n = 1) to infinity (where the energy is zero) is numerically equal to the energy of the electron in the n = 1 orbit. Now the energy of the electron in n th orbit of hydrogen (Z = 1) is given by

$$E_n = -\frac{me^4}{8\varepsilon_0^2 h^2}\left(\frac{1}{n^2}\right).$$

For the ground orbit, n = 1.

$$\therefore \quad E_1 = \frac{me^4}{8\varepsilon_0{}^2 h^2}$$

$$= -\frac{(9.11\times10^{-31})\times(1.6\times10^{-19})^4}{8\times(8.85\times10^{-12})^2\times(6.63\times10^{34})^2}$$

$$= -2.17 \times 10^{-18}\ \text{joule}$$

$$= -\frac{2.17\times10^{-18}}{1.6\times10^{-19}} = -13.6\ \text{eV}.$$

Hence, the energy required to remove the electron from n = 1 orbit to infinity is 13.6 eV. This is known as the binding energy of the hydrogen atom.

Again, for n = 8 orbit,

$$E_8 = -\frac{me^4}{8\varepsilon_0{}^2 h^2}\left(\frac{1}{8^2}\right) = \frac{E_1}{64}$$

$$= -\frac{13.6}{64} = -0.213 \text{ eV}.$$

Hence the energy required to remove an electron from n = 8 orbit to infinity is 0.213 eV.

(ii) The energy expression for

$$He^+ = -\frac{mZ^2e^4}{8\varepsilon_0^2 h^2}\left(\frac{1}{n^2}\right)$$

$$= -\frac{4me^4}{8\varepsilon_0^2 h^2}\left(\frac{1}{n^2}\right) = -4\,(E_n)_H.$$

This is, the energies for He^+ are 4 times the corresponding energies for H. Thus the results are

$$4 \times 13.6 = 54.4 \text{ eV}$$

$$4 \times 0.213 = 0.85 \text{ eV}.$$

Example 7(b):

A photon of energy 12.1 eV is completely absorbed by a hydrogen atom, originally in its ground state, so that the atom is excited. What is the quantum number of this state ? The ground state (negative) energy of hydrogen atoms is 13.6 eV.

Solution:

Let the required quantum number be n. Then

$$E_1 - E_n = -12.1 \text{ eV}$$

or $$E_1 - \frac{E_1}{n^2} = -12.1 \text{ eV}. \quad \left[\because E_n = \frac{E_1}{n^2}\right]$$

But $$E_1 = -13.6 \text{ eV}.$$

$$1 - \frac{1}{n^2} = \frac{-12.1}{-13.6} = 0.89.$$

$$E_1 - E_n = -12.1 \text{ eV}$$

or $$E_1 - \frac{E_1}{n^2} = -12.1 \text{ eV}. \quad \left[\because E_n = \frac{E_1}{n^2}\right]$$

But $$E_1 = -13.6 \text{ eV}.$$

$$1 - \frac{1}{n^2} = \frac{-12.1}{-13.6} = 0.89$$

or $$\frac{1}{n^2} = 1 - 0.89 = 0.11$$

or $$n^2 = \frac{1}{0.11} = 9.$$

$\therefore$ $$n = 3.$$

Example 7(c):

A beam of electron bombards a sample of hydrogen. Through what minimum potential difference must be electrons have been accelerated if the first line of the Balmer series is to be emitted ? The ionisation potential of hydrogen atom is 13.6 volts. Explain how many possible spectral lines can be expected if the atom finally attains the normal state.

Solution:

Normally the hydrogen atoms are in the ground state (n = 1). The first Balmer line is emitted when the atom returns from n = 3 to n = 2 state. Hence, in order to emit this line the atom must be first raised to the n = 3 state by electron bombardment. Therefore, the energy of the bombarding electrons must be equal to the difference of energy between n = 1 and n = 3 states ($E_1 \sim E_3$).

The energy of hydrogen atom in the n th state is

$$E_n = -\frac{Rhc}{n^2}.$$

If E_1 is the energy in the ground (n = 1) state, then $E_1 = -$ Rhc. Thus

$$E_n = \frac{E_1}{n^2}$$

For n = 3, we have $E_3 = \dfrac{E_1}{9}$

$$\therefore \quad E_1 - E_3 = E_1 - \frac{E_1}{9} = \frac{8}{9}E_1.$$

But $E_1 = -13.6$ eV.

$$\therefore \quad E_1 - E_3 = \frac{8}{9} \times (-13.6) = -12.1 \text{ eV}.$$

Hence the bombarding electrons must be accelerated by 12.1 volts.

If the atom finally attains the normal state, the possible number of spectral lines is 3 (3 → 2, 3, → 1, 2 → 1).

Example 7(d):

The wavelength of the first line of Balmer series of hydrogen is 6562.8 Å. Calculate (i) the ionisation potential and the first excitation potential of the hydrogen atom. ($h = 6.63 \times 10^{-34}$ J-s, $c = 3 \times 10^8 m$ s^{-1}).

Solution:

(i) The wavelengths of the Balmer lines are given by

$$\frac{1}{\lambda} = R\left(\frac{1}{2^2} - \frac{1}{n^2}\right), \qquad n = 3, 4, 5, ...$$

For the first line n = 3 and l = 6562.8 Å = 6562.8 × 10^{-10} m.

$$\therefore \quad \frac{1}{6562.8 \times 10^{-10}} = R\left(\frac{1}{2^2} - \frac{1}{3^2}\right) = \frac{5R}{36}$$

$$\text{or} \qquad R = \frac{36}{5 \times 6562.8 \times 10^{-10}\,\text{m}} = 1.097 \times 10^7 \text{ m}^{-1}.$$

This ionisation potential of an atom is numerically equal to the (ionisation) energy required to remove an electron completely from the atom in the normal state. In hydrogen atom, the electron stays in the first orbit (n = 1). Hence the energy required to remove this electron to infinity (where the energy is considered to be zero) is numerically equal to the energy of the electron in the first orbit. We know that electron energy in hydrogen atom is given by

$$E_n = -\frac{Rhc}{n^2}.$$

In the ground state, n = 1.

$$\therefore \quad E_1 = -Rhc$$

$$= -13.6 \text{ eV.} \qquad \text{(as obtained in the leas problem).}$$

Here the ionisation potential of the hydrogen atom is 13.6 volts.

(ii) The first excitation potential of the atom is the energy required to shift the electron from n = 1 to n = 2 orbit, that is, $E_1 \sim E_2$. Now,

$$E_1 = -13.6 \text{ eV}.$$

and $$E_2 = \frac{E_1}{4} = -3.4 \text{ eV}. \qquad \left[\because E_n \propto \frac{1}{n^2}\right]$$

$$\therefore \quad E_1 \sim E_2 = 13.6 - 3.4 = 10.2 \text{ eV}.$$

Hence the first excitation potential is 10.2 volts.

Example 8:

The wavelength of a yellow line of sodium is 5896 Å. Calculate its wave number and frequency.

Solution:

The wave number $\bar{v}$ is the reciprocal of wave-length λ, that is,

$$\bar{v} = \frac{1}{\lambda}.$$

Here λ = 5896 Å = 5896 × 10^{-8} cm.

$$\therefore \quad \bar{v} = \frac{1}{5896 \times 10^{-8} \text{ cm}} = 16960 \text{ per cm}.$$

The frequency v is related to the wavelength 1 by

$$c = v\lambda,$$

where c is the speed of light. Thus

$$v = \frac{c}{\lambda} = \frac{3 \times 10^8 \text{ cm/s}}{5896 \times 10^{-8} \text{ cm}} = 5.1 \times 10^{12} \text{ s}^{-1}$$

Example 9:

The ionisation potential of an atom is 14.2 eV. Calculate the series limit in its absorption spectrum. Dates as in last problem.

Solution:

The series limit in the absorption spectrum of an atom corresponds to the energy which when absorbed by a ground-state atom, ionises it. Thus, if V be the ionisation potential of an atom, the wavelength at the series limit is given by

$$\lambda = \frac{hc}{eV}$$

$$= \frac{(6.62 \times 10^{-34}\,\text{J s}) \times (3 \times 10^{8}\,\text{m s}^{-1})}{(14.2 \times 1.6 \times 10^{-19}\,\text{J})}$$

$$= 0.874 \times 10^{-7}\ \text{m}$$

$$= 874\ \text{Å}.$$

Example 10:

In a Franck-Hertz experiment the first dip in the current-voltage graph for hydrogen was observed at 10.2 volts. Calculate the wavelength of light emitted by hydrogen atom when existed to the first excitation level. ($h = 6.62 \times 10^{-34}$ J s, $c = 3.0 \times 10^{8}$ m/s, $e = 1.6 \times 10^{19}$ C.)

Hint : $\lambda = \dfrac{hc}{eV}$.

Solution:

1217 Å.

Example 11:

With Franck-Hertz type of experiment on sodium, the first spectral line to appear is the D-line, $\lambda = 5.89 \times 10^{-7}$ m. What is the first excitation potential of sodium ? Given : $h = 6.63 \times 10^{-34}$ J, s, c 3 × 108 m/s and 1 eV = 1.6×10^{-19} J.

Solution:

Let V volt be the excitation potential. Then the (excitation) energy imparted to the electron will be eV joule, where e coulomb is the charge on the electron. This energy is re-emitted as photon (radiation) when the electron returns to the normal state. If v be frequency of the emitted radiation, the photon energy will be hv. Thus

$$eV = h\nu$$

or
$$V = \frac{h\nu}{e}.$$

If l be the wavelength of the emitted radiation, then $\nu = c/\lambda$.

$$\therefore \quad V = \frac{h\nu}{e\lambda}$$

$$= \frac{(6.63 \times 10^{-34}\,\text{J s})(3 \times 10^{8}\,\text{m s}^{-1})}{(1.6 \times 10^{-19}\,\text{C})(5.89 \times 10^{-7}\,\text{m})}$$

$= 2.1\ (J/C) = 2.1\ V.$

Example 12(a):

The first two excitation potentials of atomic hydrogen in Franck-Hertz experiment are 10.2 and 12.09 volts. Draw an energy level diagram and show all possible transitions for emission and absorption along their wavelengths. ($h = 6.63 \times 10^{-34}$ J s, $c = 3.0 \times 108$ m/s, 1 eV $= 1.6 \times 10^{-19}$ J.)

Solution:

The energy level diagram, and the *three* possible transitions (a), (b) and (c) for emission are shown in the figure.

The frequency of radiation resulting from the transitions (a) is given by

$$\nu_a = \frac{E_2 - E_1}{h},$$

and the corresponding wavelength is,

$$\lambda_a = \frac{hc}{E_2 - E_1}. \qquad [\because c = \nu\lambda]$$

Now, from the Fig., $E_2 - E_1 = 10.2$ eV $= 10.2 \times (1.6 \times 10^{-19})$ J. Therefore

$$\lambda_a = \frac{(6.63 \times 10^{-34}\ \text{J s}) \times (3.0 \times 10^8\ \text{m s}^{-1})}{(10.2 \times 1.6 \times 10^{-19}\ \text{J})}$$

$$= 1.216 \times 10^{-7}\ \text{m} = 1216 \times 10^{-10}\ \text{m} = 1216\ \text{Å}.$$

$$\text{Similarly, } \lambda_b = \frac{hc}{(E_3 - E_1)}$$

$$= \frac{(6.63 \times 10^{-34}) \times (3.0 \times 10^8)}{(12.09 \times 1.6 \times 10^{-19})}$$

$$= 1.026 \times 10^{-7}\ \text{m} = 1026 \times 10^{-10}\ \text{m} = 1026\ \text{Å}.$$

$$\text{Also, } \lambda_c = \frac{hc}{(E_3 - E_2)} = \frac{hc}{(E_3 - E_1) - (E_2 - E_1)}$$

$$= \frac{hc}{(12.09 - 10.2)\ \text{eV}} = \frac{hc}{1.89\ \text{eV}}$$

$$= \frac{(6.63 \times 10^{-34}) \times (3.0 \times 10^{8})}{(1.89 \times 1.6 \times 10^{-19})}$$

$$= 6.567 \times 10^{-7} \text{ m} = 6567 \times 10^{-10} \text{ m} = 6567 \text{ Å}.$$

In absorption, only the transitions starting from n = 1 shall be observed which correspond to 1216 Å and 1026 Å.

Example 12(b):

(a) *Compare the assumptions made by Planck in discussing cavity radiation, those by Einstein in connection with the photoelectric effect and those by Bohr in connection with the hydrogen spectrum.*

(b) *Explain the implication of the fact that- the energies of the hydrogen atom orbits are negative.*

(c) *Can a hydrogen atom absorb a photon of energy greater than the binding energy of the atom ?*

Solution:

(a) *Assumptions of Planck, Einstein and Bohr* : Planck had assumed that the atoms of a hot body behave as oscillators and have discrete (quantised) energies. They do not emit radiant energy continuously, but only in 'jumps' or 'quanta'. Planck, however, still maintained that radiation propagates continuously through space as electromagnetic waves.

Einstein, in order to explain the photoelectric effect, went a step ahead. He proposed that the radiation not only is emitted as quantum at a time but also propagates as individual quanta (photons). He thus treated the propagation of radiation as particle propagation rather than wave propagation.

Bohr's in order to explain the hydrogen spectrum, adopted the Planck's quantum hypothesis that the radiation is emitted discontinuously from the atom. He started with the quantisation of the angular momentum of the electron in the orbit which ultimately results in the quantisation of energy of the atom.

(b) *Negative Energy of Hydrogen Orbits* : An electron revolving in an hydrogen orbit has a negative potential energy by virtue of its attraction towards, the nucleus, and also kinetic energy (which is positive) by virtue of its motion. The potential energy is greater in magnitude .than the kinetic energy so that the net energy is negative. The negative' energy signifies that the electron cannot escape from The atom. A positive energy for a nucleus electron combination would mean that electron is not bound to the nucleus. Such a combination cannot constitute an atom.

Example 12(c):

A μ^- meson (charge – e, mass = 207 m, where m is mass of electron) can be captured by a proton to form a hydrogen- like "mesic" atom. Calculate the radius of the first Bohr orbit, the binding energy, and the wavelength of the first line in the Layman series for such an atom. The mass of the proton is 1836 times the mass of electron. The radius of first Bohr orbit and the binding energy of hydrogen are 0.53 Å and 13.6 eV respectively. R_y = 109737 cm^{-1}.

Solution:

The reduced mass of the system is

$$m = \frac{(207\,m)(1836\,m)}{207\,m + 1836\,m} = 186\ m.$$

From Bohr theory, the radius of the first orbit (n = 1) of a hydrogen -like atom for Z = 1, is given by (taking finite mass of nucleus in consideration).

$$r_1 = 4\pi\varepsilon_0 \frac{h^2}{4\pi^2 \mu e^2} = 4\ \pi\varepsilon_0 \frac{h^2}{4\pi^2 (186m)\, e^2}$$

$$= \frac{1}{186}\left(4\pi\varepsilon_0 \frac{h^2}{4\pi^2 me^2}\right).$$

The quantity in the bracket is the first Bohr orbit of hydrogen atom which is 0.53 Å. Therefore

$$r_1 = \frac{1}{186} \times 0.54\ \text{Å} = 2.85 \times 10^{-3}\ \text{Å}.$$

Again, from Bohr's theory, the ground-state energy for a hydrogen-like atom with Z = 1 is given by

$$E_1 = -\frac{\mu e^4}{8\varepsilon_0^2 h^2} = -186 \frac{me^4}{8\varepsilon_0^2 h^2}$$

$$= -186 \times 13.6 = -2530 \text{ eV}.$$

Hence the binding energy is 2530 eV.

The wavelength of the Lyman lines are given by

$$\frac{1}{\lambda} = R_M\left(\frac{1}{1^2} - \frac{1}{n^2}\right), \; n = 2, 3, 4,...$$

where, R_M is the Rydberg constant for the music atom. For the first line, n = 2 so that

$$\lambda = \frac{4}{3R_M}.$$

Now, $$R_M = \frac{\mu e^4}{8\varepsilon_0^2 ch^3}$$

and $$R_\infty = \frac{me^4}{8\varepsilon_0^2 ch^3}, \text{ so that}$$

$$R_M = \frac{\mu}{m} R_\infty = 186 \, R_\infty = 186 \times 109737 \text{ cm}^{-1}.$$

Here $$\lambda = \frac{4}{3R_M} = \frac{4}{3 \times 186 \times 109737 \text{ cm}^{-1}}$$

Example 13(a):

A positronium atom is a system consisting of a position and an electron. Calculate the reduced mass, the Rydberg constant and the wavelength of the first Balmer line for positronium. (Give $m = 9.1 \times 10^{31}$ kg, $R_H = 1.09737 \times 10^{-3}$ Å^{-1} and $H_\alpha = 6563$ Å).

Solution:

The positron has the same mass m as the electron and has equal but positive charge. The reduced mass of the electron-positron atom is therefore

$$m = \frac{(m)(m)}{m + m} = \frac{1}{2} m = 4.55 \times 10^{-31} \text{ kg},$$

while the reduced mass of electron in hydrogen is very nearly m.

The Rydberg constant $\left(\frac{\mu e^4}{8\varepsilon_0^2 ch^3}\right)$ for positronium is therefore half that for hydrogen (with infinitely heavy nucleus). Thus

$$R_P = \frac{1}{2} R_H = 0.54868 \times 10^{-3} \text{ Å}^{-1}.$$

The wavelength of first Balmer line (H_α) for hydrogen atom is given by

$$\frac{1}{\lambda_H} = R_H \left(\frac{1}{2^2} - \frac{1}{3^2}\right),$$

while that for positronium atom is $\frac{1}{\lambda_p} = R_P \left(\frac{1}{2^2} - \frac{1}{3^2}\right)$.

$$\text{Thus} \quad \frac{\lambda_P}{\lambda_H} = \frac{R_H}{R_P} = 2.$$

$$\therefore \quad \lambda_P = 2\lambda_H = 2 \times 6563 = 13126 \text{ Å}.$$

Example 13(b):

Find the recoil speed of hydrogen atom after it emits a photon in going from n = 3 to n = 1 state. Electron mass is 9.11 × 10^{-31} kg and h = 6.626 × 10^{-34} J s.

Solution:

The energy of the hydrogen atom in the n th state is given by

$$E_n = -\frac{Rhc}{n^2},$$

where R is Rydberg constant. From this, we get

$$E_1 - E_3 = -\frac{Rhc}{1^2} + \frac{Rhc}{3^2} = -\frac{8}{9} R h c.$$

The energy of the emitted photon is

$$\Delta E = E_1 \sim E_3 = \frac{8}{9} R h c.$$

The momentum of the photon is

$$p = \frac{\Delta E}{c} = \frac{8}{9} R h.$$

By conservation of momentum, the recoil momentum of the hydrogen atom will be equal (and opposite) to the momentum of the emitted photon. The recoil speed of the atom is

$$v = \frac{\text{momentum}}{\text{mass}} = \frac{8}{9}\frac{Rh}{m_H}.$$

But $m_H = 1836$ m, where m is electron mass.

$$\therefore v = \frac{8}{9}\frac{Rh}{(1836\,m)}.$$

Putting the given values of h, m and using $R = 1.097 \times 10^7\ m^{-1}$, we get

$$v = \frac{8}{9}\frac{(1.097\times10^{7}\,m^{-1})(6.626\times10^{-34}\,Js)}{1836\,(9.11\times10^{-31}\,kg)}$$

$$= 3.86\ m/s.$$

Example 13(c):

Calculate the wavelength of light emitted by an atom excited to next level by 2 eV.

Solution:

Do yourself

Ans. : 6206 Å.

Example 14:

The ionisation potential of hydrogen tom is 13.6 volt. Find the wavelength of the Lyman series limit.

Solution:

When an atom absorbs energy so that its two stationary states becomes existed simultaneously, two superimposed sets of radiations are produced to give a 'beat' variation. If v¢ and v¢¢ represent the frequencies of two superimposed vibrations, the frequency v of the emitted beat is

$$v = v' - v''.$$

The wave mechanical vibrations are related to the energies of the corresponding state by the usual quantum expressions $E' = hv'$ and $E' - E'' = hv''$ so that

$$E' - E'' = h(v' - v'')$$

$$\Delta E = 13.61 \left(\frac{1}{(1)^2} - \frac{1}{\infty}\right) = 13.61 \text{ eV.}$$

Therefore, ionisations potential = 13.61 eV.

Example 15:

A beam of monochromatic photons of energy 9 eV is incident on hydrogen gas all of whose atoms are in the ground state. It is found that the beam is fully transmitted without absorption. Why ? The ground state energy of an electron in the hydrogen atom is $E_l = -13.6$ eV.

Solution:

The minimum energy that can be absorbed by ground-state hydrogen atom is $E_1 \sim E_2$, which would excite it to the next state (n = 2). Now,

$$E_1 \sim E_2 = E_1 \sim \frac{E_1}{4} \qquad \left[\because E_n = \frac{E_1}{n^2}\right]$$

$$= \frac{3}{4} E_1 = \frac{3}{4} \times 13.6 = 10.2 \text{ eV.}$$

Hence, photons of energy 9 eV cannot be absorbed by hydrogen atoms.

Example 16:

Calculate the ionisation potential of hydrogen from the following date.

Solution:

According to Bohr's theory, the energy of an electron in the nth orbit of hydrogen atom is

$$E = -\frac{2\pi^2 e^4 m}{n^2 h^2}$$

When the atom is in the normal state, the only electron in the hydrogen atom in the first orbit, *i.e.*, n = 1.

$$\therefore E = -\frac{2\pi^2 me^4}{h^2}$$

or $$E = \frac{2 \times (3.14)^2 \times 9 \times 10^{-28} \times (4.8 \times 10)^4}{(6.6 \times 20^{27})^2} = 2.165 \times 10^{-11} \text{ eg.}$$

To remove the electron from first orbit to infinity 2.165×10^{-11} erg of energy must be supplied. The amount of energy is called the ionisation potential of hydrogen atom.

Ionisation potential of hydrogen atom

$$= 2.165 \times 10^{-11} \text{ erg}$$

$$= \frac{2.165 \times 10^{-11}}{1.6 \times 10^{-12}} \text{ electron volts} = 13.53 \text{ eV}.$$

Example 17:

Energy in a Bohr's orbit is given to be equal to – B/n^2 *where* $B = 2.179 \times 10^{-11}$ *erg. Calculate the frequency of radiation and also the wave number when the electron jumps from the third orbit to the second* ($h = 6.62 \times 10^{-27}$ *erg sec) orbit.*

Solution:

Energy of a Bohr's orbit is given by

$$E = \frac{B}{n^2} = \frac{2.179 \times 10^{-11}}{n^2} \text{ erg.}$$

For the third orbit, n = 3

$$E_2 = \frac{2.179 \times 10^{-11}}{9} = -0.2421 \times 10^{-11} \text{ erg.}$$

For the second orbit, n = 2

$$E_2 = \frac{2.179 \times 10^{-11}}{4} = -0.54475 \times 10^{-11} \text{ erg.}$$

$$E_3 - E_2 = (0.54475 \times 10^{-11}) - (0.2421 \times 10^{-11})$$
$$= 0.30265 \times 10^{-11} \text{ erg.}$$

When electron jumps from the third to the second orbit, a photon is emitted, the energy of which is given by

$h\nu = E_3 - E_2$ where ν is the frequency

$$\nu = \frac{E_3 - E_2}{h} = \frac{0.30265 \times 10^{-11}}{6.62 \times 10^{-27}} = 4.572 \times 10^{14} \text{ sec}^{-1}.$$

The corresponding wave number is given by

$$\bar{v} = \frac{1}{\lambda} = \frac{v}{c} = \frac{4.572 \times 10^{14}}{3 \times 10^{-10}} = 15240 \text{ cm}^{-1}$$

Example 18:

Give using spectral notation for the following states of the atom.

(i) n = 4, L = 2, S = 0

(ii) n = 4, L = 1, S = 1, J = 0 and

(iii) n = 3, L = 2, multiplicity 2.

Solution:

(i) Multiplicity (2S + 1) = 1,

J = L + S = 2 ∴ State will be $4\ ^1D_2$.

(ii) Multiplicity (2S + 1) = 3 ∴ State will be $4\ ^3P_0$.

(iii) Multiplicity will be (2S + 1) = 2 or S = 1/2

As L = 2, J = 5/2 or 3/2.

The two states which are positive are $3\ ^2D_{5/2}$ or $^2D_{3/2}$. *Give using spectral notation for the following states of the atom.*

(i) n = 4, L = 2, S = 0

(ii) n = 4, L = 1, S = 1, J = 0 and

(iii) n = 3, L = 2, multiplicity 2.

Example 19(a):

Calculate the velocity of the electron in the Bohr's orbit given that

h = 6.625 × 10^{-27} erg, sec., m = 9.11 × 10^{-28} gm.

e = 4.819 × 10^{-10} e s.u.

Solution:

In Bohr's hydrogen atom, centripetal force is equal to the electrostatic force. Thus,

$$\frac{mu^2}{r} = \frac{e^2}{r^2} \text{ or } u = \frac{e}{\sqrt{(mr)}} \qquad \text{...(A)}$$

Now $r = \frac{h^2}{4\pi^2 me^2}$ or $mr = \frac{h^2}{4\pi^2 e^2}$.

Taking square root of both sides, we get

$$\sqrt{(mr)} = \sqrt{\left(\frac{h^2}{4\pi^2 e^2}\right)} \text{ or } \sqrt{(mr)} = \frac{h}{2\pi e}.$$

Substituting this value in equation (A), we get

$$u = \frac{2\pi e^2}{h} = \frac{2 \times 3.14 \times (4.819 \times 10^{-10})^2}{6.625 \times 10^{-27}}$$

$$= 2.192 \times 10^8 \text{ cm/sec.}$$

Example 19(b):

Give e= 4.78 × 10^{-10} e.s.u., h = 6.65 × 10^{-27} *ergs, sec, and m* = 9 × 10^{-33} *gm., calculate the radius of the lowest orbit of the hydrogen atom.*

Solution:

According to Bohr's theory, the radius of the n th orbit of hydrogen atom is given by

$$r = \frac{n^2 h^2}{4\pi^2 me^2} [\because Z = 1]$$

For the lowest orbit, n = 1, it means that

$$r = \frac{n^2 h^2}{4\pi^2 me^2} = \frac{(6.65 \times 10^{-27})^2}{4 \times (3.14)^2 \times (9 \times 10^{-28}) \times (4.78 \times 10^{-10})^2}$$

$$= 0.53 \times 10^{-8} \text{ cm.} = 0.53 \text{ Å.}$$

Example 20:

Given that the Rydberg constant for hydrogen is 109678 cm $^{-1}$*, find the long and short wavelength limits of the Lyman series.*

Hint : For long wavelength limit (first line) n = 2 and for short wavelength limit n = ∞.

Solution:

1216 Å, 912 Å.

Example 21(a):

Calculate the wavelengths of the first two lines of the Balmer series and the series limit. (h = 6.63 × 10 ³⁴ J s, c = 3.0 × 10⁸ ms ¹, e = 1.6 × 10 ¹⁹ C, m = 9.1 × 10 ³¹ kg, ε₀ = 8.85 × 10 ¹² C²/N m²).

Solution:

Let us first calculate the Rydberg constant from the given data. We have

$$R = \frac{me^4}{8\varepsilon_0{}^2 ch^3}$$

$$= \frac{(9.1\times10^{-31}\,\text{kg})\times(1.6\times10^{-19}\,\text{C})^4}{8\times(8.5\times10^{-12}\,\text{C}^2/\text{Nm}^2)^2\times(3.0\times10^8\,\text{ms}^{-1})\times(6.63\times10^{-34}\,\text{Js})^3}.$$

Now, the wavelengths of the spectral lines of Balmer series are given by

$$\frac{1}{\lambda} = R\left(\frac{1}{2^2}-\frac{1}{n^2}\right), \qquad n = 3, 4, 5, \ldots$$

For the first line, n = 3.

$$\therefore \qquad \frac{1}{\lambda_1} = R\left(\frac{1}{2^2}-\frac{1}{3^2}\right) = \frac{5R}{36}$$

or $$\lambda_1 = \frac{36}{5R} = \frac{36}{5\times(1.097\times10^7\,\text{m}^{-1})} = 6.563\times10^{-7}\,\text{m}$$

$$= 6563\times10^{-10}\text{m} = 6563\ \text{Å}.$$

For the second line, n = 4. We can see that l23 = 4861 Å.

For the series limit, n = ∞.

$$\therefore \qquad \frac{1}{\lambda_\infty} = \frac{R}{4}$$

or $$\lambda_\infty = \frac{4}{R} = \frac{4}{1.097\times10^7}$$

$$= 3.646\times10^{-7}\,\text{m} = 3646\ \text{Å}.$$

Example 21(b):

Calculate the wavelength of the eight line of the Balmer series of the hydrogen atom. Rydberg constant for hydrogen atom is

$1.097 \times 10^7 m^{-1}$.

Hint : For the eight line, n = 10.

Solution:

3798 Å.

Example 22:

The wavelength of the first line of Balmer series is 6563 Å. Calculate Rydberg constant.

Solution:

The wavelength of the spectral lines of Balmer series are given by

$$\frac{1}{\lambda} = R_H \left(\frac{1}{2^2} - \frac{1}{n^2}\right), \; n = 3, 4, 5, ...$$

For the first ($\lambda = 6563 \times 10^{-10}$ m), n = 3.

$$\therefore \quad \frac{1}{6563 \times 10^{-10} \text{ m}} = R_H \left(\frac{1}{2^2} - \frac{1}{3^2}\right) = \frac{5}{36} = R_H$$

or
$$R_H = \frac{36}{5 \times (6563 \times 10^{-10} \text{ m})} = 1.097 \times 10^7 \text{ m}^{-1}.$$

Example 23:

Find the wavelength of the photon emitted when the hydrogen atom goes from n = 10 state to the ground state. ($R_H = 1.097 \times 10^{-3}$ Å^{-1}).

Solution:

Since the atom drops to the ground state (n = 1), the photon emitted belongs to the Lyman series. The wavelengths of the spectral lines in this series are given by

$$\frac{1}{\lambda} = R_H \left(\frac{1}{1^2} - \frac{1}{n^2}\right), \; n = 2, 3, 4, ...$$

For n = 10, we have

$$\frac{1}{\lambda} = R_H \left(1 - \frac{1}{100}\right) = \frac{99}{100} R_H$$

or
$$\lambda = \frac{100}{99 \, R_H}$$

$$= \frac{100}{99 \times (1.097 \times 10^{-3}\ \text{Å}^{-1})} = 921\ \text{Å}.$$

Example 24(a):

The average life-time of an electron in an excited state of hydrogen atom is about 10^{-8} s. How many revolutions does an electron in the n = 2 state make before dropping to the n = 1 state? (R = 1.097 × 107 m^{-1}).

Solution:

Let v be the velocity of electron (mass m, charge e) in an orbit of radius r. This basic equations are.

$$\frac{mv^2}{r} = \frac{1}{4\pi\varepsilon_0}\frac{e^2}{r^2}$$

and $$mvr = n\frac{h}{2\pi}.$$

These equations give

$$v = \frac{nh}{2\pi mr},\ r = \frac{n^2h^2\varepsilon_0}{\pi me^2}.$$

The number of revolutions of the electron in the orbit per second is

$$f = \frac{v}{2\pi r} = \frac{2Re}{n^3} \qquad \text{(as proved in Q. 8)}$$

For n = 2 state, we have

$$f = \frac{2 \times (1.097 \times 10^7\ m^{-1}) \times (3 \times 10^8\ ms^{-1})}{8}$$

$$= 8.2 \times 10^{14}\ s^{-1}.$$

Hence the number of revolutions of the electron in its life-time of 10^{-8} second is $(8.2 \times 10^{14}) \times 10^{-8} = 8.2 \times 10^6$.

Example 24(b):

An orange photon of wavelength 600 nm is emitted from an atom. Find the difference in energy in the two atomic states involved. Find the same result for the red 6563

Å line of hydrogen. (h = 6.63 × 10^{-34} j s, c = 3 × 10^8 m s^{-1}, 1 eV = 1.6 × 10^{-19} J).

Solution:

By Bohr's postulate, the emitted frequency is given by

$$v = \frac{\Delta E}{h},$$

where ΔE is the difference in energy. But $v = c/\lambda$. Therefore

$$\frac{c}{\lambda} = \frac{\Delta E}{h} \text{ or } \Delta E = \frac{hc}{\lambda}.$$

Here λ = 600 nm = 600×10^{-9} m.

$$\therefore \quad \Delta E = \frac{(6.63 \times 10^{-34}\ \text{J s}) \times (3.0 \times 10^{8}\ \text{m s}^{-1})}{600 \times 10^{-9}\ \text{m}}$$

$$= 3.31 \times 10^{-19}\ \text{J}$$

$$= \frac{3.31 \times 10^{-19}}{1.6 \times 10^{-19}} = 2.07\ \text{eV}.$$

For λ = 6563 Å $\times 10^{-10}$ m, we can show that

$$\Delta E = 3.03 \times 10^{-19}\ \text{J} = 1.9\ \text{eV}.$$

Example 25:

What is the smallest wavelength in the spectral lines of Paschen series in the hydrogen atom ?

Solution:

The formula for Paschan series is

$$\frac{1}{\lambda} = R\left(\frac{1}{3^2} - \frac{1}{n^2}\right), \; n = 4, 5, 6, \ldots$$

For the smallest wave length, $n = \infty$.

$$\therefore \quad \lambda = \frac{9}{R} = \frac{9}{1.097 \times 10^{7}\ \text{m}^{-1}}$$

$$= 8.204 \times 10^{-7}\ \text{m} = 8204\ \text{Å}.$$

Example 26:

The series limit of Balmer series is at 3646 Å. Calculate the wavelength of the first member Ha of this series.

Solution:

The formula for Balmer series is

$$\frac{1}{\lambda} = R\left(\frac{1}{2^2} - \frac{1}{n^2}\right), \; n = 3, 4, 5,$$

For series limit n = ∞ ad $\lambda = \lambda_\infty$ = 3646 Å.

$$\therefore \quad \frac{1}{3646 \text{ Å}} = \frac{R}{4}$$

or

$$R = \frac{4}{3646 \text{ Å}}.$$

Again, for the first member n = 3.

$$\therefore \quad \frac{1}{\lambda_1} = R\left(\frac{1}{2^2} - \frac{1}{3^2}\right) = \frac{5R}{36}$$

or

$$\lambda_1 = \frac{36}{5R} = \frac{36}{5 \times \dfrac{4}{3646 \text{ Å}}} = 6563 \text{ Å}.$$

Example 27:

The first line of the Balmer series in the spectrum of hydrogen has a wavelength of 6563 Å. Calculate the wavelength of the first line of Lyman series in the same spectrum.

Solution:

The wavelength of the spectral lines of hydrogen spectrum are given by

$$\frac{1}{\lambda} = R\left(\frac{1}{n_f^2} - \frac{1}{n_i^2}\right),$$

where R is Rydberg constant. For the first member of Balmer series $n_f = 2$ and $n_i = 3$.

$$\therefore \quad \frac{1}{\lambda_1} = R\left(\frac{1}{2^2} - \frac{1}{3^2}\right) = \frac{5R}{36}.$$

For the first member of the Lyman series, $n_f = 1$ and $n_i = 2$.

$$\therefore \quad \frac{1}{\lambda_1'} = R\left(\frac{1}{1^2} - \frac{1}{2^2}\right) = \frac{3R}{4}.$$

From (i) and (ii), we have

$$\frac{\lambda_1}{\lambda_1'} = R\frac{5R}{36}\times\frac{4}{3R} = \frac{5}{27}.$$

$$\therefore \quad \lambda_1' = \frac{5}{27}\lambda_1 = \frac{5}{27}\times 6563$$

$$= 1215 \text{ Å}.$$

Example 28(a):

The first member of Balmer series of hydrogen has a wavelength of 6563 Å. Calculate the wavelengths of the second and fourth members.

Solution:

The wavelengths of the spectral lines of Balmer series are given by

$$\frac{1}{\lambda} = R\left(\frac{1}{2^2}-\frac{1}{n^2}\right),\ n = 3, 4, 5, \ldots$$

For the first member, n = 3.

$$\therefore \quad \frac{1}{\lambda_1} = R\left(\frac{1}{2^2}-\frac{1}{3^2}\right) = \frac{5R}{36} \qquad \text{....(1)}$$

For the second member, n = 4.

$$\therefore \quad \frac{1}{\lambda_2} = R\left(\frac{1}{2^2}-\frac{1}{4^2}\right) = \frac{3R}{16} \qquad \text{...(2)}$$

Dividing eq. (2) by (1), we get

$$\frac{\lambda_1}{\lambda_2} = \frac{3R}{16}\times\frac{36}{5R} = \frac{27}{20}.$$

$$\lambda_2 = \lambda_1\times\frac{20}{27}.$$

But $\lambda_1 = 6563$ Å (given).

$$\therefore \quad \lambda_2 = (6563 \text{ Å})\times\frac{20}{27} = 4861 \text{ Å}.$$

Similarly $\lambda_4 = 4102$ Å.

Example 28(b):

Calculate :

(i) Energy required to ionise hydrogen atom in the ground state,

(ii) Limit of Balmer series.

(Rydberg constant is 1.097 × 10⁷ m⁻¹,

h = 6.63 × 10⁻³⁴ J s,

c 3.0 × 10⁸ m s⁻¹,

1 eV = 1.6 × 10⁻¹⁰ J).

Solution:

(i) The energy required to ionise, that is, to remove an electron from the hydrogen atom in the ground state (n = 1) to infinity (where the energy is zero) is numerically equal to the energy of the electron in the n = 1 orbit. Now, the energy of the electron in n th orbit of hydrogen is given by

$$E_n = -\frac{me^4}{8\varepsilon_0^2 h^2}\left(\frac{1}{n^2}\right)$$

$$= -\frac{Rhc}{n^2}. \qquad \left[\because R = \frac{me^4}{8\varepsilon_0^2 ch^3}\right]$$

For the ground (first) orbit, n = 1.

$$\therefore \quad E_1 = -Rhc$$

$$= -(1.097 \times 10^7 \text{ m}^{-1})(6.63 \times 10^{-34}\text{J})(3 \times 10^8 \text{m s}^{-1})$$

$$= -21.8 \times 10^{-19} \text{ J}$$

$$= -\frac{21.8\times 10^{-19}}{1.6\times 10^{-19}} = -13.6 \text{ eV}.$$

Hence the energy required to remove the electron from n = 1 orbit to infinity is 13.6 eV.

(ii) The Balmer series is represented by

$$\frac{1}{\lambda} = R\left(\frac{1}{2^2} - \frac{1}{n^2}\right), \; n = 3, 4, 5, \ldots$$

For the series limit n = ∞ so that

$$\frac{1}{\lambda} = \frac{R}{4}.$$

$$\therefore \quad \lambda = \frac{4}{R} = \frac{4}{1.097\times 10^7 \text{ m}^{-1}}$$

$$= 3.646 \times 10^{-7} \text{ m} = 3646 \text{ Å}.$$

Example 29:

The ground-state energy of an electron in the hydrogen atom is – 13.6 eV. Compute the energy of:

(i) n = 3 state in He ,

(ii) n = 2 state in Li .

Solution:

The energy of an electron in the nth state of a hydrogen like atom is given by

$$E_n = -\frac{m Z^2 e^4}{8 \varepsilon_0^2 h^2}\left(\frac{1}{n^2}\right)$$

$$= -\frac{R Z^2 h c}{h^2}. \qquad \left[\because R = \frac{me^4}{8\varepsilon_0^2 ch^3}\right].$$

For H, Z, = 1, so that

$$(E_n)_H = -\frac{Rhc}{n^2}.$$

When n = 1 (ground state) $(E_1)_H$ = – 13.6 eV (given).

$\therefore$ Rhc = 13.6 eV.

Hence $(E_2)_H = -\frac{Rhc}{4} = -3.4$ eV

and $(E_3)_H = -\frac{Rhc}{9} = -\frac{13.6}{9} = -1.5$ eV.

(i) For He^+, Z = 2, so that

$$(E_3)_{He} = -\frac{4Rhc}{9} = 4\,(E_2)_H = -6.0 \text{ eV}.$$

(ii) For Li^{++}, Z = 3, so that

$$(E_2)\,L_i = -\frac{9Rhc}{4} = 9\,(E_2)_H = -30.6 \text{ eV}.$$

Example 30:

Calculate the approximate wave number and wavelength of the fundamental absorption peak due to the stretching vibrations of a carbonyl group > C = O. The force constant for a double bond has an approximate *value of 1 × 10^6 dynes/cm. The masses of carbon and oxygen atoms are 2 × 10^{23} and 2.6 × 10^{23}g per atom.*

Solution:

Using equation (16), we get

$$\varpi = \frac{1}{2 \times 3.14 \times 3 \times 10^{10}} \sqrt{\frac{1 \times 10^6 \times (2.0 + 2.6) \times 10^{-23}}{2.0 \times 2.6 \times 10^{-46}}}$$

$$= 1.6 \times 10^3 \text{ cm}^{-1}, \; v = \frac{10^4}{1.6 \times 10^3} = 6.3 \, \mu$$

The carbonyl stretching bond has been found experimentally to be in the region of 5.3 to 6.7 μ or 1500 – 1900 cm^{-2}.

3

MAGNETOCHEMISTRY AND MAGNETIC PROPERTIES OF SUBSTANCES

MAGNETOCHEMISTRY

Magnetochemistry is the application of magnetic susceptibilities and of closely related quantities to the solution of chemical problems. The past few years have seen magnetochemistry to take its place along which dielectric constant, X-ray diffraction, and molecular and atomic spectra. Magnetochemistry is one of the most powerful techniques at the disposal of the chemist.

Magnetochemistry began with Michael Faraday more than one hundred twenty five years ago. It enjoyed a vigorous growth under the guidance of Piere Curie, A. Pascal, Van Vleck, Stoner, Klemm, S.S. Bhatnagar, Gilbert N. Lewis and many others.

MAGNETIC PROPERTIES OF SUBSTANCES

The magnetic properties of molecules have been found to be analogous to their electric properties in the following two ways :

(i) There are some molecules, which are having permanent magnetic dipole moments.

(ii) The effect of magnetic field is to induce a further contribution to the overall magnetic moment.

The analogous of the electric polarisation is magnetisation..

The magnetic properties of substances are of less general interest to the chemist than are the electric properties. Nevertheless, since the theory of magnetic properties is analogous in several ways to the electric properties, the important information about the molecular structure can be obtained by studying the magnetic behaviour of matter.

DEFINITIONS AND UNITS

Magnetic Permeability

The force F, acting between two magnetic dipoles of pole strengths m_1 and m_2, separated by a distance r, is given by

$$F = m^1m^2/\mu r^2 \quad ...(1)$$

where μ is a constant characteristic of the medium and is called *magnetic permeability* of the medium. Magnetic permeability is the magnetic counterpart of electric permittivity.

Magnetic Permeability (μ) of a medium represents the tendency of magnetic lines of force to pass through the medium relative to the tendency of the same lines of force to pass through the air or vacuum. It is possible to classify substances according to the values of μ. For vacuum, μ is taken as 1, while for other media μ may be greater or less than unity. If $\mu < 1$ the substance composing the medium has been said to be diamagnetic.

Thus, diamagnetic substances have been less permeable to the magnetic lines of force than vacuum and the lines of force prefer to pass through a vacuum rather than through the substance.

On the other hand, when $\mu > 1$, the substance has been said to be paramagnetic and the tendency for lines of force to pass through the substance has been greater than through a vacuum. The behaviour of diamagnetic and paramagnetic substances in magnetic field has been shown in the Fig. 3.1.

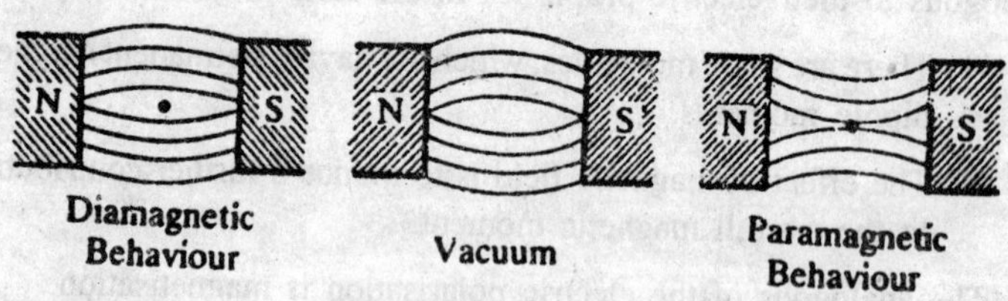

Fig. 3.1 : Behaviour of diamagnetic and paramagnetic substances.

Substances with very large values of μ *i.e.*, of the order of 10^3, have been said to be *ferromagnetic*. These include Fe, Co. Ni and their alloys.

Magnetic Susceptibility

The intensity of magnetisation (I) induced at any point in a body is directly proportional to the magnetising field (H), *i.e.*,

$$I \propto H$$

or $$I = KH \qquad ...(1)$$

where K is a proportionality constant and is known as the magnetic susceptibility of the medium. As, I the intensity of magnetisation is the magnetic moment per unit volume, K is therefore called the volume susceptibility which is defined as the ratio of intensity of magnetisation produced in the material to the magnetising force.

In some cases, it is useful to measure molar susceptibility which is defined as the susceptibility per gram of the substance, *i.e.*,

$$\chi = K/\rho \qquad ...(2)$$

where ρ is the density of the substance, χ is also known as *specific susceptibility* or *mass susceptibility*.

In some cases, it is useful to measure *molar susceptibility* which is defined as the susceptibility per gram mole, *i.e.*,

$$\chi_M =- X.M \qquad ...(3)$$

where χ_M is called molar susceptibility and M the molecular weight of the substance.

Units of Measurements

We have defined,

$$k = \frac{I}{H} = \frac{\text{Magnetic moment / volume}}{\text{Magneti sing field}}$$

$$= \frac{\text{Bohr Magneton}}{\text{Cm}^2 \text{ oersted}} = \text{B.M. cm}^{-2} \text{ oersted}^{-1}$$

and $$\chi = \frac{K}{\rho} = \frac{\text{volume Susceptibility}}{\text{density}}$$

$$= \frac{\text{B.M. cm}^{-3} \text{ oersted}^{-1}}{\text{gm cm}^{-3}} \text{B.M. gm}^{-1} \text{ oersted}^{-1}$$

The SI units of specific magnetic susceptibility and molar susceptibility are $cm^3\ g^{-1}$ and $m^3 \cdot mol^{-1}$ respectively.

Relation between Magnetic Susceptibility and Magnetic Moment

When a substance is kept in a magnetic field of strength H, the magnetic flux density B inside the medium is given by

$$B = H + 4\pi I \qquad ...(4)$$

where I represents the magnetic moment produced per unit volume of the substance. In SI system, the unit of B is kg s^2 A^{-1} called the tesla (T) and is equal to 10^4 gauss.

But
$$B = \pi H \qquad ...(5)$$

$$\mu H = H + 4\pi I$$

Dividing the above equation by H, we get

$$\mu = 1 + 4\pi \frac{I}{H} = 1 + 4\pi K \qquad ...(6)$$

In equations (4), (5) and (6) B measures the magnetic induction inside the material and μ is the permeability which is defined a the conducting power of the material for ;magnetic lines of force a compared with air.

We know from Eq. (2).

$$\chi = \frac{K}{\rho}$$

or
$$K = \chi.\rho \qquad ...(7)$$

where ρ is the density of the substance and χ is known as molar susceptibility or specific susceptibility.

On substituting Eq (7) into Eq. (6), we get

$$\mu = 1 + 4\ \pi\varepsilon\rho$$

or
$$\chi = (\mu - 1)\frac{1}{4\pi\rho} \qquad ...(8)$$

(i) In the case of diamagnetic substance, $\mu < 1$ and therefore χ is negative. Such substances are copper, graphite, antimony, bismuth, etc.

(ii) In the case of paramagnetic substances, $\mu > 1$ and therefore χ is positive. Such substances are aluminium, platinum, etc.

(iii) In the case of ferromagnetic solids, μ and χ are very high, their values are ranging between 200 and 1,00,000.

For a diamagnetic substance, χ has been found to be small, negative, independent of magnetic field intensity and also independent of temperature. For a paramagnetic substance, χ has been found to be large, positive, independent of magnetic field intensity and decreases with increase in temperature. For a ferromagnetic substance, χ has been found to be very large, positive and depends upon the magnetic field intensity, the temperature and the previous history of the sample. Then, there are the so-called antiferromagnetic substances for which χ has been found to be small and positive, shows hysteresis effects and possesses a transition temperature, called the *Neel point.*

Table 3.1 summarizes the magnitudes, signs and origins of the various forms of magnetic behaviour.

Table 3.1 : Types of Magnetic Behaviour and Their Characteristics

Types of magnetic susceptibility	*Sign*	*Approximate magnitude*	*Dependence on magnetic field intensity*	*Origin*
Diamagnetism	Negative	1×10^{-6}	Independent	Electronic Charge
Paramagnetism	Positive	$0 - 10^{-4}$	Independent	Angular momentum (electron spin)
Ferromagnetism	Positive	$10^{-2} - 10^{-4}$	Dependent	↑↑dipole exchange
Antiferromagnetism	Positive	$0 - 10^{-4}$	May depend	↑↓dipole exchange

Types of Substances

There are three types of substances :

(a) Paramagnetic Substances

These are the substances which are attracted by magnets and when placed in a magnetic field move from weaker to stronger parts of the field. Familiar examples are aluminium, manganese, oxygen, platinum, etc. Some characteristics of paramagnetic subsidences are :

(i) If a bar of paramagnetic substance is placed in a magnetic field, the lines of force tend to cumulate in it as shown in Fig. (3.2).

(ii) The susceptibility of a paramagnetic material is positive though it has a small value.

(iii) The susceptibility of a paramagnetic substance for a given magnetising force is inversely proportional to the absolute temperature and thus obeys Curie's law. If, therefore, the temperature is increased, the susceptibility decreases and at some higher temperatures it even becomes negative, *i.e.*, the substance becomes diamagnetic.

(b) Diamagnetic Substances

These are substances which are repelled by magnets and when placed in a magnetic field move from stronger to weaker part of the field. Familiar examples include bismuth, antimony, copper, water, alcohol and hydrogen. Some characteristics of diamagnetic substances are as follows :

(i) If a diamagnetic material is kept in a magnetic field, the lines of force tend to move away from the material (Fig. 3.2).

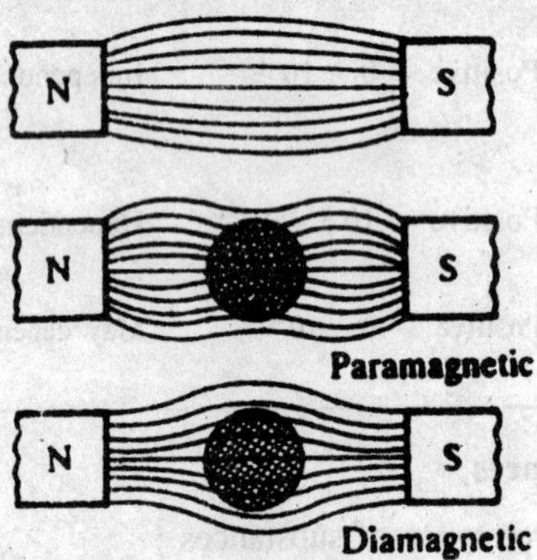

Fig. 3.2

(ii) The magnetic susceptibility of diamagnetic materials has a small value. For example, the magnetic susceptibility of bismuth is -1.4×10^{-6} e.m.u.

(iii) The magnetic susceptibility of diamagnetic substances does not change with the temperature.

(c) Ferromagnetic Substances

These are substances which are attracted by the magnets and can also be magnetised. Examples are iron, nickel, cobalt and their alloys. Some characteristics of these substances are :

(i) Ferromagnetic substances show all the properties of a paramagnetic substance to a much greater degree. The susceptibility has a positive value and the permeability is also very large.

(ii) Curie–Weiss law : The susceptibility decreases as the temperature increases and obeys Curie's law which is expressed as

$$K \propto \frac{1}{T} \quad \text{or} \quad KT = \text{constant.}$$

It is observed that as the temperature is increased, the value of susceptibility drops suddenly at a particular temperature called the *critical temperature* or *Curie point*, and the substance becomes paramagnetic. The Curie point for cobalt is 1100°C, for nickel 400°C and for steel 770°C. It is also observed that the susceptibility of a ferromagnetic substance above the Curie point is ;proportional to the difference of temperatures between the material and Curie point. This is known as *Curie-Weiss law*.

THEORIES OF PARAMAGNETISM

1. Langevin's Theory

The theory of paramagnetic substances given by Langevin (1905) has been derived on the basis of kinetic theory.

Derivation : Langevin assumed that the molecules, of the paramagnetic gas are assumed to be small permanent magnets due to rotating electrons. When a magnetic field is applied, the molecules will tend to orient themselves with their magnetic axes in the direction of the field. This tendency will, however, be opposed by the collisions due to thermal motion between the molecules. As a result, there will be an equilibrium distribution of the axes with reference to the direction of the field, which can be calculated by using the kinetic theory of gases. The principle used in this deviation are :

(i) The number of molecules, say dN, which have an orientation inclined at an angle θ with the direction of the field, H, is directly proportional to sin θ dθ , *i.e.*,

$$dN \propto \sin\theta\, d\theta \qquad ...(1)$$

(ii) In accordance with the principle of equipartition of energy, the number of molecules, dN, whose potential energy is E, is proportional to $e^{-E/kT}$, *i.e.*,

$$dN \propto e^{-E/kT} \text{ where k = Boltzmann's constant.} \qquad ...(2)$$

if μ is the molecular magnetic moment in the field of strength, H, the potential energy will be given by

$$E = -\mu H \cos\theta \qquad ...(3)$$

where θ is the angle which the axis of the magnet makes with the field direction.

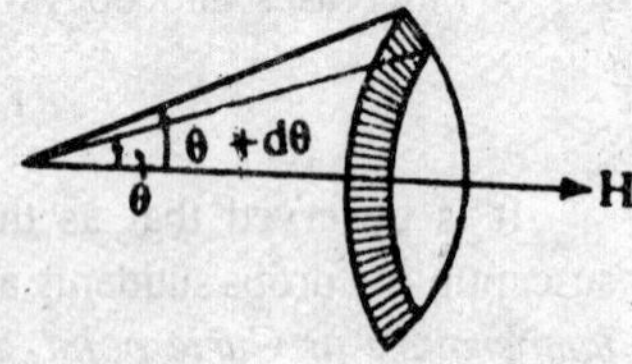

Fig. 3.3

On combining equations (1), (2) and (3), we get

$$dN = C \sin\theta\, e^{\mu H \cos\theta/kT} d\theta \qquad ...(4)$$

where C is a proportionality constant.

Since the possible orientations are contained between θ and π,

$$N = \int_0^{\pi} dN = C \int_0^{\pi} \sin\theta\, e^{\mu H \cos\theta/kT}\, d\theta$$

Put $\frac{\mu H}{kT} = \propto$ and $\cos\theta = x$ so that $-\sin\theta\, d\theta = dx$

or $$d\theta = -\frac{dx}{\sin\theta} \qquad ...(4A)$$

When θ = 0, x = + 1; π, x = – 1,

On inserting the above suppositions in eq. (4),

$$N = -C\int_{+1}^{-1} e^{\alpha x}\, dx = -C\left[\frac{e^{\alpha x}}{\alpha}\right]_{+1}^{-1} = \frac{C}{\alpha}\left[e^{\alpha} - e^{-\alpha}\right]$$

or $$C = \frac{N\alpha}{\left[e^{\alpha} - e^{-\alpha}\right]} \qquad ...(5)$$

As the resolved component of the magnetic moment of each one of the dN molecules in the field direction is μ cos θ, the resultant moment due to all the dN molecules in the field direction is μ cos θ dN. Hence, the total magnetic moment due to all the N molecules contained in a unit volume is as follows :

Total magnetic moment in unit volume

$$= \int_0^{\pi} \mu \cos\theta \, dN \qquad ...(6)$$

By definition, the intensity of magnetisation, I, is the magnetic moment per unit volume and, therefore, eq. (6) is

$$I = \int_0^{\pi} \mu \cos\theta \, dN = \int_0^{\pi} \mu \cos\theta \, C \sin\theta \, e^{\mu H \cos\theta / kT} \, d\theta \text{ [Use eq. (4)]}$$

Again applying the same suppositions,

$$I = -\mu C \int_{+1}^{-1} x \, e^{\alpha x} \, dx = \frac{\mu C}{\alpha} \left[\left\{ \frac{x \, e^{\alpha x}}{\alpha} \right\} \right]_{+1}^{-1} \frac{1 \, e^{\alpha x}}{\alpha} dx$$

$$= -\mu C \left[\frac{x \, e^{\alpha x}}{\alpha} - \frac{e^{\alpha x}}{\alpha} \right]_{+1}^{-1}$$

$$= \frac{\mu C}{\alpha} \left[\left(e^{\alpha} + e^{-\alpha} \right) - \frac{1}{\alpha} \left(e^{\alpha} - e^{-\alpha} \right) \right] \qquad ...(7)$$

Substituting eq. (5) in (7), we get

$$I = \frac{\mu C}{\alpha} \cdot \frac{N\alpha}{\left(e^{\alpha} - e^{-\alpha} \right)} \left[\left(e^{\alpha} + e^{-\alpha} \right) - \frac{1}{\alpha} \left(e^{\alpha} - e^{-\alpha} \right) \right]$$

or
$$I = \mu N \left[\frac{e^{\alpha} + e^{-\alpha}}{e^{\alpha} - e^{-\alpha}} - \frac{1}{\alpha} \right]$$

or
$$I = \mu N \left(\coth\alpha - \frac{1}{\alpha} \right) = I_o \left(\coth\alpha - \frac{1}{\alpha} \right)$$

$$= I_o \left(\frac{\alpha}{3} - \frac{\alpha^2}{45} + \frac{2\alpha^5}{945} - \ldots \ldots \right) = I_o \frac{\alpha}{3} = I_o \frac{\mu H}{3kT} \qquad ...(8)$$

As $\propto \leq I$, the higher powers in eq. (8) can be neglected. The quantity I_o is a characteristics of a paramagnetic gas and it always exceeds I. We will now calculate the expressions for paramagnetic susceptibility :

(i) *Volume susceptibility* $K = \frac{I}{H} = I_o \frac{\mu H}{3KTH} = \frac{I_o \mu}{3kT}$

or
$$K = \frac{\mu N \mu}{3kT} = \frac{\mu^2 N}{3kT} \qquad (\because I_o = \mu N) \qquad ...(9)$$

(ii) *Mass susceptibility* $x = \frac{K}{\rho} = \frac{\text{Volume susceptibility}}{\text{Density}}$

If N_m denotes the number of molecules in unit mass, the mass of one molecule is $1/N_m$, Also, the volume occupied by one molecule is 1/k where N is the Avogadro's number. Thus,

$$\rho = \frac{\text{Mass of a molecule}}{\text{Volume of molecule}} = \frac{1/N_m}{1/N} = \frac{N}{N_m} \quad ...(10)$$

$$\therefore \chi = \frac{K}{\rho} = \frac{KN_m}{N} = \frac{\mu^2 NN_m}{N.3kT} = \frac{\mu^2 N_m}{3kT} \quad ...(11)$$

(iii) Molecular susceptibility. If M is the molecular weight,

$$x_M = x.M = \frac{\mu^2 N_m}{3kT}.M = \frac{\mu^2 N}{3kT} \quad [\because N = N_m M]$$

$$= \frac{\mu^2 N^2}{3RT} = \frac{\sigma_o^2}{3R}\frac{1}{T} \quad \left[\because k = \frac{R}{N}\right] \quad ...(12)$$

where σ_o is the saturation value of the gram-molecular magnetic moment.

Consequences of Langevin's Theory

(i) *Curie's law :* According to Curie's law the susceptibility varies inversely with absolute temperature. This can be deduced from the langevin's theory by using equations (9) and (12).

$$K = \frac{\mu^2 N}{3kT} = \frac{\mu^2 N^2}{3RT} = \frac{C}{T} \quad [\because R = Nk] \quad ...(13)$$

or $K \alpha \frac{1}{T}$ which is Curie's law.

In equation (3), C is a constant replacing $\mu^2N^2/3R$. Again, equation (12) is

$$\chi_M = \frac{\sigma_o^2}{3R}.\frac{1}{T} = \frac{C_M}{T} \text{ or } \chi_M \alpha \frac{1}{T} \quad ...(14)$$

where C_M is a constant replacing $s_o^2/3R$, and known as molar Curie constant. Form equations (13) and (14), it is evident that volume susceptibility and gram-molecular susceptibility both vary inversely with the absolute temperature.

(ii) *Extensions of Langevin's theory* : Although, Langevins's theory strictly applies to gases, it can be extended to dilute solutions of paramagnetic salts which exist in the state analogous to than of gases, the ions the carriers of the magnetic moments moving in the solvent independently of each other, as experimental observations show.

Limitations of Langevin's Theory

(i) Although, this theory was quite successful to explain the behaviour of a paramagnetic gases and dilute solutions towards temperature yet it failed to explain more complicated dependence upon temperature shown by several compressed and called paramagnetic gases, concentrated solutions, crystals, etc.

(ii) It offers on explanation for the ferro-paramagnetic conversion at a certain critical temperature.

All the above complications were solved by Weiss, giving "Curie-Weiss law."

Weiss Molecular Field Theory of Paramagnetism

In Langevin, theory, an ideal gas is considered in which the mutual effect of the elementary magnets is negligible, so that the only opposing factor to the turning action of the external magnetic field is thermal agitation. But Weiss considered a real gas whose molecules are mutually influenced by their magnetic moments so that there exists an internal molecular field in a real gas produced at any point in it by all the molecules in the neigbourhood. The field produced is proportional to the intensity of magnetisation, *i.e.*,

$$H_i \propto I \text{ or } H_i = nI \qquad ...(15)$$

where n is called molecular field constant. Now the effective field, H, may be regarded as the sum of the external field H_e and internal molecular field H_i, *i.e.*,

$$H = He + Hi = He + nI \quad [\text{Use eq. (15)}] \qquad ...(16)$$

From equation (8), we have

$$I = I_0 \frac{\mu H}{3kT} = I_0 \frac{\mu (H_e + nI)}{3kT} \qquad ...(17)$$

Let us consider a gram-molecule of the substance. If σ is the density, M, the molecular weight, then σ and σ_o are the gram molecular magnetic moment and its saturation value respectively.

$$I = \frac{\sigma}{M/\rho} = \frac{\sigma\rho}{M} \text{ and } I_o = \frac{\sigma_0\rho}{M} \qquad ...(18)$$

$$\therefore \quad \frac{I}{I_o} = \frac{\sigma\rho}{M}. = \frac{M}{\sigma_0\rho} = \frac{\sigma}{\sigma_0} = \frac{\mu (H_e + nI)}{3kT} \qquad [\text{Use eq. (17)}]$$

Also $\mu = \sigma_o/N_m$ where N_m is the Avogadro's number, *i.e.*, the number of m molecules in a gram-molecule, the above reduces to

$$\frac{\sigma}{\sigma_o} = \frac{\sigma_o\,(H_e + nI)}{3N_m kT} = \frac{\sigma_o\left(H_e + \dfrac{n\sigma\rho}{M}\right)}{3RT}$$

[Use eq. (18) and $R = N_m k$]

or $$\sigma = \frac{\sigma_o^2}{3RT}\left(H_e + \frac{n\sigma\rho}{M}\right) \quad ...(19)$$

Also $$\chi_M = \frac{\sigma}{H_e} = \frac{\sigma_o^2}{3RTH_e}\left[H_e + \frac{n\sigma\rho}{M}\right] \quad \text{[Use eq. (19)}$$

or $$= \frac{\sigma_o^2}{3RT}\left[1 + \frac{n\sigma\rho}{MH_e}\right]\frac{\sigma_o^2}{3RT}\left[1 + \frac{n\rho}{M}\chi_M\right]$$

or $$\chi_M\left(T - \frac{n\rho\sigma_o^2}{3MR}\right) = \frac{\sigma_o^2}{3R} C_M \text{ (Curie molecular constant)}$$

or $$\chi_M(T - \theta) = C_M \text{ or } \chi_M = \frac{CM}{T - \theta} \quad ...(20)$$

where $$\theta = \frac{\sigma_o^2 n\rho}{3RM} \quad ...(21)$$

Equation (20) is known as Curie-Weiss law and θ is called temperature of Curie constant or Weiss constant. From equation (20), it follows that due to the existence of molecular field, the susceptibility varies inversely as the excess of temperature over a certain critical value θ, known as Curie point. The value of θ is positive in many ferromagnetic substances. The Cuire temperature (θ) corresponds to the point at which ferromagnetism becomes paramagnetism.

In case of paramagnetic substances, θ is usually small and may be positive, or negative, depending upon the experimental temperature.

From expression (20), it follows that when $T < \theta$, the paramagnetic substance is converted into diamagnetic.

Determination of Curie Point : Eq (20) may be put as

$$\frac{1}{\chi_M} = \frac{T - 0}{C_M}$$

or $$\frac{1}{\chi_M} = \frac{T}{C_M} - \frac{\theta}{C_M} \quad ...(22)$$

Equation (22) is of the form y = mx + c. Plot a graph between $1/c_M$ on the Y-axis and T on X-axis. A straight line will be obtained. If the line passes through the origin, the paramagnetic substance will obey the simple Curie's law, *i.e.*, Langevin's theory is valid. If the straight line does not pass through the origin, Curie-Weiss law is valid. The slope of this line is $1/C_M$ which consequently determines the molar constant. C_M. The intercept on the Y-axis gives q/C_M; the value of Curie point (q) can be determined if the value of C_M is substituted in the former.

Applications : For this, see the Applications of magnetic susceptibilities.

THEORIES OF DIAMAGNETISM

Langevin's Theory

In (1904-5) Langevin applied the electronic theory to explain the magnetic behaviour of the substances. According to this theory, the atom of a diamagnetic substance is such that the resultant magnetic moment of the atom as a whole is zero. When an external field is applied to such an atom, there occurs a uniform precession of electron orbits.

The precession does not alter the position of the system but simply displaces the positions of electron orbits relative to each other so that atom as a whole possesses a negative magnetic moment but not zero. This can be calculated as follows :

Let us consider an electron of charge e and mass m, moving in a circular orbit of radius r with u and ω the linear and angular velocities respectively. The revolving electron is kept in a circular orbit by the force of attraction F due to the central positive nucleus balanced against the centrifugal force. This yields,

$$F = \frac{mu^2}{r} = m\omega^2 r \qquad [\because u = r\omega] \qquad ...(1)$$

If a magnetic field H is now applied perpendicular to the plane of the orbit of this single electron, then an additional force, F_1 or H_{eu} or H_{erw} will act radially on the electron at right angles to both the orbit and the magnetic field. According to Lenz's law the direction of this force is opposite to that of the centrifugal force. Therefore, the net force, F', on the electron in the magnetic field is, *i.e.*,

$$F' + F - F_1 = m\omega^2 r - H_{erw} \qquad ...(2)$$

The effect of this force changes the angular velocity of the electron with out changing the radius of the orbit. Suppose the new angular velocity be $\omega + \Delta\omega$. Then the net force F' is given by

$$F' = m(\omega + \Delta\omega)^2 r = m(\omega^2 + \Delta\omega^2 + 2\omega\ \Delta\omega)r$$

$$= m\ (\omega^2 + 2\omega\ \Delta\omega)\ r\ [\ \Delta\omega^2 \text{ being negligible}] \quad ...(3)$$

From relations (2) and (3), w get

$$m\omega^2 r - H_{rew} = m\ (\omega^2 + 2\omega\Delta\omega)r$$

or $$\Delta\omega = -\frac{H_e}{2m} \quad ...(3A)$$

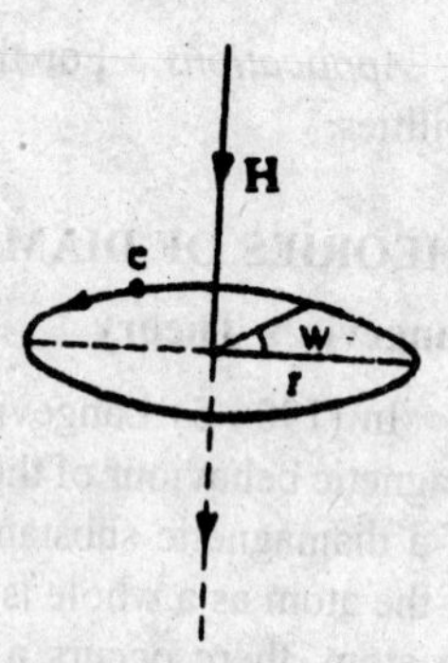

Fig. 3.4 : Orbit of a revolving electron.

The negative sign indicates that the change in angular velocity acts in a sense opposite to that of the electron. The change in angular velocity d gives rise to change in magnetic moment which may be derived as follows :

A change e revolving in a circle at the rate on n revolutions per second is equivalent to ne. Suppose a is the area of the circle.

By Ampere's theorem, the magnetic moment of the electron orbit is nea. As n $\left(= \frac{\omega}{2\pi}\right)$ is directly proportional to ω, the change in magnetic moment, $\Delta\mu$, due to a change $d\omega$ in angular velocity is given by

$$\Delta\mu = e.\ a.\ dn$$

$$= e.a.d.\left(\frac{\omega}{2\pi}\right) = \frac{ea}{2\pi} d\omega$$

$$= e\left(\pi r^2\right) \times \frac{d\omega}{2\pi} \qquad \left[\because\ a = \pi r^2\right] \quad ...(4)$$

Substituting eq. (3A) in (4), we get

$$\Delta\mu = -\frac{e\pi r^2\ He}{2\ (2\pi m)} = -\frac{H_e{}^2 r^2}{4m} \quad ...(5)$$

If the electronic orbit is not perpendicular to the applied field then r is replaced by r_1, the projection of the radius r of the orbit on the plane perpendicular to the magnetic field. Therefore, equation (5) becomes

$$\Delta\mu = -\frac{e^2 r_1^2 H}{4m} \quad ...(6)$$

If an atom of atomic number Z is considered, there are Z electronic orbits and these orbits may lie in all possible directions in space. Therefore, the total induced magnetic moment is given by

$$\sum \Delta\mu = \sum\left(-\frac{e^2 r_1^2 HZ}{4m}\right) = -\frac{e^2 HZ}{4m}\sum r_1^2 \quad ...(7)$$

If x, y, and z are the co-ordinates of the radius r of an orbit, then one can put it as

$$r^2 = x^2 + y^2 + z^2 \quad ...(8)$$

If z-axis is selected in the plane perpendicular to the applied field, then

$$r^2 = x^2 + y^2 \quad ...(9)$$

For a spherically symmetrical atom,

$$x^2 = y^2 = z^2 \quad ...(10)$$

Then, equations (8) and (9) become as

$$r^2 = x^2 + x^2 + x^2 = 3x^2 \quad ...(11)$$

and $\quad r_1^2 = x^2 + x^2 = 2x^2 = 2/3\ r^2 \quad$ [Using eq. (11)] ...(12)

Substituting the value of r_1^2 in eq. (7).

$$\sum \Delta\mu = -\frac{e^2 HZ}{4m}\sum \frac{2}{3} r^2 = -\frac{e^2 HZ}{6m}\sum r^2 \quad ...(13)$$

The magnetic moment per gram-atom

$$\sum \Delta\mu = -\frac{1}{6}\frac{Ne^2 ZH}{m}\sum r^2 \quad ...(14)$$

where N is the Avogadro's number. Now the magnetic moment per unit volume is defined as the intensity of magnetisation, *i.e.*,

$$1 = \frac{\sum \Delta\mu_A}{\text{Volume}} = \frac{\sum \Delta\mu_A}{A/\rho} = \frac{\rho \sum \Delta\mu_A}{A} = -\frac{\rho}{6A}\frac{Ne^2 ZH}{m}\sum r^2 \quad ...(15)$$

where ρ is the density of the substance and A its atomic weight. The volume susceptibility is given as

$$\therefore k = \frac{1}{H} = -\frac{\rho}{6A}\frac{Ne^2 ZH}{mH}\sum r^2 = -\frac{1}{6}\frac{\rho}{A}\frac{Ne^2 Z}{m_e}\sum r^2$$

Mass susceptibility,

$$\chi = \frac{K}{\rho} = -\frac{1}{6}\frac{\rho}{A}\frac{Ne^2Z}{m\rho}\sum r^2 = -\frac{1}{6}\frac{Ne^2Z}{mA}\sum r^2 \qquad ...(16)$$

It is usually convenient to deal with atomic susceptibility χ_A, *i.e.*, susceptibility which the substance would possess if one gram atom of it were contained in unit volume, *i.e.*,

$$\chi_A = \chi A$$

$$= -\frac{1}{6}\frac{Ne^2Z}{mA}.A\sum r^2 = -\frac{1}{6}\frac{NZe^2}{m}\sum r^2 \qquad ...(17)$$

Validity of the Langevin's Theory : The expression (17) for atomic susceptibility lends itself as a means of testing the validity of the theory ; rewriting eq. (17) as

$$\chi_A = -\frac{1}{6}\left(\frac{e}{m}\right)ZNe.R^2 \qquad \left[\because \sum r^2 = R^2\right] \qquad ...(18)$$

Now Ne = 9650 e.m.u. and e/m = 1.77×10^7 e.m.u. Substituting these values in (18), we get

$$R^2 = -6\times\frac{1}{1.77\times10^7\times9650}\times\frac{\chi_A}{Z} = -0.35\times10^{-10}\frac{\chi_A}{Z} \qquad ...(19)$$

The magnetic susceptibilities of some substances (χ_A) given in Table 3.2 are experimentally measured and substituted in eq. (19) to yield the values of R (Table 3.2).

Table 3.2

Elements	***Z***	χ_A	***R***
He	2	-1.9×10^{-6}	0.57×10^{-8} cm
N	7	-6.0×10^{-6}	0.55×10^{-8} cm
Cu	29	-18×10^{-6}	0.47×10^{-8} cm
Ag	47	-31×10^{-6}	0.48×10^{-8} cm

The atomic radii calculated by using eq. (19) are of the same order of magnitude (about 10^{-8} cm) as the value given by the kinetic theory. The fair agreement between the theoretical predictions and experimental results thus reveals that the theory of Langevin of diamagnetic substances is essentially correct.

It was P. Pascal who showed that diamagnetism in a molecule can be estimated from contributions of atoms and bonds within the molecule. Such diamagnetic contributions of atoms and bonds in a molecule are termed as pascal's constants. Pascal's constants for atoms and bonds have been included in Table 3.3.

Table 3.3

Atom of Bond	*Pascal's constant*	*Atom or Bond*	*Pascal's constant*
H	-2.9×10^{-6}	I	-44.6
C	-6.0	Si	-13.0
N (open chain)	-5.5	S	-15.1
N (ring)	-4.6	B	-7.0
N (amines)	-1.5	$C = C$	5.5
O (alcohol, ether)	-4.6	$C = C - C = C$	10.6
O (ketones)	-1.7	$N = N$	1.85
O(carboxy)	-3.3	$C = N$	8.2
F	-6.3	$C = N$	0.8
Cl	-20.1×10^{-5}	$C = C$	0.8
Br	-30.6	Benzene ring	1.4

MAGNETIC SUSCEPTIBILITY IS AN ADDITIVE AND SUBSTITUTIVE PROPERTY

P. Pascal has demonstrated that magnetic susceptibility is an additive and substitutive property and hence χ_M can also be calculated by using atomic and bond contributions (Table 3.3) which are supposed to apply to all molecules. This can be understood by taking an example of benzoic acid,

$$7C = 7(-6.00 \times 10^{-6}) = -42.00 \times 10^{-6}$$

$$6H = 6(-2.93 \times 10^{-6}) = -17.58 \times 10^{-6}$$

$$1O - = 1(-4.61 \times 10^{-6}) = -4.61 \times 10^{-6}$$

$$1O - = 1(-3.36 \times 10^{-6}) = -3.36 \times 10^{-6}$$

$$1 \text{ benezering} = 1(-1.4 \times 10^{-6}) \quad = -1.4 \times 10^{-6}$$

$$\chi_M = 68.95 \times 10^{-6}$$

The observed value cM = – 70.3 × 10^{-6} cm^3 mol^{-1}.

THEORY OF FERROMAGNETISM

Weiss' Theory of Ferromagnetism

In order to explain the special properties of ferromagnetism and also the relation between para and ferromagnetism, Weiss modified his molecular field theory by the introduction of an additional ideal of "domain" of molecules. He applied the same assumption, $H = H_e + nI$ as for paramagnetism. If the molecular field coefficient n is positive, the spontaneous magnetisation of these domains may occur. The value of such a spontaneous magnetisation due to the internal molecular field may be obtained by extrapolating the external field H to zero in the following relation which will, in turn, determine the condition for spontaneous magnetisation, *i.e.*,

$$H = H_e + nI \qquad ...(1)$$

Let us consider a gram-mole of a ferromagnetic substance of molecular weight M and density ρ. Suppose σ is the gram-molecular magnetic moment and σ_0 its saturation value. As the domains are assumed to obey the general theory of paramagnetism, it follows that

$$\frac{\sigma}{\sigma_0} = \coth \alpha - \frac{1}{\alpha} \qquad ...(2)$$

where $\propto = \mu H/kT$. If the external field H_e is zero, equation (1) becomes as

$$H = n\,I = n\frac{\sigma\rho}{M}$$

$$\therefore \quad \alpha = \frac{\mu H}{kT} = \frac{\mu n \sigma \rho}{kTM} = \frac{\sigma_0 n \sigma \rho}{N_m kTM} = \frac{\sigma_0 n \sigma \rho}{RTM}$$

This may be rewritten as

$$\frac{\sigma}{\sigma_0} = \frac{RTM}{\sigma_0^{\;2}\, n\, \rho}\,\alpha \qquad ...(3)$$

Equations (2) and (3), are two simultaneous equations which determine the conditions of spontaneous magnetisation. When curves re drawn between $\frac{\sigma}{\sigma_0}$ and ∝ corresponding to the two equations, the first gives the Langevin curve (I) while the second a straight line (II) passing through the origin whose slope equal to $RTM/\sigma_0^{\,2} n \rho$ increases with

T (Fig. 3.5). The two curves intersect at the origin and at A where $\propto = \propto'$. The slope of the curve II (RTM/ σ_o^2 n ρ) will certainly increase with the rise of temperature until a certain critical temperature is approached at which $\propto = 0$. (This temperature is known as Curie temperature). In such a situation, the curve II becomes a tangent to the Langevin's curve and the slope of the line is 1/3 (as shown in Langevin equation for paramagnetic susceptibility). This critical temperature is the temperature at with a ferromagnetic substance behaves as a paramagnetic. For temperatures lower than the critical temperature, the slope of the Weiss curve is given by

$$\frac{MRT3}{n\rho\sigma_o^2} < \frac{1}{3} \quad \text{or} \quad I\frac{n\rho\sigma_o^2}{3RM} < \theta$$

Hence, below the Curie point θ, in the absence of the external field, the domains are spontaneously magnetised, approaching the saturation value as the temperature approaches absolute zero. The value of θ is given by the intersection of the two curves. But at higher temperatures,

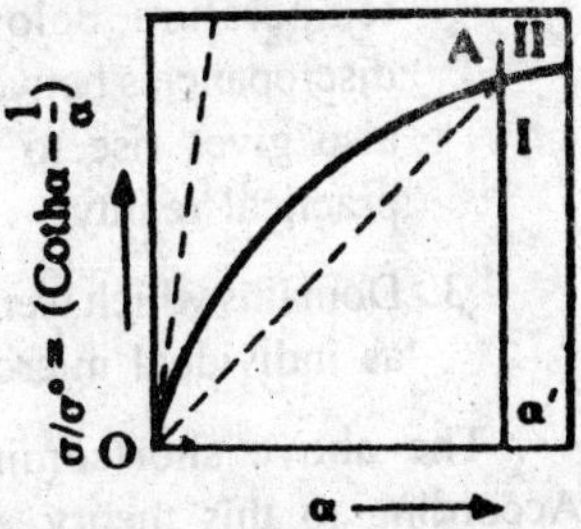

$$T > \frac{n\rho\sigma_o^2}{3RM} > \theta$$

Fig. 3.5

From the above relation it follows that, above the Curie point θ, the spontaneous magnetisation no longer occurs and the ferromagnetic properties disappear and the substance becomes paramagnetic. At temperatures, not too near the transition point, the Curie-Weiss law is obeyed. If $\propto$ is small at high temperature, one can put it as follows :

$$\frac{\sigma}{\sigma_o} = \frac{\alpha}{3} = \frac{\mu H}{3kT} = \frac{\mu(H_e + nI)}{3kT} = \frac{\sigma\left(H_e + \frac{n\sigma\rho}{M}\right)}{3RT}$$

The above relation modifies the following equation (in a similar manner as we have done in the theory of paramagnetism).

$$\chi_M = \frac{\sigma_o^2}{3R(T-\theta)} = \frac{C_M}{T-\theta}$$

From the above discussion, it follows that a substance having a permanent magnetic moment and a positive magnetic field coefficient will be a ferromagnetic substance below the Curie point whereas above it as a paramagnetic obeying Curie-Weiss law.

Criticism of Weiss Theory of Ferromagnetism

1. The classical theory adopted by Weiss fails to explain why and how internal fields between the molecules of a ferromagnetic materials can have such large values. Furthermore it does not throw any light on the nature of molecular field postulated by Weiss.

2. This theory fails to explain the breakdown of linear relationship expressed by the Curie-Weiss law close to the Curie point. Further, when the creation of intensity of magnetisation with temperature below Curie point was studied, there were discrepancies between theoretical and experimental curves. This also gives rise to anther disagreement between theory and a practical reality.

3. Domains which were postulated by Weiss could not be identified as individual microcrystals or like granules.

The above shortcomings were overcome by quantum theory. According to this theory, magnetic moment cannot have continuous values but only some discrete permissible values. Domain properties have been explained on the basis that magnetic properties are determined by the groups of the atoms which do not play any important role in determining the crystal structure. Further, Heisenberg explained the nature and origin of intramolecular fields in a satisfactory manner.

MEASUREMENT OF MAGNETIC SUSCEPTIBILITY

If a body is suspended in a magnetic field, it is under the influence of two forces :

(i) The forces acting on the body due to the permanent magnetisation.

(ii) The force acting on the body due to the induced magnetisation.

Suppose there are three principal axes x, y, and z. If we are considering the force only in x direction, the mechanical force F_x acting on a unit volume is given by Jean's equation :

$$F_x = -d\frac{\partial V'}{\partial x} - \frac{H^2}{8\pi}\frac{\partial \mu}{\partial x} + \frac{\partial}{\partial x} + \left(\frac{H^2}{8\pi}.\rho\frac{\partial \mu}{\partial \rho}\right) \qquad ...(1)$$

where d denotes the density of the magnetic force, V' the magnetic potential, ρ the density of the magnetic fluid, and μ and H the permeability and strength of field respectively.

If we are considering a body which does not have permanent magnetism, the term in equation (1) becomes zero, *i.e.*,

$$-d\frac{\partial V'}{\partial x} = 0 \qquad ...(2)$$

Therefore, equation (1) modifies to

$$F_x = -\frac{H^2}{8\pi}\frac{\partial \mu}{\partial x} + \frac{\partial}{\partial x}\left(\frac{H^2}{8\pi}\rho \cdot \frac{\partial \mu}{\partial \rho}\right) \qquad ...(3)$$

If we are considering the medium of the body to be isotropic, then

$$\frac{\partial \mu}{\partial x} = 0 \qquad ...(4)$$

Thus, equation (3) modifies to

$$F_x = \frac{\partial}{\partial x}\left(\frac{H^2}{8\pi}\rho\frac{\partial \mu}{\partial \rho}\right) \qquad ...(5)$$

We know $\mu - 1 = C\rho$ where C is a constant.

or
$$\frac{\partial \mu}{\partial \rho} = C \qquad ...(7)$$

Substituting equations (6) and (7) in (5), we get

$$F_x = \frac{\partial}{\partial x}\left(\frac{H^2}{8\pi}\frac{\mu - 1}{C} . C\right) = \frac{\mu - 1}{8\pi}\frac{\partial H^2}{\partial x} \qquad ...(8)$$

Similarly, we can write equations for the forces along y-and z-axes, *i.e.*,

$$F_y = \frac{\mu - 1}{8\pi}\frac{\partial H^2}{\partial y} \qquad ...(9)$$

$$F_z = \frac{\mu - 1}{8\pi}\frac{\partial H^2}{\partial z} \qquad ...(10)$$

In most of the instruments used for measuring magnetic susceptibilities, the field, (H_y) is considered along y-axis whereas its gradient is considered along the x – direction, keeping the field along y-and z-axes uniform. Thus, equations (9) and (10) reduce to zero and, therefore,

$$F_x = \frac{\mu - 1}{8\pi} \frac{\partial H_y^2}{\partial x} \quad ...(11)$$

Suppose we consider a body of small volume dV of length dx. Suppose μ_1 and μ_2 are the permeability of the body and surrounding medium respectively. Thus, equation (11) becomes

$$dF_x = \frac{\mu_1 - \mu_2}{8\pi} \frac{\partial H_y^2}{\partial x} \partial V \quad ...(12)$$

$\because$ $\mu_1 = 1 + 4\pi k_1$ ...(13)

and $\mu_2 = 1 + 4\pi k_2$ ...(14)

where k_1 and k_2 are the volume susceptibilities of the specimen and surrounding medium. On substituting equations (13 and (14) in (12), we get

$$dF_x = \frac{4\pi (k_1 - k_2)}{8\pi} dV \frac{\partial H_y^2}{\partial x}$$

or $$dF_x = \frac{1}{2} (k_1 - k_2) dV \frac{\partial H_y^2}{\partial x} \quad ...(15)$$

or $$F_x = (k_1 - k_2) dV H_y \frac{\partial H_y}{\partial x} \quad ...(16)$$

From the above treatment it follows that the methods of measurement of magnetic susceptibilities can be grouped under two main heads :

(a) Non-uniform field methods.

(b) Uniform field methods.

Most of the balances used fro measurement of magnetic susceptibilities are of uniform field type. Examples are Gouy method, Bhatnagar-Mathur Method, Qunicke's method, etc.

The methods using non-uniform magnetic field are Curie's balance, Curie–Cheneveau's balance, etc.

We will now discuss some of these balances.

The Gouy Method

The method, devised by a French scientist Gouy in 1899, is one of the simple and important methods and is used for the measurement of magnetic susceptibilities of various types of substances. Usually 0.5 –

1g of the samples in the form of powdered solids, liquids and of solutions of moderate concentrations are used.

Principle : The basic principle underlying this method is based upon the fact that the magnetic field exerted on the sample, when placed in the magnetic field varies directly as its mass (*i.e.*, the difference in the masses when the field is switched off and on)

Apparatus : The arrangement of the Gouy's apparatus is shown diagrammatically in Fig (3.6). The substance in taken in a long cylindrical tube which is suspended from one arm of the microbalance in such a way that its lower end lies in a strong magnetic field whereas the upper end is in the region of negligible magnetic field.

The axis of the cylindrical tube is at right angle to the magnetic field. In order to obtain maximum uniformity of the magnetic field, the two pole pieces of the magnet marked N and S are placed very close to each other. For many types of investigations it is convenient to use a magnetic field of 5000 to 15000 oersteds.

Theory : If a cylindrical sample of matter is suspended between the poles of a magnet so that one end of the sample is in a region of field intensity and the other end in a region of negligible field, then a sample will experience a force along its length. The magnitude of the vertical magnetic force acting on the specimen of length dx and volume dV is given by

$$dF_x = \frac{1}{2}(k_1 - k_2)\, dV \frac{\partial {H_y}^2}{\partial x} \quad ...(1)$$

where k_1 and k_2 are the volume susceptibilities of sample and surrounding atmosphere respectively and H_y is the maximum field to which the sample is adjusted. If A the area of cross-section of the specimen, it can be written as,

$$dV = A\, dx \quad ...(2)$$

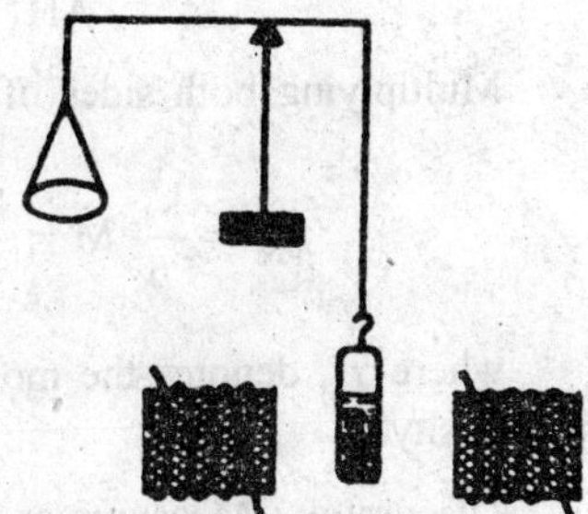

Fig. 3.6

Substituting equation (20 in (1), we get

$$dF_x = \frac{1}{2}(k_1 - k_2)\, A \frac{\partial {H_y}^2}{\partial x} dx \quad ...(3)$$

The total vertical force acting on the sample can be obtained by integrating the equation (3) within the suitable limits of the field . Thus, we can write

$$\int dF_x = \frac{1}{2}(k_1 - k_2)\, A \int_{H_0}^{H} \frac{\partial H_y^{\ 2}}{\partial x} dx$$

or $$F_x = 1/2\ (k_1 - k_2)\ A\ (H^2 - H_o^{\ 2}) \qquad ...(4)$$

As the field H_0 at the upper end of the specimen is quite negligible, it can be neglected; thus equation (4) becomes as

$$F_x = 1/2\ (k_1 - k_2)\ AH^2 \qquad ...(5)$$

The vertical force F_x can be calculated by the microbalance. The specimen is weighed first when the field is off and when the field is on. Thus,

$$F_x = \Delta\ mg \qquad ...(6)$$

where g is the gravitational constant and Δm the apparent change in mass of the sample on the application of the magnetic field. On combining equation (5) and (6), we get

$$\Delta\ mg = 1/2\ (k_1 - k_2)\ AH^2$$

or $$k_1 - k_2 = \frac{2\Delta m.g}{AH^2}$$

or $$k_1 = k_2 + \frac{2\Delta m.g}{AH^2} \qquad ...(7)$$

Multiplying both sides of equation (7) with M/ρ, we get

$$\chi_M = \frac{k_1}{\rho} M = \frac{M}{\rho}\left(k_2 + \frac{2\Delta m.g}{AH^2}\right) \qquad ...(8)$$

where χ_M denotes the molar susceptibility of the specimen and ρ its density.

Discussion : Measurements on metals or alloys are very simple by the Gouy method. The sample has only to be cast or machined into the cylindrical shape. The magnetic susceptibilities of powdered samples may be measured by packing them to cylindrical glass sample tubes. Correction should be made for the susceptibility of the glass, which is generally diamagnetic with a temperature coefficient.

The susceptibility of pure liquids is also conveniently measured in glass sample tubes. As the difficulty of packing does not arise with liquids, the accuracy may be considerably greater.

Limitation : The Gouy method is not suitable for investigation of gases although rough measurements on oxygen and on other paramagnetic gases and vapours have been made.

At temperatures 400 to 500°C, depending upon the apparatus, the method is not reliable because of convection currents.

Errors : Errors arise due to (a) amount of ferromagnetic impurity present in the specimen (b) lack of complete uniformity in the field especially just outside the pole pieces.

It is to be remembered that the spacemen should never by of rectangular cross-section because it would touch the pole pieces at once if it is slightly disturbed.

Bhatnagar-Mathur Method

In 1928, Bhatnagar and Mathur modified the Gouy balance in such a way that it could be used for the measurement of susceptibilities of, particularly, liquids (para-and diamagnetic).

Principle : The basic principle underlying the balance, is that the force acting on the sample varies directly proportional to the magnetic pull, producing a displacement of the pointer when the magnetic current is started.

Apparatus : The balance due to Bhatnagar-Mathur is depicted in Fig. (3.7). S' is a fine silver spiral which is suspended from a brass hook at the top, A is a specimen whose susceptibility is to be measured. It is suspended from the lower end of the spiral. The lower end of the specimen lies in the region of strong and uniform magnetic field, P is a very thin pointer, attached to the spiral. The displacement of P is measured by means of a microscope provided with a scale.

Theory : When a specimen is Put in between the two poles of a strong magnet as shown in Fig. (3.7), the vertical magnetic pull on the specimen is given by

$$F_x = \frac{1}{2}(k_1 - k_2)\, AH^2 \qquad ...(9)$$

In this balance, the strong magnetic field pulls the specimen which in turn pulls the spring, producing a displacement of the pointer. Reading is recorded with and without the application of magnetic field. Thus, the force F_x is defined as

$$F_x = \frac{1}{s} g \qquad ...(10)$$

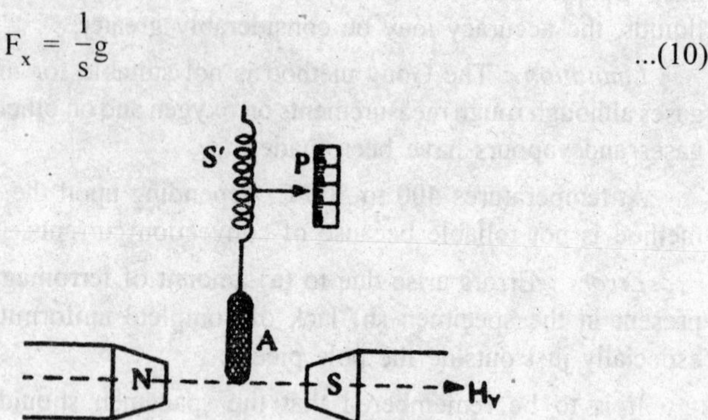

Fig. 3.7

Where l is the displacement of the pointer and s the displacement of pointer produced by 1 gm of the substance. From equations (9) and (10), we get,

$$\frac{1}{2}(k_1 - k_2) AH^2 = \frac{1}{s} g$$

or $$k_1 = k_2 + \frac{2lg}{sAH^2}$$

where k_1 and k_2 are the volume susceptibilities of the specimen and the surrounding medium.

The Quincke's Method

This method is given by G. Quincke in 1885 and strictly suitable for liquids, aqueous solutions and with some modifications for gases.

Principle : The underlying principle of this method is the same as that employed in Gouy's method except that the force acting on the liquid sample is measured interns of the hydrostatic pressure developed when the liquid is placed in a capillary tube so that the meniscus stands in a strong and uniform field. When field is applied, the meniscus will fall if the liquid is diamagnetic or will rise if the liquid is paramagnetic.

Apparatus : The apparatus used in the Quincke's method is shown diagrammatically in Fig. (3.8). This apparatus makes use of a specially designed modified U tube (ABC'CD). The limb C'D is a capillary tube which is placed in a uniform magnetic field of strength 25,000 oersted. The field near the wider end of the U tube (end C') is 50 – 100 oersted which is quite negligible. When the field is applied, the liquid meniscus rises or falls in the capillary but C' D if the liquid sample is paramagnetic or diamagnetic in nature respectively.

In most of the cases, the rise or fall of the meniscus is observed directly. Accuracy of the readings may by increased slightly by including the capillary.

Theory : As soon as the current is switched on , a strong and uniform magnetic field is established at the upper surface of the capillary tube whereas the lower surface of the capillary experiences a weak field. If the liquid is paramagnetic, the meniscus will rise whereas a fall will be observed if the liquid is a diamagnetic liquid. Suppose, Δh is the change in the height of the meniscus. Then, the corresponding hydrostatic pressure will be given by

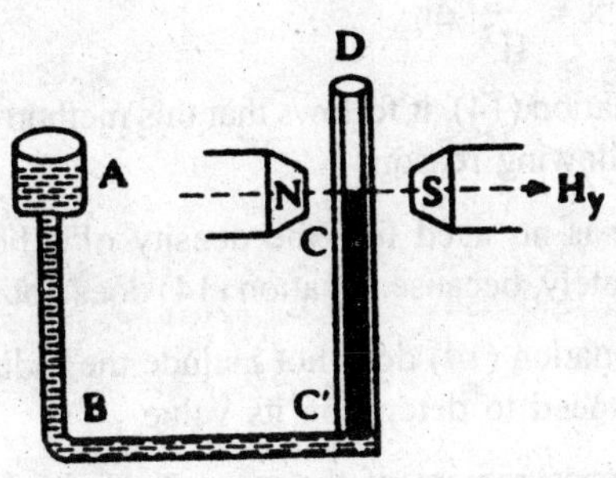

Fig. 3.8

$$p = \rho.g.\Delta h \qquad ...(11)$$

But force = pressure × area

$$\therefore \quad \text{force} = (\rho.g.\Delta h) \times A \qquad ...(12)$$

where A is the area of cross section of the tube. Equation (12) represents the vertical force ΔF_x. *i.e.*,

$$\Delta F_x = (\rho.g.\Delta h) \times A$$

The vertical force due to change in the hydrostatic pressure will be balanced by magnetic pull given by equation (5). *i.e.*,

$$(\rho.g.\Delta h) \times A = \frac{1}{2}(k_1 - k_2) AH^2$$

$$\text{or} \qquad K_1 - k_2 = \frac{2\rho g\, \Delta h}{H^2}$$

$$\text{or} \qquad \frac{k_1}{\rho} - \frac{k_2}{\rho} = \frac{2g\, \Delta h}{H^2}$$

$$\text{or} \qquad \frac{k_1}{\rho} - \frac{k_2}{\rho_o} - \frac{\rho_o}{\rho} = \frac{2g\, \Delta h}{H^2}$$

$$\text{or} \qquad \chi - \chi_o = \frac{2g\, \Delta h}{H^2} \qquad \left[\because \chi = \frac{k}{\rho}\right] \qquad ...(13)$$

where χ and χ_o are the mass susceptibilities of the liquid (or density ρ) and gas or vapour (of density ρ_o) above the liquid respectively. If a wider tube (AB) is of sufficient larger diameter, and the susceptibility (χ_o) of the vapour above the liquid meniscus is negligible, equation (13) modifies to

$$x = \frac{2g}{H^2}\, \Delta h \qquad ...(14)$$

From equation (14), it follows that this method has some advantages due to the following reasons :

(i) There is no need for the density of a liquid to be measured separately, because equation (14) does not include density term.

(ii) As equation (14) does not include the radius of capillary, there is no need to determine its value.

(iii) The measurement of the magnetic field can be avoided if the magnetic susceptibilities are measured for a sample and a reference substance under identical conditions, *i.e.*,

$$\frac{\chi_s}{\chi_r} = \frac{(\Delta h)_s}{(\Delta h)_r} \qquad ...(15)$$

where s and r are used as subscripts for a liquid (whose magnetic susceptibility is to be measured) and reference liquid (whose magnetic susceptibility is known). From equation (15), it is also evident that the magnetic susceptibility of a liquid can be measured by simply noting down the change in the height of the liquid meniscus under strong and uniform magnetic field.

Applications : The Quincke's method has been found to be very useful for the determination of magnetic susceptibilities of the following types of substances :

(i) *For solutions :* Bauer-Piccard applied the Quincke's method for the determination of the susceptibilities of solutions and deduced the following equation

$$\chi_{sol} = \frac{2g\,\Delta h}{H^2} + \chi_o \frac{\rho_o}{\rho_{sel}} \quad ...(16)$$

where χ_o and ρ_o denote the susceptibility and density of vapours above the solution respectively. The magnetic susceptibility of the solution can also be determined by applying the following relation :

$$\chi_{sol} = f_s\chi_s + (1 - f_s)\,\chi_w \quad ...(17)$$

where χ_{sol}, χ_s, χ_w are the susceptibilities of the solution, dissolved salt and solvent water respectively, and fs is the mole fraction of the solute.

(ii) *For gases :* The quincke's method may also be used for gases. If the susceptibility of the vapour over the meniscus is not negligible, the hydrostatic pressure (p) developed on the application of the field is

$$p = 1/2\,(k - k_o)\,H^2$$

where k and k_o are the volume susceptibilities of liquid and vapour respectively.

Precautions : In order to avoid the surface tension changes, the bore of the narrow limb must be strictly uniform. Further, the bore should not be very narrow otherwise on applying field, the surface of the liquid will get deformed and there will be no rise and fall of liquid column.

Errors : Errors arise due to :

(a) lack of uniformity of the field, H.

(b) lack of uniformity of narrow limb bore which gives rise to different surface tensions when the field is on and then when off.

(c) initial field, when the energising current is off, may now be zero for which a correction should be applied.

(d) deposition of solute on the walls of the narrow tube during the experiment when using solutions of salts.

(e) lack of cleanliness of glass.

Curie's Balance

The principle of this method is based on the non-homogeneous field method. The Curie's balance is a simple and useful magnetic balance, devised by Curie in 1895. Curie's balance had been used to study the variation of susceptibility with temperature, giving a clear distinction among para-, dia- and ferr-magnetic substances.

The principle underlying the Curie's balance is the same as that employed by Faraday. According to this principle, if a specimen is kept in a non-uniform magnetic field, a force is exerted on the specimen giving it a displacement which is measured in terms of torque required to bring the specimen back to the original position.

Let us consider the small volume, dV, of the specimen. The force exerted dF_x on the sample is given by equation :

$$dF_x = (k_1 - k_2)\, dV H_y \frac{\partial H_y}{\partial x} \quad \text{...(1)}$$

As the field gradient lies along the x-axis, the quantities $H_x \frac{\partial H}{\partial x}$ and $H_z \frac{\partial H_x}{\partial x}$ are neglected. The total force exerted on the specimen may be integrating equation (1). Thus,

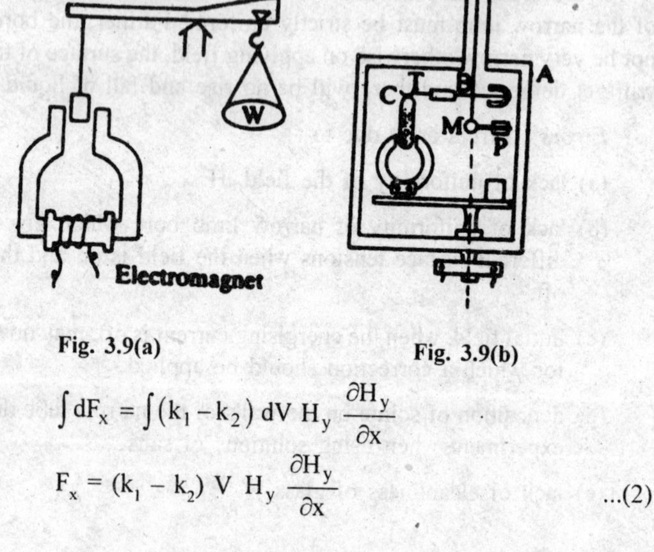

Fig. 3.9(a) Fig. 3.9(b)

$$\int dF_x = \int (k_1 - k_2)\, dV\, H_y \frac{\partial H_y}{\partial x}$$

or

$$F_x = (k_1 - k_2)\, V\, H_y \frac{\partial H_y}{\partial x} \quad \text{...(2)}$$

But $W = V\rho$ ...(3)

and $\chi_1 = \dfrac{k_1}{\rho}$ and $\chi_2 = \dfrac{k_2}{\rho}$ ...(4)

Substituting equations (3) and (4) in (2), we get

$$F_x = (\chi_1 - \chi_2)\ WH_y \frac{\partial H_y}{\partial x} \quad ...(5)$$

where W is the weight of the sample. From equation (5), it follows that also the quantities except $\dfrac{\partial H_y}{\partial x}$ can be measured very easily. In order to measure $\dfrac{\partial H_y}{dx}$, a test coil and a ballistic galvanometer is employed. In the Curie's method, the non-uniform field can be generated by inclining the two pole-pieces of an electromagnet at a certain angle. The Curie balance using an electromagnet is shown in Fig. (3.9a).

The Curie method has the advantage that only a small sample is required and that the specimen can be of irregular shape. The only disadvantage of this method is that $\dfrac{\partial H_y}{\partial x}$ cannot be determined very accurately at a point and, hence, the precision is usually not better than a few per cent. The method is of great sensitivity if the sample is held on a horizontal arm which is suspended from a torsion fibre.

In 1930, Cheneveau improved the Curie's balance by using a permanent magnet of the shape shown in Fig. (3.9b). It consists of a torsion arm TA suspended by a fine wire from the torsion head J. One end of the torsion arm (C) supports the sample which is free to move between the poles of a small permanent magnet . The magnet may be moved forward or backward with respect to the sample, and , because of its magnetic susceptibility the sample is either repelled or attracted. Movement of the torsion arm may be followed by a pointer or by a mirror (M), and lamp and scale. The magnet is slowly turned so as to recede from the specimen till the deflection reaches a maximum. Readings for the maximum deflection are recorded for both sides of zero. For calibration, distilled water is utilised as a standard substance.

First of all the observations are recorded fro the empty glass tube and then with the substance in it. The susceptibility is calculated from the following equation :

$$\frac{k}{k_1} = \frac{m_1}{m} \frac{x \pm x_2}{x_1 \pm x_2}$$

where k_1 and m_1 are the susceptibility and mass respectively of known substance of known susceptibility, m is the mass of them sample of unknown susceptibility k, x, and x_1 are the deflections for the unknown substance and reference substance respectively, and x_2 the deflection for empty tube. The positive sign (+) for x_2 is to be utilised if the empty tube possesses a susceptibility opposite to the test substance.

The Curie-Cheneveau balance is of most use in magnetochemical analyses such as are required in rare-earths work Useful elaborations of the Curie-Cheneveau balance are described by Wilson, Oxley and Vaidyanathan.

Advantage : The method is suitable for studying the substances in all three states, *i.e.*, solid, liquid and gas. Furthermore, it provides a wide range temperature study of susceptibilities extending from 25° to 137° C.

Errors : The errors arise if :

(a) The value of $H_y \dfrac{\partial H_y}{\partial x}$ is not constant over at least 1 cm region to maintain the specimen to a standard position.

(b) The replacement is not precise due to the three dimensional freedom to a specimen.

APPLICATIONS OF MAGNETIC SUSCEPTIBILITIES

Paramagnetism in Atoms, Ions and Molecules

Single, uncoupled electrons behave in a magnetic field as if each were a magnet. This is due to the spin of the electron. The magnetic moment due to the spinning electron is called the *spin moment*. The coupled electrons have equal and opposite spins and the resultant spin moment due to them is zero.

An atom, ion or molecule in which all the electrons are coupled and the resultant spin moment is zero, is not attracted to the magnetic field and is said to be *diamagnetic.*

An atom, ion or molecule which has one on more uncoupled electrons and has a resultant spin moment, is attracted into a magnetic field and is said to be paramagnetic. The spin moment is directly proportional to the number of uncoupled electrons. The molar magnetic susceptibility is related to the effective magnetic moment by

$$\mu_{eff} = \left(\frac{3k\ \chi_M\ T}{N\beta^2} \right)^{\frac{1}{2}} \quad ...(1)$$

where β is the Bohr's magneton and k, the Boltzmann's constant.

The magnetic moment on the basis of classical theory may be calculated from

$$\mu_{eff} = g\ \sqrt{J\ (J + 1)} \quad ...(2)$$

where g is the Land splitting factor and J is the resultant angular momentum which is given by

$$J = L + S \quad ...(3)$$

where L is the total angular momentum of the orbital motion of the electron and S, the corresponding spin angular momentum. From equation (30 it follows that both the orbital motions of the electron and also the spinning of the electron about its own axis should contribute to the magnetic moment of an atom or ion or molecule. But in a polyatomic system the orbital motions are so rigidly oriented with respect to nuclei that the applied magnetic field cannot affect the total angular momentum. This is also true in case of monatomic species in liquid or solid states in which the changes in the orbital motion due to an external magnetic field are prevented by the influence of solvent molecules or their m mutual cohesion.

So in most cases (excepting the rare-earth ions) the orbital angular moment is not operating and one has to consider the contribution from the spinning of the electron. Thus, equation (3) becomes as

$$J = S \quad ...(4)$$

Substituting equation (4) in (2), we get

$$\mu_{eff} = g\sqrt{S\ (S + 1)} \quad ...(5)$$

The value of g is 2 thus.,

$$\mu_{eff} = 2\ \sqrt{S\ (S + 1)} = \sqrt{4S\ (S + 1)} = \sqrt{2S\ (2S\ (2S + 2)} \quad ...(6)$$

But S, the total spin moment, is equal to the total number, n, of the unpaired electrons multiplied by s, the spin quantum number, *i.e.*,

$$S = ns = 1/2 \qquad [\because s = 1/2]$$

Thus, equation (6) becomes

$$\mu_{eff} = \sqrt{2} \times 1/2n\ (2 \times 1/2n + 2) = \sqrt{n}\ (n + 2) \quad ...(7)$$

In equation (7), n represents the number of unpaired electrons. From equations (1) and (2), we have

$$\sqrt{n\,(n+2)} = \left(\frac{3k\ \chi_M\ T}{N\beta^2}\right)^{\frac{1}{2}}$$

or $$\chi_M = \frac{\beta^2 N}{3kT} n\,(n+2) \quad ...(8)$$

From equation (8), one can determine the number of unpaired electrons. This is evident from the Table 3.4.

Table 3.4

No. of unpaired electrons	*S*	μ_{eff} *in Bohr*	$\chi_m = \beta^2 N/3kT\ n(n+p)$
1	1/2	1.732	1260×10^{-6} cm³ mol⁻¹
2	2/2	2.828	3360
3	3/2	3.873	6293
4	4/2	4.899	10100
5	5/2	5.916	14700
6	6/2	6.928	20120

Evidently, determination of molar magnetic susceptibility will lead to the knowledge of the number of unpaired electrons present in the molecule or the ion concerned. Following are the examples which illustrate this point :

(i) The molar susceptibility of oxygen at ordinary temperature has been found to be 3360×10^{-6}, *i.e.*, a moment of about 2.8 Bohr magnetons . Thus, in oxygen molecule there will be two unpaired electrons and the bonding will be represented by

$$:\dot{O} : \dot{O}:$$

(ii) The mercurous chloride is purely diamagnetic and does not show paramagnetic susceptibility. This confirms that the molecule must be Hg_2Cl_2 and not HgCl, since the latter would have made it paramagnetic.

(iii) In $[Mn(CN)_6]^{3-}$, there should be two unpaired 3d elections and the magnetic moment should be 2.83 in this case. But the observed value of magnetic moment is 3.0 Bohr magnetons which is in good agreement with the above result, confirming that there are two unpaired electrons.

(iv) In $[Cr(CN)_6]^{3-}$, there should be two unpaired 3d elections. But the observed magnetic moment is 3.0 Bohr magnetons which confirms that there are two unpaired electrons.

Magnetic Titrations

A chemical reaction in which the magnetic moment of certain atoms or ions changes appreciably may be followed by means of magnetic measurements. In this way it becomes possible in many cases to known whether the reaction has gone to completion or not. In some other cases it is also possible to know that what quality of one of the reacting substances was initially present. Conversely, if the amounts of reacting materials are known, it is possible to conclude what the reaction has been through knowing the point at which the reaction is complete. An illustration of the latter procedure is shown in Fig. (3.10) where the change in susceptibility of a solution of ferrihaemoglobin after various additions of KCN is plotted. From these data it was concluded that the reaction involved one cyanide per haeme. Some other examples of magnetic titrations are :

(i) When the paramagnetic hexammine nickel nitrate is titrated which postassium cyanide, the ultimate product is diamagnetic $Ni(CN_4)^{2-}$ with the formation of no intermediate products.

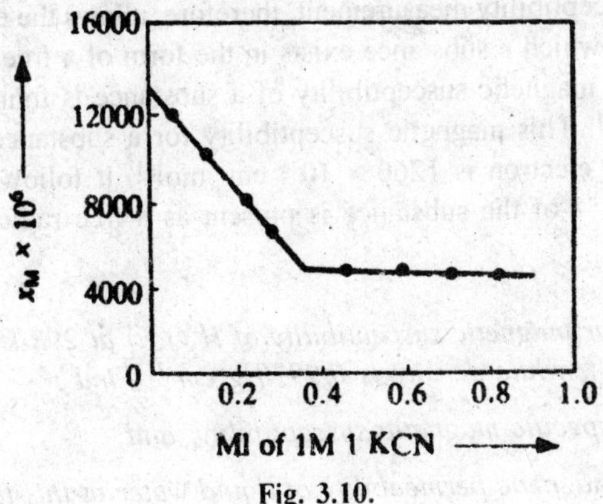

Fig. 3.10.

(ii) When Fe^{2+} ions are titrated with bipyridine, mono –, c is – and tris-complexes are formed which are diamagnetic but the inter mediate complexes are paramagnetic.

Oxygen in Air

Another application of magnetochemistry is the determination of oxygen in gases. The method is based on the fact that the volume susceptibility of oxygen is greater by a factor of 100 to 1000 than the susceptibility of all other gases except nitric oxide, nitrogen dioxide and chlorine dioxide. Hence if three gases are absent, it is easy to determine oxygen in amounts greater than a few per cent.

Determination of Structure of Coordination Compounds

Magnetic susceptibility measurements have been used successfully in determining the number of unpaired electrons in the coordination compounds. Therefore, it becomes possible to assign suitable structures to these coordination compounds. For example, the magnetic moment for the complex $[Co(NH_3)_6]^{3+}$ is zero BM and while that of $[CoF_6]^{3-}$ is 4.26. This shows that $[Co(NH_3)_6]^{3+}$ is having no unpaired electrons while $[CoF_6]^{3-}$ is having four unpaired electrons.

Similarly, it can be proved that $[Fe(CN)_6]^{3-}$ ion is having one unpaired electron, $(FeF_6)^{3-}$ ion is having five unpaired electrons and $[Ni(CN)_4]^{2-}$ is having no unpaired electrons.

Evaluation of Free Radicals

As a free radical is always associated with an unpaired electron, magnetic susceptibility measurement, therefore, allows the evaluation of the extent to which a substance exists in the form of a free radical. For example, the magnetic susceptibility of a substance is found to 1.40×10^{-6} m^3 mol^{-1}. This magnetic susceptibility for a substance containing one unpaired electron is 1260×10^{-6} cm^3 mol^{-1}. It follows, therefore, that about 11% of the substance is present as a free radical.

Example:

The molar magnetic susceptibility of H_2O (l) at 298 k is -13.0×10^{12} m^3 mol^{-1}, while density is 0.9970 g cm^3. Find :

(a) The specific magnetic susceptibility, and

(b) The magnetic permeability of liquid water at this temperature.

Solution:

(a) Molar magnetic susceptibility, χ_M is related to the specific magnetic susceptibility, χ as

$$\chi_M = M\chi$$

$$-13.0 \times 10^{-12}\ m^3\ mol^{-1} = 18 \times 10^{-3}\ kg\ mol^{-1} \times \chi$$

$$\therefore \quad \chi = \frac{-13.0 \times 10^{-12}\ m^3\ mol^{-1}}{18 \times 10^{-3}\ kg\ mol^{-1}}$$

$$= 0.722 \times 10^{-9}\ m^3\ kg^{-1}$$

(b) Specific magnetic susceptibility χ is given by

$$\chi = \frac{\mu - 1}{4\pi\rho}$$

$$0.722 \times 10^{-9} = \frac{k-1}{4 \times 3.14 \times \left(0.9970 \times 10^{-3}\right)}$$

$$\therefore \mu - 1 = 0.722 \times 10^{-9} \times 4 \times 3.14 \times (0.9970 \times 10^{-3})$$

$$\mu - 1 = 9.041 \times 10^{-12}$$

$$\mu - 1 = 1 + 9.041 \times 10^{-12}$$

$$= 0.9999.$$

4

ELECTRIC PROPERTIES OF MOLECULES AND DIPOLE MOMENT

INTRODUCTION

It is possible to obtain considerable structural information about a molecule by placing it in an electrostatic field. If a molecule is kept in an electric field, the field is able to distort the electronic structure and alters the equilibrium positions of the nuclei, thereby causing the separation of the centres of the positive and negative charges. This can be expressed by stating that the electric field has been able to induce a dipole moment in a molecule. The induced dipole moment μ_{ind} has been directly proportional to the strength of the applied electrostatic field, E *i.e.*,

$$\mu_{ind} \propto E$$

$$\mu_{ind} = \alpha E$$

where α is a constant of proportionality and is termed as polarisability of the molecule.

This constant can be regarded as a measure of the ease with which the molecule can get polarised.

If E is equal to unit electric field strength, then α is equal to the induced dipole moment.

DIPOLE MOMENT

In some of the covalently linked molecules there exists a slight separation of charges due to the bond forming shared pair of electrons being pulled more towards one of the two bonded atoms. Substances having molecules of this type are said to have polar bonds and such molecules themselves are said to be polar molecules. For example, in a molecule of hydrogen fluoride, the fluorine atom is more electronegative, *i.e.*, it possesses a much greater tendency than hydrogen to attract the electron pair shared between them, to itself. Hence, the electron pair lies closer to the fluorine atom than to hydrogen, resulting a slight positive charge on the hydrogen atom and a slight negative charge on fluorine atom. Thus, this leads to the formation of permanent dipole in which two oppositely charged poles (H^+ and F^-) are separated by a rigid link or bond as shown below :

$$\overset{+\delta}{H} - \overset{-\delta}{F}$$

Thus, the molecule having permanent dipoles are said to have polarity. The extent of polarity is generally expressed in terms of *dipole moment* which is defined as

"It is the product of the magnitude of the charges (positive or negative) and the distance between them, i.e., bond length.

If q is the magnitude of the charge at each end of the dipole molecule and d is the distance between the two charges, the dipole moment is given by

$$\vec{\mu} = q.d$$

The dipole moment is a vector quantity, *i.e.*, the dipole has ;magnitude as well as direction. It is usually regarded to the directed from the negatively charged end towards the positively charged end (Fig. 1).

$$-q \xrightarrow{d} +q$$

Fig. 4.1 : The definition of dipole moment $\vec{\mu} = q.d$

The dipole moment for a group of point charges q_i is given by

$$\vec{\mu} = \sum_i q_i \vec{d}_i \qquad ...(1)$$

where $\vec{d}_i$ is the vector from the origin to charge q_i.

For a distribution of zero total charge, the dipole moment is equal to the product of the absolute charge of either the positive or the positive or negative distribution and the vector distance d between the centres of the two distributions. Thus,

$$\vec{\mu} = q^+\vec{d} = q^-\vec{d} \qquad ...(2)$$

When the molecule consists of three or more atoms, the net dipole moment of the molecule is given by the vectorial sum of the individual dipole moments, *i.e.*,

$$\vec{\mu} = \vec{\mu}_1 + \vec{\mu}_2 + \dots \dots$$

The arrows on $\vec{\mu}, \vec{\mu}_1, \vec{\mu}_2, \dots \dots$ indicate that dipole moments are vectorial quantities

Units

As the electric charge is of the order of 10^{-10} e.s.u. unit and the distance is of the order of 10^{-8} cm the value of dipole moment of the molecule will be of ;the order of $10^{-10} \times 10^{-18}$. This quantity is called a Debye unit and is denoted by the symbol D.

$$\mu = (4.8 \times 10^{-10} \text{ e.s.u.}) \times (10^{-8} \text{ cm}) = 4.8 \times 10^{-10} \text{ e.s.u. cm} = 4.8\text{D}$$

where 1 D = 1×10^{-18} e.s.u. cm. For example, the dipole moment of HCl is 1.03 × 10–18 e.s.u. cm and is written as 1.03D.

The dipole moment is expressed in units of debyes (D), after Peter Debye who made significant contributions to the understanding of polar molecules. In the SI system, the electronic charge e = 1.602×10^{-19} C and d = 10^{-10} m; hence,

$$m = (1.602 \times 10^{-19} \text{ C}) \times (10^{-10} \text{ m}) = 1.602 \times 10^{-29} \text{ C m} = 4.8 \text{ D}$$

where one debye is defined as 1 D = 3.336×10^{-30} C m

INDUCED OR DISTORTION POLARISATION

A molecule, though neutral, is made up of positively charged nuclei and negatively charged electrons. When this molecule is kept in an electric field between two charged plates, the positively charged nuclei will be directed towards the negative plate and the negatively charged electrons towards the positive plate. This results a change in the molecule, causing a ;positive charge at one end and negative charge at the other, as shown in Fig. 4.2. *The creation of positive and negative charges (called electric dipoles) in a neutral molecule under the influence of electric field is termed as electrical dissociation or polarisation of the molecule.*

The electrical distortion of polarisation of the molecule is temporary and is only taking place under the influence of electric field. It means that as soon as the electric field is removed the positive and negative charges disappear and the molecule reverts to its original state, *i.e.*, the nuclei and the electrons are restored to their original positions. This type of polarisation of the molecule is known as *induced polarisation* and the dipole formed is known as induced dipole.

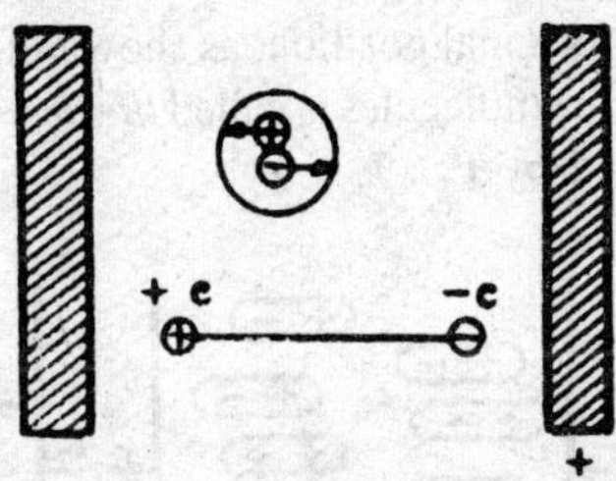

Fig. 4.2

Types of Induced Polarisation

As all the molecules consist of positively charged nuclei surrounded by negatively charged electrons, the electric field may polarise the molecules in the following two different ways :

(i) *Electric polarisation : In this, the electrons will be distorted with respect to the nuclei towards the positive plate of the capacitor.* It is denoted by P_e.

(ii) *Atomic polarisation* : In this, *the nuclei will be distorted with respect to each other*. It is denoted by P_a.

Thus, induced polarisation = $P_e + P_a$

or $$P_i = P_e + P_a$$

ORIENTATION POLARISATION

Each of the polar molecules possesses a positive and a negative end. Normally in the absence of an electric field, the molecules would be oriented in all directions due to the thermal effect, Fig. 4.3. But when such molecules are placed in the electric field between two charged plates, two distinct effects will arise :

(i) Firstly, there will be usual distortion of positive and negative charges which would give rise to *induced polarisation.*

(ii) Secondly, the field would tend to orient all the molecules in the direction of the field because they are polar Fig. 4.4. If the molecule are stationary, the electric field would tend to orient them at an angle of 180° to the direction of the field or at 90° to the condenser plates, as shown in Fig. 4.4. But the molecules themselves are in constant motion, known as *thermal agitation.* This thermal agitation opposes such orientation and the molecules would occupy some mean position between the direction of the field and their original position, as shown in Fig. 4.5. This effect of the field on the molecules is called *the orientation polarisation.* This is denoted by P_o.

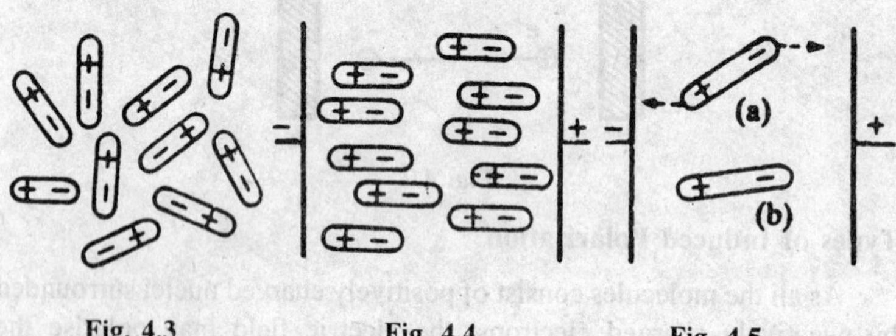

Fig. 4.3 Fig. 4.4 Fig. 4.5

TOTAL MOLAR POLARISATION

The total molar polarisation P_M, will be given by

$$P_M = P_i + P_o$$

where P_i is the induced polarisation and P_o, the orientation polarisation. The induced molar polarisation of a substance is independent of temperature. It depends on the ratio of specific properties of the substances like dielectric constant and density. Both these properties are easily measurable.

Orientation polarisation of molecules depends upon the temperature. For example, the molar polarisation of substances like O_2, CO_2, N_2, which do not have permanent dipole moments and hence no orientation polarisation, are independent of temperature changes. It is not so with substances like HCl, $C_6H_5NO_2$ or CH_3Cl which possess permanent dipole moments. Their molar polarisations change with change in temperature. Thus, one can distinguish between polar and nonpolar molecules,

POLARISABILITY

Molecules acquire dipole moment when they are placed in an electric field. The magnitude of this induced dipole moment, μ_i is proportional to the strength of the field, F, and it is written as follows :

$$\mu_i = \alpha F \quad ...(1)$$

In general, it depends on the orientation of the molecule. If the field becomes very strong μ_i also varies as F^2 and coefficient of proportionality β in βF^2 is called hyperpolarizability. The effectiveness of an applied field in making a molecule polar can be determined by polarizability of the molecule. The polarizability is defined as the dipole moment induced by an electric field of unit strength.

Dielectric Constant

It is a property of any given medium. For a vacuum $\in = 1$, but for any other medium $\in$ is greater than unity. The dielectric constant of a substance can be determined by measuring first the capacity of a condenser with a vacuum between the plates, C_0, and next, the capacity of the same condenser when filled with the given substance, C. Then, the dielectric constant follows :

$$\in = \frac{C}{C_0} \quad ...(2)$$

Capacities can be measured by various electronic circuits available which use alternating currents with frequencies of 10^6 to 10^7 per second.

MOSOTTI-CLAUSIUS EQUATION

The relationships between the dielectric constant an induced polarisation was deduced by Mosotti (1850) and Clausius (1879) and generally known as the Mosotti-Clausius equation.

In the derivation of the Mosotti equation it was assumed that each molecule occupies a small sphere (a macroscopic quantity). Furthermore, the charges within spherical molecules are so uniformly distributed that before the application of the external field the permanent moment in the molecule is zero

Let us suppose that the distortion or induced polarization, P_i produced due to an electric field does not contain permanent dipoles.

The electron and nuclei in any molecule are, to some extent, mobile and so when the molecule, whether it is polar or non-polar, is placed in an electric field there will be a small displacement of the electrical centres; with the result that a dipole, in addition to one which may already be present, will be induced in the molecules. If μ_i is the electrical moment of the induced dipole produced by a field of intensity F acting on a single molecule, then

$$\mu_i = \alpha_D F \quad ...(1)$$

The constant α_D is called the polarisability of the molecule; ;it is measure of the ease with which the molecule can be polarised, that is the ease of displacement of positive and negative with respect to each other, in an electric field shown in Fig. 4.6.

Suppose the strength of a uniform electric field produced by two charged plates is E_o : then in any non-polar medium the field strength is reduced to E because the dipole induced in the molecules acts in opposition to the applied field. The ratio E_o/E is the dielectric constant $\in$ of the medium. It can be shown by electrostatics that

$$E_o = E + 4\pi I \quad ...(2)$$

or $$E_o - E = 4\pi I$$

or $$E\left(\frac{E_o}{E} - 1\right) = 4\pi I$$

or $$E (\in - 1) = 4\pi I \quad ...(3)$$

where $\in$ (E_o/E) is called the dielectric constant of the medium, I being the induced electric moment per unit volume, *i.e.*,

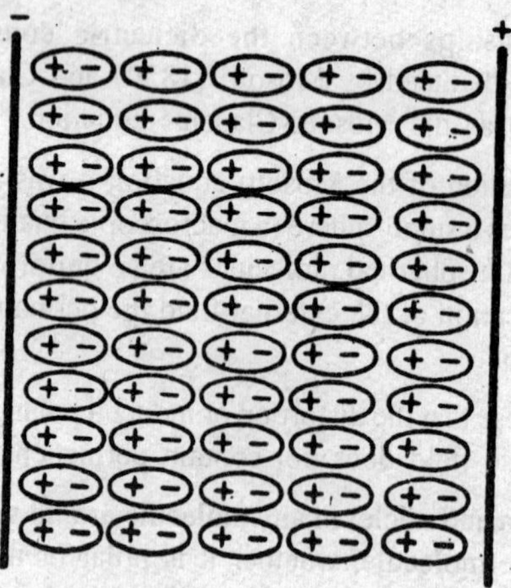

Fig. 4.6 : Orientation of polar molecules in an electric field.

$$I = \mu_i n \quad ...(4)$$

where n is the number of molecules per m_l, and μ_i is the electrical moment of the induced dipole.

The electrical intensity F is made up of several parts:

(i) Field of charge on the plates, *i.e.*, E_o,

(ii) Force -4π I due to the charge induced in the surfaces of the dielectric in contact with the plates.

(iii) 4/3 π I resulting from the charge induced on the surface of the spherical cavity.

(iv) The field caused by the molecules within the cavity.

In the case of liquids or gases, the molecules are oriented in the absence of an external field. Therefore, the field caused by the molecules within the cavity may be neglected and, thus, the value of F may be written as

$$F = E_o + \frac{4}{3}\pi I - 4\pi I \quad ...(5)$$

Introducing equation (2) in (5), we get

$$F = E + 4\pi I + \frac{4}{3}\pi I - 4\pi I = E + \frac{4}{3}\pi I \quad ...(6)$$

introducing equation (3) in (6), we get

$$F = E + \frac{1}{3}E(\in - 1) = E + \frac{E_\in}{3} - \frac{E}{3}$$

$$F = \frac{2E}{3} + \frac{E_\in}{3} = \frac{E}{3}(2 + \in) \quad ...(7)$$

Substituting equation (7) in (1), we get

$$\mu_i = \alpha_D \cdot \frac{E}{3}(2 + \in) \quad ...(8)$$

Substituting equation (8) in (4), we get

$$I = n\alpha_D \frac{E}{3}(2 + \in) \quad ...(9)$$

Substituting equation (9) in (3), we get

$$E(\in - 1) = 4\pi n\alpha_D \frac{E}{3}(2 + \in)$$

or

$$\frac{\in - 1}{\in + 1} = \frac{4}{3}\pi n\alpha_D \quad ...(10)$$

If ρ is the density of the medium between the charged plates and M is the molecular weight, then the number of molecules n in unit volume is Nρ/M, where N is the Avogadro's number *i.e.*,

$$n = N\rho/M \qquad ...(11)$$

On substituting equation (110 in (10), we get

$$\frac{\epsilon-1}{\epsilon+1} = \frac{M}{\rho} = \frac{4}{3}\pi N\alpha_D \qquad ...(11A)$$

The left hand side of the equation is given by the symbol P_M and is called molar polarisation of the material, *i.e.*,

$$P_M = \frac{\epsilon-1}{\epsilon+1} = \frac{M}{\rho} = \frac{4}{3}\pi N\alpha_D \qquad (11B)$$

Eq. (11 B) is known as Clausius-Mosotti equation.

Since the applied field produced an induced change in the molecule by the relative displacement or distortion of electrons and nuclei, P_M is then referred to as the 'induced' or 'distorted' polarisation. Thus, equation (11 B) represents the induced or distorted polarisation, *i.e.*,

$$P_i = P_M = \frac{4}{3}\pi N\alpha_D \qquad ...(12)$$

From equation (12), it follows that the induced molar polarisation of a substance is independent of temperature and depends upon the nature of the substance, *i.e.*, its dielectric constant, molecular weight and density.

Eq. (12) has been verified experimentally for *non-polar* molecules such as H_2, CO_2, CH_4, CCl_4, etc., which do not have a permanent dipole moment. The Clausius-Mosotti equation is not obeyed by *polar* molecules such as HCl, H_2O, BH_3, etc., which possess a permanent dipole moment.

DEBYE EQUATION : ORIENTATION POLARISATION

In 1912 Debye regarded the behaviour of the polar molecules kept between the plates of a condenser. In the absence of the electric field at somewhat higher temperatures, because of thermal motion, the molecules get randomly oriented so that there occurs no net dipole moment in any direction Fig. 4.7a). However, on applying the electric field across plates of the condenser the molecules would orient themselves in the direction of the field as depicted in Fig. 4.7.

Debye (1912) deduced the molar polarisation by taking into account the orientation of the molecular dipole in an electric field and was able to draw important results in connection with the variation of polarisation with temperature.

He pointed out that the molar polarisation is not only due to the distortion of the electrons and nuclei in the molecule, but if the molecule is assumed to have a permanent moment, μ, in absence of the electric field, the temperature coefficient can be readily explained an thus the electric moment per molecule is no longer given by equation (1) but by

$$\mu_M = \alpha F = (\alpha_D + \alpha_o) F = \alpha_D F + \alpha_o F \qquad \text{...(12A)}$$

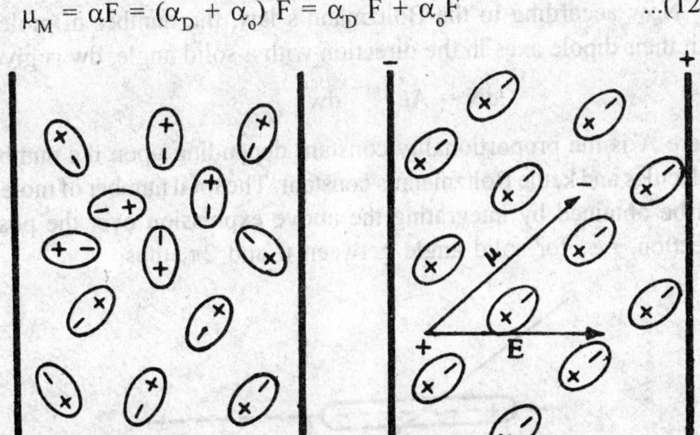

Fig. 4.7 : Orientation of molecules : (a) in the absence of an electric field (b) in the presence of an electric field.

or $$\mu_M = \mu_i + m \qquad \text{...(13)}$$

where μ_i is the moment that the induced dipole would acquire during the application of field, *i.e.*, the induced polarisation $_m$ is the moment due to orientation polarisation, and α is the polarisability, which is the sum of distortion polarisability α_D and α_o, a part due to the orientation of the molecular dipoles in the field direction.

Debye, in order to calculate the orientation moment, assumed that an equilibrium is established between two forces acting on a dipole molecule, viz. :

(i) the action of the measuring field, tending to orientate the dipole in one direction, and

(ii) the thermal agitation, which produces a random distribution of the dipole axis. Further, he regarded the molecules as a system of charges and disregarded the moment induced in the molecules by the field F.

If μ is the permanent molecular moment and the molecular dipoles are so oriented that their axes make an angle θ with the direction of the filed, the potential energy U of a molecule is given by

$$U = -\mu F \cos\theta \qquad \text{... (14)}$$

and the permanent moment in the direction of the field is $\mu F \cos\theta$.

Now according to the Boltzmann's law, the number of molecules with their dipole axes in the direction with a solid angle, dw is given by

$$dN = Ae^{-U/kT}\,dw$$

where A is the proportionality constant depending upon the number of molecules and k, the Boltzmann's constant. The total number of molecules can be obtained by integrating the above expression over the possible direction, *i.e.*, for solid angle between 0 and 2π; thus,

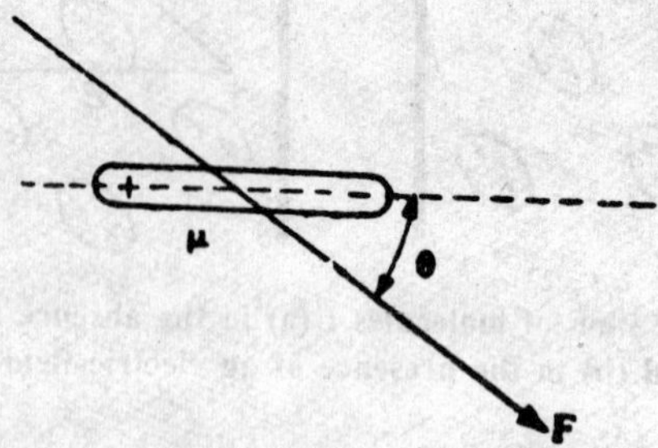

Fig. 4.8 : Dipole in electric field.

Total number of molecules

$$\int_0^{2\pi} Ae^{-U/kT}\,dw = \int_0^{2\pi} Ae^{\mu F\cos\theta/kT}\,dw$$

and the total moment in the direction of the field if given by

$$\text{total moment} = \int_0^{2\pi} Ae^{\mu F\cos\theta/kT}\,.\mu\cos\theta\,dw$$

Therefore, the average moment per molecule $\overline{m}$ in the direction of the field is

$$\overline{m} = \frac{\int_0^{2\pi} Ae^{\mu F \cos\theta/kT} . \mu \cos\theta \, dw}{\int_0^{2\pi} Ae^{\mu F \cos\theta/kT} . dw}$$

Substituting $\mu F/kT = x$ and $\cos\theta = t$ and making use of the relation $dw = 2\pi \sin\theta, d\theta$, we get

$$\overline{m} = \frac{\mu \int t e^{tx} dt}{\int e^{tx} dt}$$

$$\overline{m} = \frac{e^x + e^{-x}}{e^x - e^{-x}} - 1/X = (\text{COTH } X - 1/X) = l\,(X) \qquad ...(15)$$

where L (x) represents the function derived by P. Langevin (1905) in connection with his work on the paramagnetism of gaseous molecules. If x is small particularly for small field strength, when saturation effects do not arise. It is found that

$$L(x) \approx \frac{1}{3} x$$

and then equation (15) gives

$$\overline{m} = \frac{1}{3} \mu x = \frac{\mu^2}{3kT} F \qquad ...(16)$$

On comparing equations (12 A) and (13), we get

$$\overline{m} = \alpha_o F \qquad ...(17)$$

Substituting equation (16) in (17), we get

$$\alpha_o F = \frac{\mu^2}{3kT} F \quad \text{or} \quad \alpha_o = \frac{\mu^2}{3kT} \qquad ...(18)$$

But the total polarisation, α is given by

$$\alpha = \alpha_D + \alpha_o = \alpha_D + \frac{\mu^2}{3kT} \qquad ...(19)$$

The molar polarisation, which is assumed from dielectric constant $\frac{\epsilon - 1}{\epsilon + 2} \cdot \frac{M}{\rho}$ is the sum of the two effects, induced and orientation polarisation *i.e.*,

$$P_M = P_i + P_o$$

But $$P_M = \frac{\epsilon - 1}{\epsilon + 2} \cdot \frac{M}{\rho} \qquad ...(20)$$

If α represents the total polarisability, the equation (11 A) can be written by replacing with α , i.e,

$$\frac{\in-1}{\in+2}\cdot\frac{M}{\rho}=\frac{4\pi N}{3}\alpha \qquad ...(21)$$

Substituting eq. (19) in (21) ; we get

$$\frac{\in-1}{\in+2}\frac{M}{\rho}=\frac{4\pi N}{3}\left(\alpha_D+\frac{\mu^2}{3kT}\right) \qquad ...(22)$$

For gases at low pressure, $\in \to 0$ (So that $\in + 2 = 3$)

$$\therefore \quad \frac{\in-1}{3}\frac{M}{\rho}=\frac{4\pi N}{3}\left(\alpha_D+\frac{\mu^2}{3kT}\right) \qquad ...(23)$$

The equation (22) is known as the Debye equation. It is clear from the equation that

$$P_M = P_i + P_o$$

such that Pi = $(4\pi N\ \alpha_D)/3$, the induced polarisation and $P_o = \frac{4}{3}\pi N.\frac{\mu^2}{3kT}$ is the orientation polarisation.

Total Polarisation and Temperature

As the temperature is raised, the thermal movement of the molecule increases and this tends to oppose the orienting effect of the applied electric field on the permanent molecular dipoles ; the orientation polarisation should thus decrease with increasing temperature, in agreement with equation (11A). The dependence of the total polarisation on temperature may be seen by writing the Debye equation (11A) in the form

$$P_M = \frac{4}{3}\pi N\ \alpha_D + \frac{4}{3}\pi N\left(\frac{\mu^2}{3kT}\right)$$

$$\text{or} \quad P_M = a + (b/T) \qquad ...(24)$$

$$\text{where} \quad a = \frac{4}{3}\pi N\alpha_D \quad \text{and} \quad b = \frac{4}{3}\pi N\left(\frac{\mu^2}{3k}\right) = P_oN$$

According to equation (24), therefore, the plot of the total molar polarisation of a gas or vapour, calculated from the dielectric constant, against the reciprocal of the absolute temperature should be a straight line. If the substance is non-polar then μ, hence b, is zero and the line will be parallel to the 1/T axis, but if it is polar the slope, equal to b, is determined by μ, the permanent dipole moment of the molecule. The experimental data for carbon tetrachloride, hydrogen chloride and methyl chloride are recorded in (Fig. 4.8); the results are in agreement with expectation, since the first of these is symmetrical and non-polar whereas the other are polar molecules.

Deviation From the Debye Equation

Debye equation affords a good linear relationship between the molar polarisation, P_M and 1/T in case of solutions of polar substances, some polar liquids etc. It could not be applied correctly in case of dilute solution and pure polar liquid.

An account of the above deviation could be made by quantum theory given by Van Vleck and large number of other theories suggested by Onsagar, Kirkwood, etc.

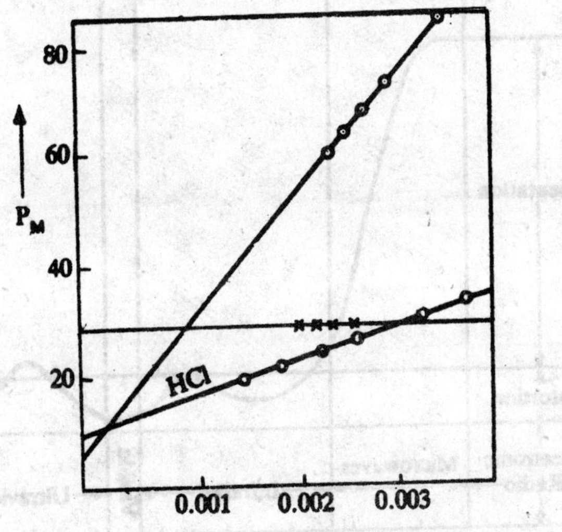

Fig. 4.9. Variation of total polarisation with temperature.

DEPENDENCE OF POLARIZABILITY ON FREQUENCY

The treatment given above for the derivations of Debye equation is based on subjecting molecules to static electric fields. When, however the applied field is oscillating, the Debye equation cannot be applied as when the frequency is very high, the permanent dipole moments of molecules would not orient themselves sufficiently rapidly to align themselves with the ever-changing field direction. The result is that at such high frequencies the permanent dipole moment is not making any contribution to the polarization of the molecules.

As a molecule needs nearly one pico-second (10^{-12} second) to rotate in a liquid, the dipolar contribution gets vanished when the dielectric constant (or relative permittivity) is measured at frequencies above 10^{11} Hz (*i.e.*, in the microwave region). Accordingly it becomes possible to measure the rate of molecular orientation in liquids by carrying out observations on the variation of polarization with frequency. Because of this, the technique of the dielectric relaxation is based on this principle.

As microwave frequencies, the orientation polarization gets decreased but thereafter the molecular distortion causes the stretching and the bending of the molecule by the electric field Fig. 4.10.

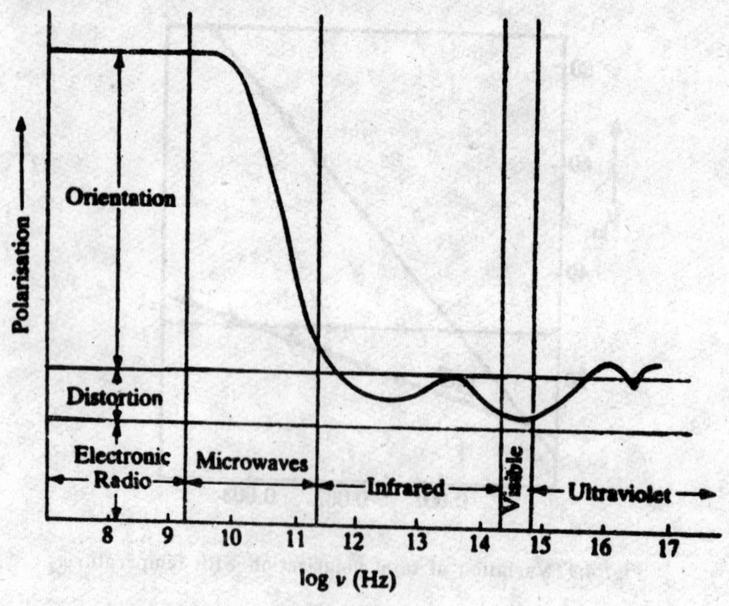

Fig. 4.10 : The dependence of P_M on frequency.

This takes place in the infrared region. At still higher frequencies when the radiation frequency becomes so high that it changes more quickly than the time needed for the molecule to change its shape, which refers to the time required for a molecule to vibrate, the *distortion polarization* do not make any contribution to P_M.

At the higher optical frequencies (UV/visible region) only the *electronic contribution* to P_M remains. The reason is that electrons are light enough to respond to the rapidly changing direction of the applied electric field. In the infrared (IR) region, both nuclei and the electrons follow the electric field but in the UV visible region only electronic contribution is left.

It is known that a light wave is electromagnetic in nature, *i.e.*, it has rapidly changing electric and magnetic fields associated with it. The electric field of ;the light wave is able to polarize the medium through which it passes, pulling the electrons back and forth rapidly.

Maxwell in his electromagnetic theory of light demonstrates that at optical frequencies the dielectric constant (*i.e.*, relative permittivity) and the refractive index of the medium are related as follows :

$$\epsilon_r = n_r^2 \quad \text{...(1)}$$

This *Maxwell relation* is only true if the permanent dipole moment of a molecule does not contribute to ϵ_r and hence to P_M.

Thus, at optical frequencies, substitution of the Maxwell relation into the Clausius-Mosotti equation gives rise to

$$P_M = R_M = \frac{n_r^2 - 1}{n_r^2 + 2} \frac{M_M}{\rho} = \frac{N_A \alpha_d}{3 \epsilon_0} \quad \text{...(2)}$$

where R_M refers to the molar refraction or molar refractivity. Eq (2) is known as the *Lorentz equation.* Using this equation, it impossible to determine induced polarizations from the refractive index measurements. If these data are combined with dielectric data, the dipole moment can be determined from the Debye equation.

Eq. (2) can also be inverted to yield an expression for the refractive index of a medium :

$$n_r = \left(\frac{V_M + 2R_M}{V_M - R_M} \right)^1 / 2 \text{ where } V_M = M_M/\rho \quad \text{...(3)}$$

Eq. (2) can be rearranged to give an expression for distortion polarizability, α_d :

$$\alpha_d = 3\epsilon_o\, R_M/N_A \qquad ...(4)$$

so that the polarizability volume becomes as follows :

$$\alpha'_d := \alpha_d/4\pi\epsilon_o = 3R_M/4\pi N_A \qquad ...(5)$$

It may be remarked that the Maxwell relation $\epsilon_r = n_r^2$ is only true for nonpolar molecules. For non-polar molecules ϵ_r and n_r need not be measured at the same frequency. Even when ϵ_r and n_r are measured at different frequencies the orientation polarizability would be absent from non-polar molecules and the nuclear distortion polarizability becomes negligible in comparison with the electronic distortion polarizability.

This implies that the measurement of ϵ_r and n_r at different frequencies includes only the electronic polarization. Thus, for benzene, a nonpolar molecule, $n_r = 1.50$ $\epsilon_r = 2.27$ so that $\epsilon_r \approx n^2$. However, for water, a polar molecule, $n_r = 1.33$ and $\epsilon_r = 80$ so that $\epsilon_r \neq n_r^2$. For polar molecules $\epsilon_r = n_r^2$ only if both ϵ_r and n_r have been measured at the same frequency.

Molar Refractivity

R_M, of a molecule is additive, i.e, it is equal to the sum of the molar refractivities of individual atoms or of individual bonds. Table 4.1, lists the molar refractivities of some atoms, ions and bonds.

Table 4.1 : Molar Refractivities at 589 nm

Bond	*Molar refractivity R_M (cm^3 mol^{-1})*	*Atom (or) I*	*Molar refractivity R_M (cm^3 mol^{-1})*
C – H	1.65	H	1.1
O – H	1.85	He	0.5
C – C	1.20	C	2.42
C = C	2.79	Li^+	0.07
C = C	4.79	Na^+	0.46
C – O	1.41	K^+	2.12
C = O	3.34	Rb^+	3.57
C = N	4.69	Be^+	0.20
		Mg^{2+}	0.24
		Ca^{2+}	1.19

It is possible to calculate the molar refraction R_M from the measured value of the refractive index of the liquid or the solid. Eq. (2) can be combined with the Debye equation to obtain the following expression for the relative permittivity at low frequencies:

$$P_M = R_M + \frac{N_A \mu^2}{9 \in_o kT} \quad ...(6)$$

i.e.,

$$\frac{\in_r - 1}{\in_r + 1} \frac{M_M}{\rho.} = R_M + \frac{N_A \mu^2}{9 \in_o kT} \quad ...(7)$$

In the c.g.s. system, Eq (6) may be written as follows :

$$P_M = R_M + \frac{4\pi N_A \mu^2}{9 kT} \quad ...(8)$$

DETERMINATION OF DIPOLE MOMENT

Here we will describe only few methods for the dipole moment measurements which are normally employed.

1. Ebert's Method

This method is based on the fact that *in the gases and to a smaller extent in liquids, the molecules have free movements.*

But in the cases of solids, the molecules remain almost stationary in space lattice. So polar molecules in liquid or gaseous state will suffer orientation as well as distortion polarisation, when subjected to electric field.

But, in solids no orientation polarisation even in polar substances is possible, *i.e.*, p_o is zero (it is not always true). It means that in a polar molecule the total polarisation in gaseous (or liquid) and solid states will be markedly different.

The total molar polarisation,

$$P_{gas} = P_i + P_M$$

$$P_{solid} = P_i \text{ or } P_M = P_{gas} - P_{solid}$$

or

$$\frac{4\pi N}{3}\left(\frac{\mu^2}{3kT}\right) = P_{gas} - P_{solid}$$

$\therefore$

$$\mu = \sqrt{\left(\frac{9kT}{4\pi N} P_{gas} - P_{solid}\right)}$$

All the quantities except μ are known in the above expression.

2. Vapour-Temperature Method

As we known that Debye equation is accurately applicable to gases, therefore, this method is perhaps most accurate.

The Debye equation is :

$$P_M = \frac{\epsilon-1}{\epsilon+2}\cdot\frac{M}{\rho} = \frac{4\pi N}{3}\left(\alpha_D + \frac{\mu^2}{3kT}\right)$$

or $\quad P_M = A + (B/T)$ where $A \; \frac{4\pi N\alpha_D}{3}$

and $\quad B = \frac{4\pi N}{3}\cdot\frac{\mu^2}{3kT}$ ...(1)

$\mu = \sqrt{\left(\frac{9kT}{4\pi N}\right)}.\sqrt{B}$: Substituting the values of k, N and π, we get

$$\mu = 0.0128 \times \sqrt{B} \times 10^{-18} \text{ e.s.u.}$$

From the Debye equation, it is obvious that if a graph is plotted between total polarisation (P_M) and 1/T, a straight line will be obtained Fig. 4.11.

In the case of polar molecules, the slope of line II will give the value of constant B which consequently determines μ.

If the molecules are non-polar, the curve will be a horizontal line as curve I (μ = 0). The intercept of the lines gives the value of induced polarisation, P_i.

An alternative application of the same principle is to calculate the total molar polarisation, say P_1 and P_2 at two different temperatures T_1 and T_2 and then using eq. (1) B can be calculated as follows :

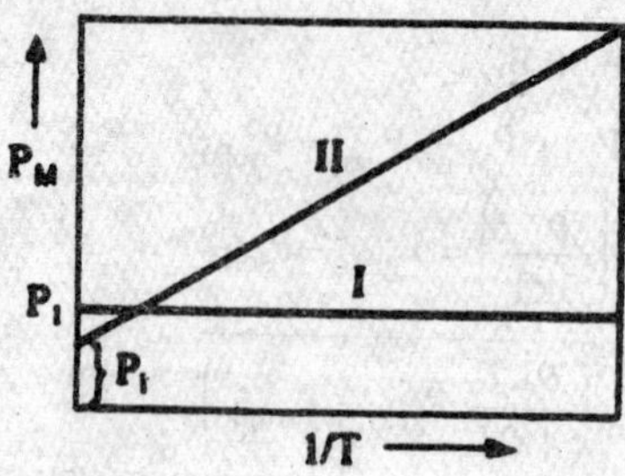

Fig. 4.11 : Plot of PM and 1/T.

$\because \quad (PM)_1 = A + B/T_1$ and $(PM)_2 = A + B/T_2$

$$\therefore \quad (PM)_1 = (PM)_2 = B\left(\frac{1}{T_1} - \frac{1}{T_2}\right) = B\left(\frac{T_2 - T_1}{T_1 T_2}\right)$$

$$\therefore \quad B = [(PM)_1 - (PM)_2]\left(\frac{T_1 T_2}{T_2 - T_1}\right) \qquad ...(2)$$

The value of the dipole moment is then calculated by using Eq. (2).

The method has been widely used for a large number of gases.

Limitations

Since the range of temperature required for accurate results is large, the method is applicable only to those substances which do not decompose within the temperature used. This condition does not hold in liquids because of the interaction of adjacent molecules. The temperature method just described, which is capable of giving accurate values for the dipole moment, is thus applicable to gases and vapours where the molecules are relatively far apart.

The method involves complicated experimentation by involving the measurement of dielectric constant and density of the vapour at a series of temperatures.

3. Refraction method

This method involves the measurement of dielectric constant of the vapours in terms of refractive index only at one temperature. We have,

$$P = P_i + \frac{4\pi N}{3}.\frac{\mu^3}{3kT} P = P_i + \frac{4\pi N}{3}.\frac{\mu^3}{3kT}$$

$$\text{or} \qquad \mu = \sqrt{\left(\frac{9k}{4\pi N}\right)}.\sqrt{\{(p - p_i)\}} = 0.0128\sqrt{[(p - p_i)T]}$$

where P_i is the distortion polarisation. Thus, for the determination of μ, the total polarisation p, and the distortion P_i must be known accurately. We know,

$$P = \frac{\epsilon - 1}{\epsilon + 1}.\frac{M}{d}$$

Maxwell showed that for the same electromagnetic vibrations, the refractive index (μ) is related to the dielectric constant by the expression.

$$\in = n^2.$$

So far non-polar molecules,

$$P = \frac{n^2 - 1}{n^2 + 1} \cdot \frac{M}{d} = \frac{4\pi\mu\alpha_i}{3}$$

The refractive index, n is measured with visible light and then converted for long infrared region by the relation

$$n = n_\infty + (a/\lambda^2)$$

where 'a' is a constant; n is the refractive index at wave length λ, and n_∞ that for long wavelength.

4. Dilute Solution Method

The above methods which have been discussed are only strictly applicable when mutual interaction of the molecules can be regard as negligible. For gases at high pressures and with liquids, this condition is certainly not fulfilled.

If a polar substance is dissolved in a non-polar solvent, the tendency for dipole interaction to occur is generally comparatively small. Hence the total polarisation of the dilute solution can be regarded as equal to the sum of the polarisation of the constituents. On the basis of this assumption, Debye deduced the following relationship :

$$P_{1,2} = \frac{\in_m - 1}{\in_m + 2} \times \frac{x_1 M_1 + x_2 M_2}{\rho}$$

where $P_{1,2}$ = total polarisation of mixture of two constituents 1, 2.

$\in_m$ = dielectric constant of the mixture.

x_1, x_2 = mole fractions of the non-polar solvent and polar solute respectively.

M_1, M_2 = molecular weights of the constituents 1, 2.

ρ = density of the solution.

Since $p_{1,2}$ is the sum of the polarisations of two constituents, we have,

$$P_{1,2} = x_1 p_1 + x_2 P_2$$

or $$\frac{(\in_m - 1)}{(\in_m + 2)} \frac{(x_1 M_1 + x_2 M_2)}{\rho} = \frac{\in_{m1} - 1}{\in_{m2} + 2} \cdot \frac{M_1}{\rho_1} x_1 + \frac{\in_{m2} - 1}{\in_{m2} + 2} \cdot \frac{M_2}{\rho_2} x_2$$

where P_1 and P_2 are the molar polarisations of solvent and solute respectively. The value of P_1 can be evaluated from the measurements of dielectric constant of pure solvent.

The value of P_2 is determined by knowing the dielectric constant at known values of x_1 and x_2. Then μ can be determined as usual by plotting p_2 against 1/T.

In this method the solvent for low dielectric constant is used some corrections are, however, necessary sometimes to eliminate the solvent effects if any.

Modification :

Van Vleck (1934) modified the Debye equation by introducing $bN\mu^2$. He found that equation is applicable to concentrated solution also

$$P = \frac{\epsilon - 1}{\epsilon + 1} = \frac{4\pi N}{3}\left[\alpha_D + \frac{\mu^2}{3kT + bN\mu^2}\right]$$

where b is a constant and N is the number of molecules per unit volume. Obviously zero concentration (very dilute solution) reduces above equation to Debye equation.

Molecular Beam Method

This method is based on the direct measurement of the behaviour of molecules in non-homogeneous electric field.

A thin molecular beam of vapourised molecules of the substance is passed through a non-homogeneous electric field and is finally received on a brass plate cooled with liquid air. The trace left as result of condensation of molecules can be seen through a microscope.

In case of non-polar molecules, the trace left by the molecules is laterally shifted. But in case of polar molecules there will be addition a broadening of the trace. This broadening under constant working conditions has been found to be proportional to the dipole moment. From the broadening of a trace from a substance of known dipole moment, we can calculate that of an unknown substance.

This method has proved to be of great value in the determination of dipole moment of several compounds of low volatility, and substances which are insoluble in non-polar solvents. Some results for the permanent dipole moment of various substances are shown in Table 4.2.

Table 4.2 : Dipole moments of various molecules (In Debye Units)

Organic molecules	μ	*Inorganic molecules*	μ
CH_4, C_3H_6, C_2H_4, C_2H_2	0	H_2, Cl_2, Br_2, I_2, N_2	0
C_6H_6 naphthalene, diphenyl	0	CO_2, CS_2, $SnCl_4$, SnI_4	0
CCl_4, CBr_4,	0	HCl	1.03
CH_3OH	1.70	HBr	0.78
CH_3Cl	1.86	HI	0.38
CH_3Br	1.80	H_2O	1.84
C_2H_5Br	2.03	HCN	2.23
CH_3NH_2	1.24	NH_3	1.46
CH_3COOH	1.74	SO_2	1.63
0-dichlorobenzene	2.50	N_2O	0.17
m-dichlorobenzene	1.72	CO	0.12
p-dichlorobenzene	0	PH_3	0.55
0-chloronitrobenzene	4.64	PCl_3	0.78
m-chloronitrobenzene	3.73	$AsCl_3$	1.59
p-chloronitrobenzene	2.83	$AgClO_4$	4.7

BOND MOMENTS AND THE MOLECULAR DIPOLE MOMENT

It is more convenient to consider the permanent dipole moment of a molecule in terms of its *bond moments*. It is known that the bond moment is a *vector quantity* and can be associated with a chemical bond. The bond moments for some bonds are given in Table 4.3

Table 4.3 : Bond Moments of Some Chemical Bonds

Bond	*Bond Moment (D)*	*Bond*	*Bond Moment (D)*	*Bond*	*Bond Moment (D)*
C – H	0.4	C – N	0.2	S – H	0.8
C – F	1.4	C = O	2.3	F – Cl	0.9
C – Cl	1.5	C – NO_2	3.5	F – Br	1.3
C – Br	1.4	H – I	0.4	Br – Cl	0.6
C – I	1.2	O – H	1.6	H – F	1.9
C – O	0.9	N – H	1.3	H – Cl	1.0
				H – Br	0.8

The bond moments for a given bond have been found to vary from molecule to molecule, though the variation is generally very small. For a molecule having only one bond, the bond moment is equal to the molecular dipole moment. However, for a polyatomic molecule, the molecular dipole moment is equal to the resultant of the bond moments and can be obtained by vectorial addition of the bond moments. Hence, there exists an intimate relationship between dipole moment and molecular geometry. It is also possible to resolve the molecular dipole moment into bond moments. The ionic character of a bond can also be estimated from the knowledge of bond moment and the bond moment can be calculated assuming that the bond is completely ionic.

APPLICATIONS OF DIPOLE MOMENT

Dipole moment has proved to be a powerful and invaluable weapon in our attack on molecular structure. Mainly two types of information about the molecular structure are provided by dipole measurements.

(i) The percentage ionic character, *i.e.*, the extent to which a bond is permanently polarised.

(ii) An insight into the geometry of the molecules, especially the bond angles, bond length and symmetry of the molecules.

Inorganic Compounds

Main applications of dipole moment are as follows :

1. Study of Per cent Ionic Character

Dipole moment measurement is employed in the study of percent ionic character. For example, the distance between H and Cl in HCl determined by infra-red spectrum is 1.27Å . If H and Cl are completely separated, the HCl molecule would be expected to have dipole moment as :

$$\mu_{oal}\ (HCl) = 4.80 \times 10^{-10} \times 1.27 \times 10^{-8} \text{ e.s.u.}$$
$$= 6.4 \times 10^{-18} = 6.4 \text{ D}$$

But the experimental value of HCl has been found to be

$$\mu_{obs} = 1.03 \text{ D}$$

The large difference between the two values is due to the fact that two charges are not separated completely but they are partially separated, *i.e.*, the bond distance between H and Cl atoms is less than the expected value of 1.27A°.

The percent ionic character P of the bond AX is given by :

$$P = \frac{\mu_{obs}}{\mu_{cal}} \times 100\%$$

Thus, $$P_{HCX} = \frac{1.03}{6.14} \times 100 = 17\%$$

the result does not tell which end of the molecule is positive and which is negative.

2. Bond Moment

For polyatomic, it is customary to understand the molecular dipole moment in terms of the contribution of the individual bonds of the molecules. The contribution of individual bond is called its bond moment. The method of calculating bond moment is given below. The measured dipole moment of water is 1.85 D. This dipole moment is the vectorial sum of the individual bond moments, of two –OH bonds directed at an angle of 104.5° with respect to each other Fig. 4.12. Thus

$$\mu_{obs} = 2_{\mu OH} \cos 52.25° \text{ or } 1.85 = 2_{\mu OH}\ 0.6129$$

$$\therefore \quad \mu OH = 1.51 \text{ D.}$$

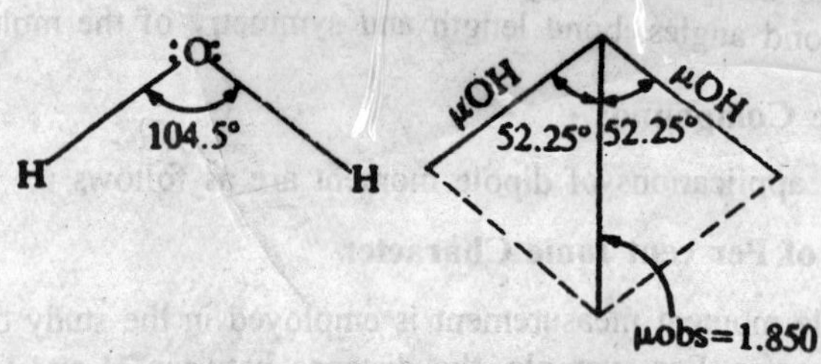

Fig. 4.12 : Calculation of bond moments

SOLVED EXAMPLES

Example 1:

The dielectric constant of nitrogen at N.T.P. is 1.00485 and the density also at N.T.P. is 1.25 gm per litre. Calculate the induced molar polarization and polarizability of the molecule (N = 6.023 × 10^{23}).

Solution:

As nitrogen is non-polar, it means that

$$P = P_i, \text{ or } P_i = \frac{E_m - 1}{E_m + 2} \cdot \frac{M}{\rho}$$